Biodiversity in
HORTICULTURAL CROPS
– Volume 6 –

The Editor

Prof. (Dr.) K.V. Peter is basically a horticulturist, a plant breeder and a University Professor. He is an acknowledged and decorated scientist and science manager. A post graduate from G.B. Pant University of Agriculture and Technology, he did post doctoral research at BARC Beltsville Maryland USA and worked at Laboratories at AVRDC Taiwan and Guadeloupe (French West Indies). He was associated with development and release of improved and high yielding varieties/hybrids in tomato, brinjal, chilli, bittergourd, melons, amaranths and cowpea. Sources of resistance to bacterial wilt in tomato, chilli and brinjal; aphids in cowpea and viral leaf curl in chilli were located by him and are now used in breeding programmes. A vegetable seed production complex established by him first at Pantnagar and later at Kerala Agricultural University in 1980 continue to supply quality seeds to farmers even today. Prof. Peter provided managerial support to the ICAR Indian Institute of Spices Research Calicut to possess the World's largest collection of black pepper and cardamom germplasm. The technology package for protected cultivation of bush pepper would make available green pepper throughout the year. A prolific writer and academic editor he authored/edited 80 books published both in India and abroad, 16 technical bulletins, 110 research papers, 22 short research notes, 15 scientific reviews, 111 papers in symposia/seminars, 85 chapters in books, 307 popular articles and 17 radio/TV programmes. He received 7 scholarships and fellowships at various stages of education. He was member in the Board of Examiners in 12 Universities, Chairman/member in 122 committees of Government/research institutes, and membership in 18 scientific societies. His students occupy important positions in civil services, research institutes and other sectors. His publishers include Taylor and Francis USA, Elsevier USA, National Book Trust, I.C.A.R., New India Publishing Agency, Astral International Pvt. Ltd; Studium Press(USA) and Universities Press Hyderabad. Now Commissioning Editor New India Publishing Agency New Delhi, he is Director, World Noni Research Foundation, Chennai since 2008. He was Director, ICAR IISR Calicut; Director of Research KAU and the Vice-Chancellor KAU. The awards received include Rafi Ahmed Kidwai Award 1996-1998 for outstanding research in Horticulture-ICAR New Delhi; Recognition Award 2000–National Academy of Agricultural Sciences, New Delhi; Dr. M.H. Marigowda National Award for the Best Horticulturist-2000; Silver Jubilee Medal for outstanding contributions in Vegetable Research, Indian Society of Vegetable Science, Varanasi 1998; Dr. Harbhajan Singh Award 1993 instituted by the Indian Society of Vegetable Science, Varanasi for the Best Paper in the Journal Vegetable Science 1993; Silver Jubilee Memonto awarded in appreciation of services rendered to the Indian Society of Vegetable Science 1998; Biotech Product and Process Development Award 2003 awarded by Department of Biotechnology, Ministry of Science and Technology, Government of India; ICAR, New Delhi awarded a set of books for being one of the Best UG Students during 1966-1969; Awarded National Scholarship (1966-69), Junior Research Fellow, ICAR (1969-1971)-only one Jr. Research Fellowship in Horticulture in India during 1969, Senior Research Fellow, ICAR (1971-1975), Scholarship for Study Abroad(1981-1982); Shiva Sakthi-HSI National Award for Life Time Achievement in Horticulture-2008; NAAS Dr. K. Ramiah Memorial Award for outstanding contribution to Crop Improvement-2009; Best Institution Award to IISR, Calicut for 1994-1999 instituted by ICAR, New Delhi; Sardar Patil Award 2003 for the best ICAR institution to Kerala Agricultural University conferred on 19/10/2004; KRLCC Award for contributions to Education and Science-2015 and Suganda Bharati award by Indian Society of Spices, Calicut. He is an elected Fellow of National Academy of Sciences, Allahabad; National Academy of Agricultural Sciences New Delhi and National Academy of Biological Sciences, Chennai.

Family: Vimala is wife, Anvar and Ajay sons, Anu and Cynara Daughers-in-law and Kuruppacharil Antony Ajay Peter, the grand son and the grand daughters Anna Vimala Anvar and Annie Terese Anvar. Parents are Late Kuruppacharil Devassey Varkey and Late Rosa Varkey.

Biodiversity in
HORTICULTURAL CROPS
– Volume 6 –

– Editor –

Prof. (Dr.) K.V. Peter

2018

Daya Publishing House®

A Division of

Astral International Pvt. Ltd.

New Delhi – 110 002

Published by : **Daya Publishing House®**
A Division of
Astral International Pvt. Ltd.
– ISO 9001:2015 Certified Company –
4736/23, Ansari Road, Darya Ganj
New Delhi-110 002
Ph. 011-43549197, 23278134
E-mail: info@astralint.com
Website: www.astralint.com

Dedication

The series BIODIVERSITY OF HORTICULTURAL CROPS published by Astral International Pvt Ltd, New Delhi is devoted to my parents Late Kuruppacharil Devassey Varkey (25/12/1911- 3/11/92) and Late Rosa Varkey (31/5/13-20/11/95) who toiled hard to educate their three sons to the highest levels possible. They inculcated in us discipline, integrity and above all "To worship God Almighty" and live with out fear, jealousy and egoism. Being the youngest in the family, I had the freedom to question his decisions and to write to him. My mother Rosa Varkey was a "saint" in all its meanings-she took food only at the end when all others and even selected neighbours had their dinner. I have never heard her speaking ill of any one- a great soul. Both were lovers of plants and maintained a homested garden.

डा.(श्रीमती) बि. मीनाकुमारी

अध्यक्ष

Dr. (Ms) B. Meenakumari
Chairperson

राष्ट्रीय जैव विविधता प्राधिकरण
NATIONAL BIODIVERSITY AUTHORITY
भारत सरकार
Govt. of India

Foreword

Biodiversity is a key factor in maintaining balance of nature and it also provides a large number of goods and services which sustain human lives. Rich biodiversity is the wealth of any nation and India is a mega biodiversity nation and we owe it to the future generations. For better conservation and sustainable management of global bioresources, biodiversity hotspots are identified. Biodiversity hotspots are biogeographic regions that possess pools of biodiversity which presently face the danger of extinction. So far 34 biodiversity hotspots with high endemic biodiversity are identified in the world as per the concept of Norman Myers (1988). Species richness, endemism and threat to extinction are the three factors determining a hot spot. These are sensitive zones and need protection from natural disasters like floods, hurricanes and drought. It is a herculean task as it involves management of both human and monetary resources.

The current series, brought out as the 6th volume, is based on "Biodiversity of Horticultural Crops" has 14 chapters, authored by 36 scientists from 15 various research institutes of ICAR and State Agricultural Universities, covers various aspects of biodiversity of more than 20 crops. The first chapter deals with the importance of biological diversity and the role of CBD and Biodiversity Act 2002, concerning conservation and management aspects of bioresources. The much quoted statement of Mahatma Gandhi *"Earth provides enough to satisfy every man's needs, but not every man's greed"* denotes the need for conserving bioresources, its sustainable

5th मंज़िल टैसल बयो पार्क, सी एस आई आर रोड, तरमणि, चेन्नई - 600 113, तमिलनाडु, भारत
5th Floor, TICEL Bio Park, CSIR Road, Taramani, Chennai – 600 113, Tamil Nadu, India
+91 44 22541805; +91 44 22541073 chairman@nba.nic.in Web: www.nbaindia.org

management and the importance in educating present and future generations in this regard.

India is the centre of origin of many horticultural crops like brinjal, citrus, banana, beans, gourds, cardamom, black pepper etc. However, over a period of time many new economic crops got introduced and domesticated such as European cauliflower, pear, kale, onion, cacao, African oil palm etc. Introduction of such new crops and their domestication resulted in altering and modifying the biodiversity of various regions of the country. The Protection of Plant Varieties and Farmers Rights Authority (PPV & FRA), New Delhi has notified Distinctiveness, Uniformity and Stability (DUS) guidelines for several crops including oil palm, coconut and cacao. NBA along with UNEP, GEF, MoEFCC, ABS, Biodiversity Finance Initiative (BIOFIN) under the GOI-UNDP, Centre for Biodiversity Policy and Law (CEBPOL) and State Biodiversity Boards, regularly monitor the status and facilitate the implementation of biodiversity policies and laws for better conservation, possession, valuation and benefit sharing of bioresources.

National Biodiversity Authority (NBA) was established in 2003, to implement Biological Diversity Act 2002 of India, which came into effect after India signed the Convention on Biological Diversity (CBD) of 1992. Over these years, the Authority has taken several initiatives for the successful implementation of biodiversity policies and laws in the country. The initiatives taken by NBA aid in conservation, sustainable management and in ensuring fair and equitable sharing of the benefits arising out of bioresources to various stake holders *i.e.,* farmers, rural and forest habitants, women conservators etc.

Biological diversity is included in the syllabi of various academic programmes such as Agricultural Botany, Horticulture, Agro-forestry, Ecology, Environmental Sciences and other allied disciplines. I believe the present series "Biodiversity of Horticultural Crops" edited by Prof. K V Peter, Former Vice-Chancellor of Kerala Agricultural University is a useful compendium to one and all concerned. Dr. K. V. Peter, the contributors and the publisher, Astral International Pvt Ltd New Delhi deserve our appreciation for completing the publication in a time bound manner.

B. Meenakumari

Horticultural Science Division

INDIAN COUNCIL OF AGRICULTURAL RESEARCH

Krishi Anusandhan Bhawan-II, Pusa, New Delhi-110 012

Ph: 011-25846490; E-mail: adghortsci@gmail.com

Dr. T. JANAKIRAM

Assistant Director General (HS)

Preface

"Biodiversity is the wealth of any nation" to quote the introduction to Biodiversity Act-2002. Enactment of Biodiversity Act followed up by the establishment of National Biodiversity Authority-2003 are landmarks in the movements for survey, collection, maintenance, description, characterization and uses of biodiversity in plants-terrestrial, aquatic and aerobic. The Division of Horticultural Sciences, ICAR manages 23 National Research Institutes-Indian/ Central Institutes, Directorates, National Research Centres-,20 Regional Stations and 13 All India Co-ordinated Research Projects/All India Network Research Projects where national crop germplasms are maintained. In fact Institutes like ICAR-IISR Calicut possesses the world's largest collection of black pepper, cardamom and ginger. ICAR-CPCRI Kasaragod maintains largest collection of coconut. Descriptors of horticultural crops are available useful for locating genes for crop improvement.

The ICAR-NBPGR, New Delhi, its ten regional stations and the ICAR-AICRN on potential crops are mandated with collection, conservation, cataloguing and collaboration with National Biodiversity Authority (NBA), Chennai and Protection of Plant Variety and Farmers Right Authority (PPV and FR Authority) New Delhi for protecting the biological wealth of India. "Vision -2050" of ICAR-NBPGR speaks the futuristics and goal in protecting the biodiversity so vital to India's short and long term interests-a hunger free India habituated by healthy people through nutritional security-.

Horticultural crops-fruits, vegetables, ornamentals, spices, medicinal and aromatic plants, plantation crops, tuber crops, bamboos and mushrooms-play a major role in the nutritional security. Prof. M S Swaminathan stated "there is a horticultural remedy for every nutritional malady". Total production of fruits and vegetables in 2017 has exceeded the food grain production overtaking" green revolution by nutrition revolution"-a dream of Gandhiji getting materialised. Modern tools of identification like DNA Finger printing, bar coding and molecular markers are making piracy in plant genetic resources rather a matter of impossibility.

The series "Biodiversity of Horticultural Crops" is entering into 6th volume carrying 14 chapters and two preambles authored by 35 active scientists from 15 research institutes. Crops covered are underutilized fruits of hot arid region, acid lime, banana, pear, cluster bean, kale, onion, ridge gourd, large cardamom, cocoa and African oil palm. The Volume is edited by Prof. K V Peter Former Vice-Chancellor Kerala Agricultural University and Former Director ICAR-IISR Calicut. The series is published by the Astral International Pvt Ltd, New Delhi.

T. Janakiram

Contents

List of Contributors

Chapter 1

N. Sivaraj, S.R. Pandravada and V. Kamala
ICAR-National Bureau of Plant Genetic Resources, Regional Station,
Hyderabad – 500 030, Telengana
E-mail: sivarajn@gmail.com

Chapter 2

Aparna Banerjee and Rajib Bandopadhyay*
UGC-Centre of Advanced Study, Department of Botany,
The University of Burdwan, Golabag, Burdwan – 713 104, West Bengal
*E-mail: rajibindia@gmail.com

Chapter 3

R.S. Singh, A.K. Singh, Sanjay Singh and Vikas Yadav
ICAR-Central Institute for Arid Horticulture,
Beechwal, Bikaner – 334 006, Rajasthan
E-mail: aksbicar@gmail.com

Chapter 4

Mahantesh Kamatyanatti[1] and Murlimanohar Baghel[2]
[1]Department of Horticulture (Fruit Science),
CCS Haryana Agricultural University, Hisar – 125 004, Haryana,
E-mail: mahanteshkk23@gmail.com
[2]Division of Fruits and Horticultural Technology,
IARI, New Delhi – 110 012

Chapter 5

Sukhen Chandra Das and K.S. Thingreingam Irenaeus
Department of Horticulture,
College of Agriculture, Tripura – 799 210,
E-mail: sukhenchandra@rediffmail.com

Chapter 6

Amit Kumar[1], Tawseef Rehman Baba[1], Nirmal Sharma[2] and Manmohan Lal[1]
[1]Division of Fruit Sciences,
SKUAST-Kashmir, Shalimar, Srinagar (J&K)
[2]Division of Fruit Sciences
SKUSAT-Jammu, Chatha, Jammu (J&K)
E-mail: khokerak@rediffmail.com

Chapter 7

Vikas Kumar, R.B. Ram and Chhatarpal Singh
Department of Applied Sciences (Horticulture) and
Department of Environmental Microbiology,
Babasaheb Bhimrao Ambedkar University, Lucknow – 226 025, U.P.
E-mail: vs1744@gmail.com

Chapter 8

Chander Parkash, S.S. Dey and Vijay Bhardwaj
ICAR IARI, Regional Station,
Katrain (Kullu Valley) – 175 129, H.P.
E-mail: cp1968@gmail.com

Chapter 9

M.A. Vaddoria[1] and Ganesh Kulkarni[2]
[1]Vegetable Research Station,
[2]Department of Genetics and Plant Breeding,
JAU Junagadh, Gujarat
E-mail: mavaddoria@jau.in

Chapter 10

D.K. Singh[1] and M. Sarkar[2]
[1]Department of Vegetable Science,
G.B.Pant University of Agriculture and Technology,
Pantnagar – 263 145, Uttarkhand
E-mail: dks1233@gmail.com
[2]Hill Millet Research Station
Navsari Agricultural University, Waghai, Navsari – 394 730, Gujarat

Chapter 11

S. Sreekrishna Bhat
Indian Cardamom Research Institute, Spices Board,
Myladumpara, Idukki – 685 553, Kerala
E-mail: sk9bhat@yahoo.co.in

Chapter 12

S. Prasannakumari Amma, V.K. Mallika and E.K. Lalitha Bai and J.S. Minimol
Cadbury-KAU Co-operative Research Project,
College of Horticulture, Kerala Agricultural University,
Vellanikkara, P O KAU, Thrissur – 680 656 , Kerala
E-mail: sprasannakumari@gmail.com

Chapter 13

P. Murugesan and G.M. Aswathy
ICAR-Indian Institute of Oilpalm Research,
Regional Station, Palode, Thiruvanandapuram – 695 562
E-mail: gesan70@gmail.com

Chapter 14

B.K. Babu and R.K. Mathur
ICAR-Indian Institute of Oilpalm Research,
P O Pedavegi – 534 450, West Godavari District, A.P.
E-mail: rkmathur1967@gmail.com

Introduction

The Vol. VI "Biodiversity of Horticultural Crops " is continuation of earlier 5 volumes focussed on biological diversity, its management including description, survey, collection, conservation, bioprospecting, bar coding and futuristics. Biological diversity depicts variety of life on planet earth including oceans, mountains, terrestrial and under terrestrial living bodies and micro-organisms. About 1. 75 million species are identified, mostly of insects, mammals and reptiles. Scientists report 13 million species though estimates range from 3-100 million. The preample to the Biological Diversity Act-2002 reports existence of 7 billion types of plants, animals and fungi inhabitating along with 7 billion people. In 1992, at the first Earth Summit in Rio de Janeiro, Brazil the vast majority of the world's nations declared that human actions were dismantling the Earth's ecosystems eliminating genes, species and biological traits at alarming rate. To quote Mahatma Gandhiji "Earth provides enough to satisfy every man's needs, but not every man's greeds. " Biodiversity is what we need; conserving biodiversity is a good deed" according to Shreemayi, School Student, New Delhi. Biological diversity is also the wealth of any nation percolated to states, regions, communities, farmers, users and future generations. It is crucial to the survival of all ecosystems, life forms and microorganisms on planet earth. Biodiversity provides a large number of goods and services which sustain our lives-write Sivaraj *et al.* (2017) in Chapter 1. Global scenario of biodiversity is changing as land, vegetation, climate, atmospheric green house gas *etc.* are changing fast. Biodiversity is considered by diversity of life on earth and calculated at different levels of biological set up together with genes, species and ecosystems along with their interactions. The Rio de Janeiro world summit held in 1992 highlighted importance of biodiversity as the core subject resulting in the Convention on Biological Diversity (CBD). The CBD aims at "conservation of biological diversity, sustainable uses of its components and fair and equitable sharing of the benefits from the use of genetic resources". Human interference (anthropogenic) impacts resulted in gigantic biodiversity loss including ecosystem disturbance, global warming, sea level increase, habitat destruction and loss of species genetic diversity (Aaaparna Banerjee and Rajib Bandopadhyay, 2017). They described a "biodiversity hotspot" as a biogeographic stretch which should be both an important pool of biodiversity and also in danger of extinction. Species richness, endemism and threat to extinction are the three factors determining a hot spot. There are a total of 34 biodiversity hotspots in 10 different zones around the world.

Singh *et al* (2017) elaborate biodiversity of underutilized fruits of hot arid region. In India, arid zones occupy 38. 70 million ha area in the states of Rajasthan, Gujarat, Haryana, Punjab, Karnataka, Andhra Pradesh and Telengana. Arid zone is characterized by high and low temperature and erratic rainfall limiting high crops productivity. Arid fruits like date palm, mulberry, phalsa, ber, aonla, bael, kinnow, lasoda and gooseberry are traditional fruit crops of the zone. Existence of biodiversity in these fruit crops makes it possible to select genotypes suited to locations. With advent of post harvest technology and products development, these arid fruits are now available throughout India and import bills getting reduced to the best interest of farmers of India. In contrast to arid fruits, pear is a temperate fruit with reported origin in China and Asia minor. It is the fifth most widely produced fruit in the world, being grown on large scale in China, Europe and the USA. Being a close "cousin" of apple, pear possesses a delicate flavour (Amit Kumar *et al.*, 2017). Protein hunger leading to stunted growth and wasting among children is very prevalent in India and consumption of pulses is encouraged. Being low in indigenous production and imported, cost/kg is unaffordable even to middle income groups. Cluster bean (guar) grown in arid and semi arid areas in South Asian continents meets two purposes-industrial gum and immature pod vegetable-rich in vit. c, vit. K, Vit. A, dietary fibre, folate, iron, manganese and potassium. Biodiversity of cluster bean is elaborated by Vikas Kumar *et al.* (2017). Kale is a leafy vegetable similar to non heading cabbage or collard. A native of Mediterranean or of Asia minor, it belongs to cole group of vegetables cultivated in Jammu and Kashmir and on a minor scale in Himachal Pradesh. Under protected cultivation, kale is grown in peri-urban areas near markets. Very rich in carotenoids, kale is an essential component of nutritional security. Being free from major pests and diseases, use of plant protection chemicals is negligible (Chander Prakash *et al.*, 2017). Onion is a spice cum vegetable and an integral part of Indian dietry. The word onion takes its name from the city built by Onias (B. C. 173) near the Gulf of Suez in Egypt. Onion was domesticated a very long time and is one of the earliest cultivated plants. Hybrid onion making use of mail sterility is preferred by farmers for higher productivity, long storage life and resistance to abiotic and biotic stresses. There are all sorts of onion-red, yellow, white, pink and extra large bulb onion. ICAR-Directorate of Onion and Garlic Research, Pune, Nasik posseses the largest germplasm collection in India (Vaddoria and Ganesh Kulkarni, 2017). Among gourds, ridge gourd has an All India likeness as a vegetable and the dried pod as dry flower and body cleanser. It is grown in Sunderbans in West Bengal and in all the districts of Kerala, Tamil Nadu and Karnataka. There are both monoecious and hermaphrodite forms, fruit length from a few centimetres to more than one meter. Tamil Nadu Agricultural University, Coimbatore has released several ridge gourd varieties popular among growers (Singh and Sarkar, 2017). Large cardamom grown in Sikkim, Darjeeling *etc.* (India), Nepal and Bhutan is the oldest of spices used by man. The genus *Ammomum* to which large cardamom belongs is the second largest genus of the family Zingiberaceae with 150 species. Large cardamom has several medicinal properties and used in Persian food delicacies. It is also used to flavour residential habitats during winter (Sreekrishna Bhatt, 2017). With improvement in living standards and quality of life, use and consumption of confectioneries –candy or pastry-are going up. Cacao (cocoa) based confectionaries are more popular among children. Multinational companies like Cadbury and Nestle are promoting research, extension and products development in cocoa. The Kerala Agricultural University-Cadbury Research project is a success story in private-public research collaboration resulting in formulation of package of practices for growing cacao, development of varieties and hybrids and development of many products within the reach of common man. West African countries like Cote d'Ivoire, Ghana, Nigeria and Cameroon possess rich germplasm and their economy depend on export of cocoa products. The fruit of cocoa is a pod carrying several beans which on fermentation give the much valued kernel, the raw material for confectionary manufacture. In India cocoa was imported from Ghana and Nigeria and now cultivated in the states of Kerala, Karnataka, Tamil Nadu and North Eastern States especially Tripura. Squirrels are major pests of the plantation crop and traps are used to catch the pests. Biodiversity in cocoa is maintained by *in situ* conservation and descriptors are available (Prasanna Kumari Amma *et al.,* 2017). Edible oil industry is faced with shortage of seeds-mustard, sunflower, rape seed, soybean-, low productivity-coconut- and reported high content of cholesterol. Palm oil grown in Malaysia and Indonesia yields oils above 5 tones/ha and is adopted to soil and climatic conditions of Tropical Kerala, Tamil Nadu, Telengana, Andhra Pradesh, Odisha and West Bengal. The Pedavegi (Andhra Pradesh) based ICAR-Indian Institute of Oil Palm Research and its regional station, Palode Thiruvanandapuram maintain a working germplasm of African Oil Palm. Descriptors on African oil palm are available (Murugesan and Aswathy, 2017). Genomics is an emerging science, a part of molecular biology and genetic engineering applied for crop improvement and quality improvement of products. Babu and Mathur (2017) at ICAR-Indian Institute of Oil Palm Research, Pedavegi narrate biodiversity and genomics of oil palm.

The Protection of Plant Varieties and Farmers Rights Authority (PPV and FR), New Delhi has notified Distinctiveness, Uniformity and Stability (DUS) guidelines for several crops including oil palm, coconut and cacao. The Chennai based National Biodiversity Authority (NBA) and its arms like UNEP, GEF, MOEF and CC ABS Project, Biodiversity Finance Initiative (BIOFIN) under the GOI-UNDP, Centre for Biodiversity Policy and Law (CEBPOL) and State Biodiversity Boards are mandated with conservation, possession, valuation, benefit sharing and identification of genome saviours. The ICAR-National Bureau of Plant Genetic Resources and its Regional Stations are depositories of plant germplasm. The ICAR-Crop based institutes also maintain working germplasm. The Botanical Survey of India (BSI), State Agricultural Universities, selected CSIR Institutes and Life Science institutes of Defence Research and Development Organization (DRDO) are also involved in survey, collection, evaluation, description, maintenance and conservation of plant biodiversity.

K.V. Peter

Preamble 1

About the Biological Diversity Act-2002 and National Biodiversity Authority-2003**

The most unique feature of earth is the existence of life and the most extraordinary feature of life is its diversity. Approximately 9 million types of plants, animals and fungi inhabit the Earth along with 7 billions of people. In 1992, at the first Earth Summit in Rio de Janeiro, Brazil, the vast majority of the world's nations declared that human actions were dismantling the Earth's ecosystems, eliminating genes, species and biological traits at an alarming rate.

Increasing concern about dwindling biological resources led to the establishment and adoption of the Convention on Biological Diversity (CBD) in 1992.This international convention for the first time recognised sovereign rights of the nations over their biological resources and emphasized that access to genetic resources should be only for environmentally sound purposes and should be subject to national legislations. India is one of the mega biodiversity countries and is also the signatory to the CBD.

In India,measures for conservation and sustainable uses of biodiversity did not start with CBD. India has a long history of conservation and sustainable use of natural resources. Environment protection is enshrined in our constitution (Article 48 A and 51A (g).Over a period of time, a stable organizational structure has been developed for protection of the environment. Numerous wide ranging policies, programmes and projects are in place serving to protect, conserve and regulate sustainable use of the biological resources of the country.

The Biological Diversity Act-2002 was enacted to adopt the objectives enshrined in the United Nation's Convention on Biological Diversity (CBD). The National Biodiversity Authority (NBA) was established in 2003 to implement India's Biodiversity Act-2002.

The Biological Diversity Act-2002 in a Nutshell

★ To regulate access to biological resources (BR) of the country with the purpose of securing equitable share in benefits arising out of the use of Biological Resources (BR's) and associated knowledge relating to BR's.;

★ To conserve and sustainably use the biological diversity;

★ To respect and protect knowledge of local communities related to biodiversity;

★ To secure fair and equitable sharing of benefits arising out of the use of biological resources, associated knowledge relating to biological resources and share the benefits with local communities who are conservers of biological resources or holders of knowledge;

★ Conservation and development of areas of importance from the stand point of biological diversity by declaring them as Biodiversity Heritage Sites (BHS);

★ Protection and rehabilitation of threatened species;

★ Involvement of institutions of State Governments in the broad scheme of the implementation of the Biological Diversity Act through constitution of Biodiversity Management Committees (BMC's);

★ Measures to conserve and sustainable use of biological resources, including habitat and species protection, environmental impact assessments (EIAs) of projects, integration of biodiversity into the plans and policies of various departments/sectors;

★ Provisions for local communities to have say in the use of their resources and knowledge, and to charge fees for this;

★ Setting up of National, State and Local Biodiversity Funds to support conservation and benefit sharing (BS).

About National Biodiversity Authority (NBA)

The National Biodiversity Authority was established in 2003 to implement India's Biological Diversity Act-2002 and has its headquarters in Chennai, Tamil Nadu, India. The NBA is a statutory and autonomous body that performs facilitative, regulatory and advisory functions on issues of biodiversity. The NBA delivers its mandates through a structure that comprises of Authority, State Biodiversity Boards (SBBs) at the state level, Biodiversity Management Committees (BMC's) at local level and various expert committees. The main functions of NBA include the following:

★ Advise the Government of India on matters relating to conservation of biodiversity, sustainable use of its components and fair and equitable sharing of benefits arising out of utilization of BRs

★ Regulate activities and issue guidelines for access to biological resources and for fair and equitable sharing in accordance with sections 3, 4 and 6 of the Biological Diversity Act-2002.

☆ Take necessary measures to oppose the grant of intellectual property rights in any country outside India on any BR obtained from India or knowledge associated with such BR derived from India illegally.

☆ Advise the state governments in selection of areas of biodiversity importance to be notified as heritage sites and suggest measures for their management.

☆ Along with SBBs provide guidance and technical support to BMCs for documenting People's Biodiversity Registers.

Milestones

☆ The NBA has processed over 1500 application forms submitted by the users of biological resources and associated traditional knowledge.

☆ About 900 applications have been cleared by the Authority, out of which,over 400 approvals were transformed into ABS agreements on Mutually Agreed Terms (MAT).

☆ An amount of Rs 360 million of Benefit Sharing has been released from the ABS agreements and the NBA is in the process of disbursing this amount to the local community or Benefit Claimers from where the biological resources were accessed till 2017.

☆ All 29 states in India have established the State Biodiversity Boards (SBBs) and 23 states have framed State Biological Diversity Rules.

☆ Sixteen states and two union territories have notified a list of species which are on the verge of extinction or likely become extinct in the near future as a threatened species under the Section 38 of the Biological Diversity Act 2002.

☆ Fifteen institutions have been designated as national repositories to receive and store voucher specimen and new taxon by the researchers.

☆ Facilitated the constitution of 41180 BMCs at the local level in 27 states.

☆ The NBA provided financial assistance of Rs 16 million for the documentation of Peoples Biodiversity Registers (PBRs) during 2015-16.A total of 2889 PBRs have been documented.

The NBA serves as the launch pad for several international projects and initiatives on biodiversity policy implementation and governance like United Nations Environment Project (UNEP),Biodiversity Finance Initiative (BIOFIN) under the GOI-UNDP, United Nations Environment Programme-Division of Environmental Law and Conventions (UNEP/DELC),United Nations University –Institute for the Advanced Study of Sustainability (UNU-IAS) and Global Environment Fund (GEF).

The ICAR-NBPGR has now developed an online Germplasm Registration Information System (GRIS) to make the entire process of germplasm registration from submission of application to evaluation by experts and decision by Plant

Germplasm Registration Committee (PGRC) simple and fast. The link for submission of proposal is given:

भाकृअप – राष्ट्रीय पादप आनुवंशिक संसाधन ब्यूरो
ICAR – National Bureau of Plant Genetic Resources
A nodal organization in India for the management of plant genetic resources
(An ISO 9001:2008 Certified Institute)
NBPGR
ICAR

जननद्रव्य पंजीकरण सूचना प्रणाली
Germplasm Registration Information System (GRIS)

Home   About Us   Announcements/Proceedings   Guidelines   Help   Contact Us

- Enter your User name and Password in the given text boxes
- Click on "Login" Button to log in to the system
- If you are a new user then Click "Sign up" and fill the registration form
- If you forgot your User name and Password, you can click Forgot User name or Password link to reset your Password
- "Remember me on this computer" box will allow your computer to remember your User name and Password if you don't wish to enter it again

User Login

User name [   ]
Password [   ]

New user? Sign up   Forgot User name or Password?
☐ Remember me on this Computer

SUBMIT   RESET

Note
For any queries, please contact us at:
Email: nbpgr.registration[AT]icar.gov.in
Phone: 91-11-25846268

Developed in ICAR National Fellow Project
Copyright © 2017 All Rights Reserved, ICAR-National Bureau of Plant Genetic Resources,
Indian Council of Agricultural Research, Ministry of Agriculture and Farmers Welfare (Govt. of India), Pusa Campus, New Delhi-110012, INDIA.

*** Source: NBA 2017.Biodiversity matters Newsletter.Vol 1 (01):1-8*

Preamble 2

Delhi Declaration on Agrobiodiversity Management*

The first International Agrobiodiversity Congress held in New Delhi, India, from 6 to 9 November, 2016 was attended by over 900 participants from 60 countries. Delegates discussed various aspects of conservation, management, access and use of agrobiodiversity in 16 technical sessions, four satellite sessions, a gene bank roundtable, a public forum, a farmers' forum and poster sessions. Based on detailed deliberations, the delegates unanimously adopted the following declaration in the concluding session on 9 November 2016.

1. We call upon nations to accord top priority to the shared vision of agrobiodiversity conservation and sustainable use towards achieving the Sustainable Development Goals (SDGs) and the Aichi Targets of the Convention on Biological Diversity addressing poverty alleviation, food, nutritional and health security, gender equity and global partnership.

2. We recognize the importance of traditional agrobiodiversity knowledge available with farm men and women, pastoralists, tribal and rural communities and its central role in the conservation and use for a food secure and climate resilient world. We, therefore, call upon countries to develop the necessary legal, institutional and funding mechanisms to catalyse their active participation.

3. We urge researchers and policymakers to initiate, strengthen and promote complementary strategies to conserve agrobiodiversity through use, including greater emphasis on using crop wild relatives. We call upon them to ensure a continuum between *ex situ, in situ,* on-farm, community-based and other conservation methods with much greater and equal emphasis on each.

4. We propose that researchers employ modern technologies including, but not limited to, genomics, biotechnology, space, computational, and nanotechnologies for genetic resources characterization, evaluation and trait discovery. The aim must be to achieve efficiency, equity, economy and environmental security through diversified agricultural production systems and landscapes.

5. We re-emphasize the necessity of global exchange of plant, animal, aquatic, microbial and insect genetic resources to diversify agriculture as well as our food basket and to meet the ever-growing food and nutritional needs of all countries. To ensure this, nations need to be catalysed to adopt both multi-lateral (as envisaged in the International Treaty on Plant Genetic Resources for Food and Agriculture) and bilateral (as per the Nagoya Protocol) instruments to facilitate the exchange of genetic resources, while ensuring equitable access and benefit sharing opportunities.

6. Countries are also expected to harmonize their existing biosecurity systems, including phytosanitary and quarantine, and enhance their capacities to facilitate safe trans-boundary movement of germplasm.

7. We also expect that the governments and civil societies lay much greater emphasis on public awareness and capacity enhancement programmes on agrobiodiversity conservation in order to accelerate its effective and efficient use.

8. We recommend the development and implementation of an Agrobiodiversity Index to help monitor on-going genetic resource conservation and management efforts, with particular emphasis on agrobiodiversity hot spots.

9. It is also urged that public and private sectors and civil societies henceforth actively invest in and incentivize the utilization of agrobiodiversity to mitigate malnutrition, increase the resilience and productivity of farms and farming households and enhance ecosystem services. Such efforts should lead to equitable benefits and opportunities, with particular emphasis on women and youth.

10. We urge countries to reprioritize their research and extension with increased investments to support the conservation and use of agrobiodiversity. Furthermore, we strongly recommend creation of an International Agrobiodiversity Fund as a mechanism to assist countries and communities in scientific *in situ* and *ex situ* conservation and enhanced use of agrobiodiversity.

11. We urge the United Nations to consider declaring a 'Year of Agrobiodiversity' in order to draw worldwide attention and catalyse urgent actions for effective management of genetic resources by the global community.

12. Finally, we recommend that the International Agrobiodiversity Congress be held every four years, with Biodiversity International playing the facilitator's role, to maintain the momentum gained in 2016 and continue emphasizing the need to implement the 'Delhi Declaration on Agrobiodiversity Management' and monitor the progress so made by the different stakeholders and countries.

* *Source: Current Science 112 (1): 10 January, 2017 p.14*

Chapter 1

Convention on Biological Diversity vis-à-vis Biological Diversity Act of India

N. Sivaraj, S.R. Pandravada and V. Kamala

*ICAR-National Bureau of Plant Genetic Resources, Regional Station,
Hyderabad – 500 030, Telengana
E-mail: sivarajn@gmail.com*

Biological diversity or biodiversity is the term attributed to the variety of life on Earth and the natural patterns it forms. The biodiversity, what we see today is the fruit of billions of years of evolution, shaped by natural processes and increasingly, by the influence of humans. It forms the web of life of which we are an integral part and upon which we so fully depend upon. So far, about 1.75 million species have been identified, mostly small creatures such as insects and Scientists reckon that there are actually about 13 million species, though estimates range from 3 to 100 million (Angelo Cropper, 1993). Biodiversity also includes genetic differences within each species - for example, between varieties of crops and breeds of livestock. Yet another aspect of biodiversity is the variety of ecosystems such as those that occur in deserts, forests, wetlands, mountains, lakes, rivers and agricultural landscapes.

"Earth provides enough to satisfy every man's needs, but not every man's greed"

– Mahatma Gandhi

"Biodiversity is what we need; conserving biodiversity is a good deed"

– Shreemayi, School student, New Delhi

Biodiversity is crucial to the survival of all ecosystems, life forms and micro-organisms on our earth (Tiwari, 2006). It is an interdependent, closely interlinked life system that binds all life on planet together in a delicate and intricate web where each strand plays a crucial role and depends upon the other for survival. In each ecosystem, living creatures, including humans, form a community, interacting with one another and with the air, water, and soil around them. It is the combination of life forms and their interactions with each other and with the rest of the environment that has made Earth a uniquely habitable place for humans. Biodiversity provides a large number of goods and services that sustain our lives.

The Earth's biological resources are vital to humanity's economic and social development. As a result, there is a growing recognition that biological diversity is a global asset of tremendous value to present and future generations (FA0, 1995). At the same time, the threat to species and ecosystems has never been as great as it is today. Species extinction caused by human activities continues at an alarming rate. The Convention on Biological Diversity is one of the three 'Rio Conventions', emerging from the UN Conference on Environment and Development, also known as the Earth Summit, held in Rio de Janeiro in 1992. The Convention on Biological Diversity (CBD) opened for signature at the Earth Summit in Rio de Janeiro in 1992 for entering into force from December 1993. The Convention on Biological Diversity is an international treaty for the conservation of biodiversity, the sustainable use of its components and the fair and equitable sharing of the benefits arising out of the utilization of genetic resources, including by appropriate access to genetic resources and by appropriate transfer of relevant technologies, taking into account all rights over those resources and to technologies and by appropriate funding (Varaprasad and Sivaraj, 2010). It represents a dramatic step forward in the conservation of biological diversity, the sustainable use of its components and the fair and equitable sharing of benefits arising from the use of genetic resources. The Convention on Biological Diversity was inspired by the world community's growing commitment to sustainable development. With 196 Parties up to now, the Convention has near universal participation among the countries (Anonymous, 2000).

The CBD covers biodiversity at all levels: Ecosystem, species and genetic levels. It also covers biotechnology through the Cartagena Protocol on Biosafety. In fact, it covers all possible domains that are directly or indirectly related to biodiversity and its role in development, ranging from science, politics and education to agriculture, business, culture and much more. The governing body of the CBD is the Conference of the Parties (COP). This ultimate authority of all governments (or Parties) that have ratified the treaty meets every two years to review progress, set priorities and commit to work plans. The Convention seeks to address all threats to biodiversity and ecosystem services, including threats from climate change, through scientific assessments, the development of tools, incentives and processes, the transfer of technologies and good practices and the full and active involvement of relevant stakeholders including indigenous and local communities, youth, NGOs, women and the business community. The Cartagena Protocol on Biosafety and the Nagoya Protocol on Access and Benefit Sharing are supplementary agreements to the Convention. The Cartagena Protocol, which entered into force on 11 September, 2003, seeks to protect biological diversity from the potential risks posed by living modified organisms resulting from modern biotechnology and to date, 170 Parties have ratified the Cartagena Protocol. The Nagoya Protocol aims at sharing the benefits arising from the utilization of genetic resources in a fair and equitable way, including by appropriate access to genetic resources and by appropriate transfer of relevant technologies. It entered into force on 12 October 2014 and to date has been ratified by 75 Parties.

History and Decades Journey of the Convention on Biological Diversity

In response, the United Nations Environment Programme (UNEP) convened the Ad Hoc Working Group of Experts on Biological Diversity in November 1988 to explore the need for an international convention on biological diversity. Soon after, in May 1989, it established the Ad-Hoc Working Group of Technical and Legal Experts to prepare an international legal instrument for the conservation and sustainable use of biological diversity. The experts were to take into account "the need to share costs and benefits between developed and developing countries" as well as "ways and means to support innovation by local people".

By February 1991, the Ad Hoc Working Group had become known as the Intergovernmental Negotiating Committee. Its work culminated on 22 May 1992 with the Nairobi Conference for the Adoption of the Agreed Text of the Convention on Biological Diversity.

The Convention was opened for signature on 5 June, 1992 at the United Nations Conference on Environment and Development (the Rio "Earth Summit"). It remained open for signature until 4 June, 1993, by which time it had received 168 signatures. The Convention entered into force on 29 December, 1993, which was 90 days after the 30th ratification. The first session of the Conference of the Parties was scheduled from 28 November - 9 December, 1994 in the Bahamas. Two decades of journey of the CBD is provided in Table 1.1.

The adoption in 2010, the International Year of Biodiversity, of a new protocol on Access and Benefit Sharing (ABS) to promote international co-operation with regard to objective (iii), will open a new era for international co-operation on biodiversity policy and practice to demonstrate that conservation and use of biodiversity are necessary conditions for sustainable development.

Table 1.1: Convention on Biological Diversity: Two Decades of Journey

CBD Milestone Activity	Day/Year
Adoption of CBD Text	22 May, 1992
CBD opened for Signatures at UNCED, Rio de Janeiro	5 June, 1992
Interim SCBS at Geneva, GEF financial mechanism of CBD	1992-1993
CBD enters into force	29 December, 1993
Conference of Parties (COP-1) - Bahamas	1994
COP-2 (Indonesia), Marine and Coastal	1995
SCBD at Montreal, CoP-3 (Argentina), Ecosystem, Agriculture, Forest and Traditional knowledge	1996
CoP-4 (Slovakia), Inland Water, Global Taxonomy, Invasive Alien Species, Expert panel on ABS set up	1998
ExCoP-1 in Colombia	1999
Cartagena Protocol on Bio-safety (CPB) adopted. CoP-5 (Kenya). CPB opened for signatures	2000
GBO1 launched	2001

CBD Milestone Activity	Day/Year
CPB enters into force	11 September, 2003
CoP-MoP-1 and CoP-7 (Malaysia). Decision to negotiate ABS taken	2004
CoP-MoP-2 (Canada)	2005
CoP-MoP-3 and CoP-8 (Brazil). GBO2 launched. Decision to complete ABS negotiations by 2010	2006
UN GA proclaim 2010 as the International Year of Biodiversity	20 December 2007
CoP-MoP-4 and CoP-9 (Germany)	2008
CoP-MoP-5 and CoP-10 (Japan). Strategic plan on Biodiversity 2020. Strategy for Resource Mobilization and Nagoya Protocol on ABS adopted. GBO3 launched	2010
UN Decade on Biodiversity proclaimed	2011
Nagoya Protocol on ABS open for signatures	1 February, 2011
Nagoya-Kula Lumpur Supplementary Protocol open for signatures	7 March, 2011
CoP-MoP-6 and CoP-11, Hyderabad, India	1-19 October, 2012
CoP-MoP-7, CoP-12, Nagoya Protocol CoP-MoP-1- Republic of Korea	2014
The Nagoya Protocol on Access to Genetic Resources and the Fair and Equitable Sharing of Benefits Arising from their Utilization to the Convention on Biological Diversity enters into force.	12 October, 2014
CoP-MoP-8, CoP-13, Nagoya Protocol CoP-MoP-2 [Cancun, Mexico (confirmed to be held in December 2016)]	2016

Preamble and Objectives of the CBD

CBD re-affirms in its preamble, the sovereign rights of the States (countries) over their own biological resources (that earlier were the common heritage of humankind).

The Convention calls on the parties to establish conditions and frame appropriate national systems to protect as well as facilitate access to genetic resources incorporating the instruments of "Mutually Agreed Terms" and "Prior Informed Consent". Articles of CBD can be fruitfully used to promote international exchange and trade in genetic resources and also to preserve global diversity (Putterman, 1994).

Article 1 Objectives

The Convention on Biological Diversity (CBD) promotes collaboration for attaining three important global objectives: (i) conservation of biodiversity; (ii) sustainable use of its components; and (iii) equitable sharing of benefits arising out of the access and utilization of genetic resources and transfer of relevant technologies, taking into accounts all rights over these resources and to technologies, and by appropriate funding.

Article 2 Use of Terms

It describes the use of terms for the purposes of this Convention:

"Biological diversity" means the variability among living organisms from all sources including, inter alia, terrestrial, marine and other aquatic ecosystems and the ecological complexes of which they are part; this includes diversity within species, between species and of ecosystems.

"Biological resources" includes genetic resources, organisms or parts thereof, populations, or any other biotic component of ecosystems with actual or potential use or value for humanity.

"Biotechnology" means any technological application that uses biological systems, living organisms, or derivatives thereof, to make or modify products or processes for specific use.

"Country of origin of genetic resources" means the country which possesses those genetic resources in *in-situ* conditions.

"Country providing genetic resources" means the country supplying genetic resources collected from *in-situ* sources, including populations of both wild and domesticated species, or taken from *ex-situ* sources, which may or may not have originated in that country.

"Domesticated or cultivated species" means species in which the evolutionary process has been influenced by humans to meet their needs.

"Ecosystem" means a dynamic complex of plant, animal and micro-organism communities and their non-living environment interacting as a functional unit.

"Ex-situ conservation" means the conservation of components of biological diversity outside their natural habitats.

"Genetic material" means any material of plant, animal, microbial or other origin containing functional units of heredity.

"Genetic resources" means genetic material of actual or potential value.

"Habitat" means the place or type of site where an organism or population naturally occurs.

"In-situ conditions" means conditions where genetic resources exist within ecosystems and natural habitats, and, in the case of domesticated or cultivated species, in the surroundings where they have developed their distinctive properties.

"In-situ conservation" means the conservation of ecosystems and natural habitats and the maintenance and recovery of viable populations of species in their natural surroundings and, in the case of domesticated or cultivated species, in the surroundings where they have developed their distinctive properties.

"Protected area" means a geographically defined area which is designated or regulated and managed to achieve specific conservation objectives.

"Regional economic integration organization" means an organization constituted by sovereign States of a given region, to which its member States have transferred competence in respect of matters governed by this Convention and which has been duly authorized, in accordance with its internal procedures, to sign, ratify, accept, approve or accede to it.

"Sustainable use" means the use of components of biological diversity in a way and at a rate that does not lead to the long-term decline of biological diversity, thereby maintaining its potential to meet the needs and aspirations of present and future generations.

"Technology" includes biotechnology.

Article 3 Principle

It states that States have, in accordance with the Charter of the United Nations and the principles of international law, the sovereign right to exploit their own resources pursuant to their own environmental policies, and the responsibility to ensure that activities within their jurisdiction or control do not cause damage to the environment of other States or of areas beyond the limits of national jurisdiction.

Article 4 Jurisdictional Scope

Subject to the rights of other States, and except as otherwise expressly provided in this Convention, the provisions of this Convention apply, in relation to each Contracting Party:

a) In the case of components of biological diversity, in areas within the limits of its national jurisdiction; and

b) In the case of processes and activities, regardless of where their effects occur, carried out under its jurisdiction or control, within the area of its national jurisdiction or beyond the limits of national jurisdiction.

Article 5 Cooperation

Each Contracting Party shall, as far as possible and as appropriate, cooperate with other Contracting Parties, directly or, where appropriate, through competent international organizations, in respect of areas beyond national jurisdiction and on other matters of mutual interest, for the conservation and sustainable use of biological diversity.

Article 6 General Measures for Conservation and Sustainable Use

Each Contracting Party shall, in accordance with its particular conditions and capabilities:

a) Develop national strategies, plans or programmes for the conservation and sustainable use of biological diversity or adapt for this purpose existing strategies, plans or programmes which shall reflect, inter alia, the measures set out in this Convention relevant to the Contracting Party concerned; and

b) Integrate, as far as possible and as appropriate, the conservation and sustainable use of biological diversity into relevant sectoral or cross-sectoral plans, programmes and policies.

Article 7 Identification and Monitoring

Each Contracting Party shall, as far as possible and as appropriate, in particular for the purposes of Articles 8 to 10:

a) Identify components of biological diversity important for its conservation and sustainable use having regard to the indicative list of categories set down in Annex I;

b) Monitor, through sampling and other techniques, the components of biological diversity identified pursuant to subparagraph (a) above, paying particular attention to those requiring urgent conservation measures and those which offer the greatest potential for sustainable use;

c) Identify processes and categories of activities which have or are likely to have significant adverse impacts on the conservation and sustainable use of biological diversity, and monitor their effects through sampling and other techniques; and

d) Maintain and organize, by any mechanism data, derived from identification and monitoring activities pursuant to subparagraphs (a), (b) and (c) above.

Article 8 *In-situ* Conservation

Each Contracting Party shall, as far as possible and as appropriate:

a) Establish a system of protected areas or areas where special measures need to be taken to conserve biological diversity;

b) Develop, where necessary, guidelines for the selection, establishment and management of protected areas or areas where special measures need to be taken to conserve biological diversity;

c) Regulate or manage biological resources important for the conservation of biological diversity whether within or outside protected areas, with a view to ensuring their conservation and sustainable use;

d) Promote the protection of ecosystems, natural habitats and the maintenance of viable populations of species in natural surroundings;

e) Promote environmentally sound and sustainable development in areas adjacent to protected areas with a view to furthering protection of these areas;

f) Rehabilitate and restore degraded ecosystems and promote the recovery of threatened species, inter alia, through the development and implementation of plans or other management strategies;

g) Establish or maintain means to regulate, manage or control the risks associated with the use and release of living modified organisms resulting from biotechnology which are likely to have adverse environmental impacts that could affect the conservation and sustainable use of biological diversity, taking also into account the risks to human health;

h) Prevent the introduction of, control or eradicate those alien species which threaten ecosystems, habitats or species;

i) Endeavour to provide the conditions needed for compatibility between present uses and the conservation of biological diversity and the sustainable use of its components;

j) Subject to its national legislation, respect, preserve and maintain knowledge, innovations and practices of indigenous and local communities embodying traditional lifestyles relevant for the conservation and sustainable use of biological diversity and promote their wider application with the approval and involvement of the holders of such knowledge, innovations and practices and encourage the equitable sharing of the benefits arising from the utilization of such knowledge, innovations and practices;

k) Develop or maintain necessary legislation and/or other regulatory provisions for the protection of threatened species and populations;

l) Where a significant adverse effect on biological diversity has been determined pursuant to Article 7, regulate or manage the relevant processes and categories of activities; and

m) Cooperate in providing financial and other support for in-situ conservation outlined in subparagraphs (a) to (l) above, particularly to developing countries

Article 9 *Ex-situ* Conservation

Each Contracting Party shall, as far as possible and as appropriate, and predominantly for the purpose of complementing *in-situ* measures:

a) Adopt measures for the *ex-situ* conservation of components of biological diversity, preferably in the country of origin of such components;

b) Establish and maintain facilities for ex-situ conservation of and research on plants, animals and micro- organisms, preferably in the country of origin of genetic resources;

c) Adopt measures for the recovery and rehabilitation of threatened species and for their reintroduction into their natural habitats under appropriate conditions;

d) Regulate and manage collection of biological resources from natural habitats for ex-situ conservation purposes so as not to threaten ecosystems and in-situ populations of species, except where special temporary ex-situ measures are required under subparagraph (c) above; and

e) Cooperate in providing financial and other support for ex-situ conservation outlined in subparagraphs (a) to (d) above and in the establishment and maintenance of ex- situ conservation facilities in developing countries.

Article 10 Sustainable Use of Components of Biological Diversity

Each Contracting Party shall, as far as possible and as appropriate:

a) Integrate consideration of the conservation and sustainable use of biological resources into national decision-making;

b) Adopt measures relating to the use of biological resources to avoid or minimize adverse impacts on biological diversity;

c) Protect and encourage customary use of biological resources in accordance with traditional cultural practices that are compatible with conservation or sustainable use requirements;

d) Support local populations to develop and implement remedial action in degraded areas where biological diversity has been reduced; and

e) Encourage cooperation between its governmental authorities and its private sector in developing methods for sustainable use of biological resources.

Article 11 Incentive Measures

Each Contracting Party shall, as far as possible and as appropriate, adopt economically and socially sound measures that act as incentives for the conservation and sustainable use of components of biological diversity.

Article 12 Research and Training

The Contracting Parties, taking into account the special needs of developing countries, shall:

a) Establish and maintain programmes for scientific and technical education and training in measures for the identification, conservation and sustainable use of biological diversity and its components and provide support for such education and training for the specific needs of developing countries;

b) Promote and encourage research which contributes to the conservation and sustainable use of biological diversity, particularly in developing countries, inter alia, in accordance with decisions of the Conference of the Parties taken in consequence of recommendations of the Subsidiary Body on Scientific, Technical and Technological Advice; and

c) In keeping with the provisions of Articles 16, 18 and 20, promote and cooperate in the use of scientific advances in biological diversity research in developing methods for conservation and sustainable use of biological resources.

Article 13 Public Education and Awareness

The Contracting Parties shall:

a) Promote and encourage understanding of the importance of, and the measures required for, the conservation of biological diversity, as well as its propagation through media, and the inclusion of these topics in educational programmes; and

b) Cooperate, as appropriate, with other States and international organizations in developing educational and public awareness programmes, with respect to conservation and sustainable use of biological diversity.

Article 14 Impact Assessment and Minimizing Adverse Impacts

1. Each Contracting Party, as far as possible and as appropriate, shall:

   a) Introduce appropriate procedures requiring environmental impact assessment of its proposed projects that are likely to have significant adverse effects on biological diversity with a view to avoiding or minimizing such effects and, where appropriate, allow for public participation in such procedures;

   b) Introduce appropriate arrangements to ensure that the environmental consequences of its programmes and policies that are likely to have significant adverse impacts on biological diversity are duly taken into account;

   c) Promote, on the basis of reciprocity, notification, exchange of information and consultation on activities under their jurisdiction or control which are likely to significantly affect adversely the biological diversity of other States or areas beyond the limits of national jurisdiction, by encouraging the conclusion of bilateral, regional or multilateral arrangements, as appropriate;

   d) In the case of imminent or grave danger or damage, originating under its jurisdiction or control, to biological diversity within the area under jurisdiction of other States or in areas beyond the limits of national jurisdiction, notify immediately the potentially affected States of such danger or damage, as well as initiate action to prevent or minimize such danger or damage; and

   e) Promote national arrangements for emergency responses to activities or events, whether caused naturally or otherwise, which present a grave and imminent danger to biological diversity and encourage international cooperation to supplement such national efforts and, where appropriate and agreed by the States or regional economic integration organizations concerned, to establish joint contingency plans.

2. The Conference of the Parties shall examine, on the basis of studies to be carried out, the issue of liability and redress, including restoration and compensation, for damage to biological diversity, except where such liability is a purely internal matter.

Article 15 Access to Genetic Resources

Recognizing the sovereign rights of States over their natural resources, the authority to determine access to genetic resources rests with the national governments and is subject to national legislation.

1. Each Contracting Party shall endeavour to create conditions to facilitate access to genetic resources for environmentally sound uses by other Contracting Parties and not to impose restrictions that run counter to the objectives of this Convention.

2. For the purpose of this Convention, the genetic resources being provided by a Contracting Party, as referred to in this Article and Articles 16 and 19 are only those that are provided by Contracting Parties that are countries of origin of such resources or by the Parties that have acquired the genetic resources in accordance with this Convention.

3. Access, where granted, shall be on mutually agreed terms and subject to the provisions of this Article.

4. Access to genetic resources shall be subject to prior informed consent of the Contracting Party providing such resources, unless otherwise determined by that Party.

5. Each Contracting Party shall endeavour to develop and carry out scientific research based on genetic resources provided by other Contracting Parties with the full participation of, and where possible in, such Contracting Parties.

6. Each Contracting Party shall take legislative, administrative or policy measures, as appropriate, and in accordance with Articles 16 and 19 and, where necessary, through the financial mechanism established by Articles 20 and 21 with the aim of sharing in a fair and equitable way the results of research and development and the benefits arising from the commercial and other utilization of genetic resources with the Contracting Party providing such resources. Such sharing shall be upon mutually agreed terms.

Article 16 Access to and Transfer of Technology

1. Each Contracting Party, recognizing that technology includes biotechnology, and that both access to and transfer of technology among Contracting Parties are essential elements for the attainment of the objectives of this Convention, undertakes subject to the provisions of this Article to provide and/or facilitate access for and transfer to other Contracting Parties of technologies that are relevant to the conservation and sustainable use of biological diversity or make use of genetic resources and do not cause significant damage to the environment.

2. Access to and transfer of technology referred to in paragraph 1 above to developing countries shall be provided and/or facilitated under fair and most favourable terms, including on concessional and preferential terms where mutually agreed, and, where necessary, in accordance with the financial mechanism established by Articles 20 and 21. In the case of technology subject to patents and other intellectual property rights, such access and transfer shall be provided on terms which recognize and are consistent with the adequate and effective protection of intellectual property rights. The application of this paragraph shall be consistent with paragraphs 3, 4 and 5 below.

3. Each Contracting Party shall take legislative, administrative or policy measures, as appropriate, with the aim that Contracting Parties, in

particular those that are developing countries, which provide genetic resources are provided access to and transfer of technology which makes use of those resources, on mutually agreed terms, including technology protected by patents and other intellectual property rights, where necessary, through the provisions of Articles 20 and 21 and in accordance with international law and consistent with paragraphs 4 and 5 below.

4. Each Contracting Party shall take legislative, administrative or policy measures, as appropriate, with the aim that the private sector facilitates access to, joint development and transfer of technology referred to in paragraph 1 above for the benefit of both governmental institutions and the private sector of developing countries and in this regard shall abide by the obligations included in paragraphs 1, 2 and 3 above.

5. The Contracting Parties, recognizing that patents and other intellectual property rights may have an influence on the implementation of this Convention, shall cooperate in this regard subject to national legislation and international law in order to ensure that such rights are supportive of and do not run counter to its objectives

Article 17 Exchange of Information

1. The Contracting Parties shall facilitate the exchange of information, from all publicly available sources, relevant to the conservation and sustainable use of biological diversity, taking into account the special needs of developing countries.

2. Such exchange of information shall include exchange of results of technical, scientific and socio-economic research, as well as information on training and surveying programmes, specialized knowledge, indigenous and traditional knowledge as such and in combination with the technologies referred to in Article 16, paragraph 1. It shall also, where feasible, include repatriation of information.

Article 18 Technical and Scientific Cooperation

1. The Contracting Parties shall promote international technical and scientific cooperation in the field of conservation and sustainable use of biological diversity, where necessary, through the appropriate international and national institutions.

2. Each Contracting Party shall promote technical and scientific cooperation with other Contracting Parties, in particular developing countries, in implementing this Convention, inter alia, through the development and implementation of national policies. In promoting such cooperation, special attention should be given to the development and strengthening of national capabilities, by means of human resources development and institution building.

3. The Conference of the Parties, at its first meeting, shall determine how to establish a clearing-house mechanism to promote and facilitate technical and scientific cooperation.

4. The Contracting Parties shall, in accordance with national legislation and policies, encourage and develop methods of cooperation for the development and use of technologies, including indigenous and traditional technologies, in pursuance of the objectives of this Convention. For this purpose, the Contracting Parties shall also promote cooperation in the training of personnel and exchange of experts.

5. The Contracting Parties shall, subject to mutual agreement, promote the establishment of joint research programmes and joint ventures for the development of technologies relevant to the objectives of this Convention.

Article 19 Handling of Biotechnology and Distribution of its Benefits

1. Each Contracting Party shall take legislative, administrative or policy measures, as appropriate, to provide for the effective participation in biotechnological research activities by those Contracting Parties, especially developing countries, which provide the genetic resources for such research, and where feasible in such Contracting Parties.

2. Each Contracting Party shall take all practicable measures to promote and advance priority access on a fair and equitable basis by Contracting Parties, especially developing countries, to the results and benefits arising from biotechnologies based upon genetic resources provided by those Contracting Parties. Such access shall be on mutually agreed terms.

3. The Parties shall consider the need for and modalities of a protocol setting out appropriate procedures, including, in particular, advance informed agreement, in the field of the safe transfer, handling and use of any living modified organism resulting from biotechnology that may have adverse effect on the conservation and sustainable use of biological diversity.

4. Each Contracting Party shall, directly or by requiring any natural or legal person under its jurisdiction providing the organisms referred to in paragraph 3 above, provide any available information about the use and safety regulations required by that Contracting Party in handling such organisms, as well as any available information on the potential adverse impact of the specific organisms concerned to the Contracting Party into which those organisms are to be introduced.

Article 20 Financial Resources

1. Each Contracting Party undertakes to provide, in accordance with its capabilities, financial support and incentives in respect of those national activities which are intended to achieve the objectives of this Convention, in accordance with its national plans, priorities and programmes.

2. The developed country Parties shall provide new and additional financial resources to enable developing country Parties to meet the agreed full incremental costs to them of implementing measures which fulfil the obligations of this Convention and to benefit from its provisions and which costs are agreed between a developing country Party and the

institutional structure referred to in Article 21, in accordance with policy, strategy, programme priorities and eligibility criteria and an indicative list of incremental costs established by the Conference of the Parties. Other Parties, including countries undergoing the process of transition to a market economy, may voluntarily assume the obligations of the developed country Parties. For the purpose of this Article, the Conference of the Parties, shall at its first meeting establish a list of developed country Parties and other Parties which voluntarily assume the obligations of the developed country Parties. The Conference of the Parties shall periodically review and if necessary amend the list. Contributions from other countries and sources on a voluntary basis would also be encouraged. The implementation of these commitments shall take into account the need for adequacy, predictability and timely flow of funds and the importance of burden-sharing among the contributing Parties included in the list.

3. The developed country Parties may also provide, and developing country Parties avail themselves of, financial resources related to the implementation of this Convention through bilateral, regional and other multilateral channels.

4. The extent to which developing country Parties will effectively implement their commitments under this Convention will depend on the effective implementation by developed country Parties of their commitments under this Convention related to financial resources and transfer of technology and will take fully into account the fact that economic and social development and eradication of poverty are the first and overriding priorities of the developing country Parties.

5. The Parties shall take full account of the specific needs and special situation of least developed countries in their actions with regard to funding and transfer of technology.

6. The Contracting Parties shall also take into consideration the special conditions resulting from the dependence on, distribution and location of, biological diversity within developing country Parties, in particular small island States.

7. Consideration shall also be given to the special situation of developing countries, including those that are most environmentally vulnerable, such as those with arid and semi- arid zones, coastal and mountainous areas.

Article 21 Financial Mechanism

1. There shall be a mechanism for the provision of financial resources to developing country Parties for purposes of this Convention on a grant or concessional basis the essential elements of which are described in this Article. The mechanism shall function under the authority and guidance of, and be accountable to, the Conference of the Parties for purposes of this Convention. The operations of the mechanism shall be carried out by such institutional structure as may be decided upon by the Conference

of the Parties at its first meeting. For purposes of this Convention, the Conference of the Parties shall determine the policy, strategy, programme priorities and eligibility criteria relating to the access to and utilization of such resources. The contributions shall be such as to take into account the need for predictability, adequacy and timely flow of funds referred to in Article 20 in accordance with the amount of resources needed to be decided periodically by the Conference of the Parties and the importance of burden-sharing among the contributing Parties included in the list referred to in Article 20, paragraph 2. Voluntary contributions may also be made by the developed country Parties and by other countries and sources. The mechanism shall operate within a democratic and transparent system of governance.

2. Pursuant to the objectives of this Convention, the Conference of the Parties shall at its first meeting determine the policy, strategy and programme priorities, as well as detailed criteria and guidelines for eligibility for access to and utilization of the financial resources including monitoring and evaluation on a regular basis of such utilization. The Conference of the Parties shall decide on the arrangements to give effect to paragraph 1 above after consultation with the institutional structure entrusted with the operation of the financial mechanism.

3. The Conference of the Parties shall review the effectiveness of the mechanism established under this Article, including the criteria and guidelines referred to in paragraph 2 above, not less than two years after the entry into force of this Convention and thereafter on a regular basis. Based on such review, it shall take appropriate action to improve the effectiveness of the mechanism if necessary.

4. The Contracting Parties shall consider strengthening existing financial institutions to provide financial resources for the conservation and sustainable use of biological diversity

Article 22 Relationship with Other International Conventions

1. The provisions of this Convention shall not affect the rights and obligations of any Contracting Party deriving from any existing international agreement, except where the exercise of those rights and obligations would cause a serious damage or threat to biological diversity.

2. Contracting Parties shall implement this Convention with respect to the marine environment consistently with the rights and obligations of States under the law of the sea.

Article 23 Conference of the Parties

1. A Conference of the Parties is hereby established. The first meeting of the Conference of the Parties shall be convened by the Executive Director of the United Nations Environment Programme not later than one year after the entry into force of this Convention. Thereafter, ordinary meetings

of the Conference of the Parties shall be held at regular intervals to be determined by the Conference at its first meeting.

2. Extraordinary meetings of the Conference of the Parties shall be held at such other times as may be deemed necessary by the Conference, or at the written request of any Party, provided that, within six months of the request being communicated to them by the Secretariat, it is supported by at least one third of the Parties.

3. The Conference of the Parties shall by consensus agree upon and adopt rules of procedure for itself and for any subsidiary body it may establish, as well as financial rules governing the funding of the Secretariat. At each ordinary meeting, it shall adopt a budget for the financial period until the next ordinary meeting.

4. The Conference of the Parties shall keep under review the implementation of this Convention, and, for this purpose, shall:

   a) Establish the form and the intervals for transmitting the information to be submitted in accordance with Article 26 and consider such information as well as reports submitted by any subsidiary body;

   b) Review scientific, technical and technological advice on biological diversity provided in accordance with Article 25;

   c) Consider and adopt, as required, protocols in accordance with Article 28;

   d) Consider and adopt, as required, in accordance with Articles 29 and 30, amendments to this Convention and its annexes;

   e) Consider amendments to any protocol, as well as to any annexes thereto, and, if so decided, recommend their adoption to the parties to the protocol concerned;

   f) Consider and adopt, as required, in accordance with Article 30, additional annexes to this Convention;

   g) Establish such subsidiary bodies, particularly to provide scientific and technical advice, as are deemed necessary for the implementation of this Convention;

   h) Contact, through the Secretariat, the executive bodies of conventions dealing with matters covered by this Convention with a view to establishing appropriate forms of cooperation with them; and

   i) Consider and undertake any additional action that may be required for the achievement of the purposes of this Convention in the light of experience gained in its operation.

5. The United Nations, its specialized agencies and the International Atomic Energy Agency, as well as any State not Party to this Convention, may be represented as observers at meetings of the Conference of the Parties. Any other body or agency, whether governmental or non-governmental, qualified in fields relating to conservation and sustainable use of biological diversity, which has informed the Secretariat of its wish to be

represented as an observer at a meeting of the Conference of the Parties, may be admitted unless at least one third of the Parties present object. The admission and participation of observers shall be subject to the rules of procedure adopted by the Conference of the Parties.

Article 24 Secretariat

1. A secretariat is hereby established. Its functions shall be:

   (a) To arrange for and service meetings of the Conference of the Parties provided for in Article 23;

   (b) To perform the functions assigned to it by any protocol;

   (c) To prepare reports on the execution of its functions under this Convention and present them to the Conference of the Parties;

   (d) To coordinate with other relevant international bodies and, in particular to enter into such administrative and contractual arrangements as may be required for the effective discharge of its functions; and

   (e) To perform such other functions as may be determined by the Conference of the Parties.

2. At its first ordinary meeting, the Conference of the Parties shall designate the secretariat from amongst those existing competent international organizations which have signified their willingness to carry out the secretariat functions under this Convention.

Article 25 Subsidiary Body on Scientific, Technical and Technological Advice

1. A subsidiary body for the provision of scientific, technical and technological advice is hereby established to provide the Conference of the Parties and, as appropriate, its other subsidiary bodies with timely advice relating to the implementation of this Convention. This body shall be open to participation by all Parties and shall be multidisciplinary. It shall comprise government representatives competent in the relevant field of expertise. It shall report regularly to the Conference of the Parties on all aspects of its work.

2. Under the authority of and in accordance with guidelines laid down by the Conference of the Parties, and upon its request, this body shall:

   (a) Provide scientific and technical assessments of the status of biological diversity;

   (b) Prepare scientific and technical assessments of the effects of types of measures taken in accordance with the provisions of this Convention;

   (c) Identify innovative, efficient and state-of-the-art technologies and know-how relating to the conservation and sustainable use of biological diversity and advise on the ways and means of promoting development and/or transferring such technologies;

(d) Provide advice on scientific programmes and international cooperation in research and development related to conservation and sustainable use of biological diversity; and

(e) Respond to scientific, technical, technological and methodological questions that the Conference of the Parties and its subsidiary bodies may put to the body.

3. The functions, terms of reference, organization and operation of this body may be further elaborated by the Conference of the Parties.

Article 26 Reports

Each Contracting Party shall, at intervals to be determined by the Conference of the Parties, present to the Conference of the Parties, reports on measures which it has taken for the implementation of the provisions of this Convention and their effectiveness in meeting the objectives of this Convention.

Article 27 Settlement of Disputes

1. In the event of a dispute between Contracting Parties concerning the interpretation or application of this Convention, the parties concerned shall seek solution by negotiation.

2. If the parties concerned cannot reach agreement by negotiation, they may jointly seek the good offices of, or request mediation by, a third party.

3. When ratifying, accepting, approving or acceding to this Convention, or at any time thereafter, a State or regional economic integration organization may declare in writing to the Depositary that for a dispute not resolved in accordance with paragraph 1 or paragraph 2 above, it accepts one or both of the following means of dispute settlement as compulsory:

(a) Arbitration in accordance with the procedure laid down in Part 1 of Annex II;

(b) Submission of the dispute to the International Court of Justice.

4. If the parties to the dispute have not, in accordance with paragaph 3 above, accepted the same or any procedure, the dispute shall be submitted to conciliation in accordance with Part 2 of Annex II unless the parties otherwise agree.

5. The provisions of this Article shall apply with respect to any protocol except as otherwise provided in the protocol concerned.

Article 28 Adoption of Protocols

1. The Contracting Parties shall cooperate in the formulation and adoption of protocols to this Convention.

2. Protocols shall be adopted at a meeting of the Conference of the Parties.

3. The text of any proposed protocol shall be communicated to the Contracting Parties by the Secretariat at least six months before such a meeting.

Article 29 Amendment of the Convention or Protocol

1. Amendments to this Convention may be proposed by any Contracting Party. Amendments to any protocol may be proposed by any Party to that protocol.

2. Amendments to this Convention shall be adopted at a meeting of the Conference of the Parties. Amendments to any protocol shall be adopted at a meeting of the Parties to the Protocol in question. The text of any proposed amendment to this Convention or to any protocol, except as may otherwise be provided in such protocol, shall be communicated to the Parties to the instrument in question by the secretariat at least six months before the meeting at which it is proposed for adoption. The secretariat shall also communicate proposed amendments to the signatories to this Convention for information.

3. The Parties shall make every effort to reach agreement on any proposed amendment to this Convention or to any protocol by consensus. If all efforts at consensus have been exhausted, and no agreement reached, the amendment shall as a last resort be adopted by a two-third majority vote of the Parties to the instrument in question present and voting at the meeting, and shall be submitted by the Depositary to all Parties for ratification, acceptance or approval.

4. Ratification, acceptance or approval of amendments shall be notified to the Depositary in writing. Amendments adopted in accordance with paragraph 3 above shall enter into force among Parties having accepted them on the ninetieth day after the deposit of instruments of ratification, acceptance or approval by at least two thirds of the Contracting Parties to this Convention or of the Parties to the protocol concerned, except as may otherwise be provided in such protocol. Thereafter the amendments shall enter into force for any other Party on the ninetieth day after that Party deposits its instrument of ratification, acceptance or approval of the amendments.

5. For the purposes of this Article, "Parties present and voting" means Parties present and casting an affirmative or negative vote.

Article 30 Adoption and Amendment of Annexes

1. The annexes to this Convention or to any protocol shall form an integral part of the Convention or of such protocol, as the case may be, and, unless expressly provided otherwise, a reference to this Convention or its protocols constitutes at the same time a reference to any annexes thereto. Such annexes shall be restricted to procedural, scientific, technical and administrative matters.

2. Except as may be otherwise provided in any protocol with respect to its annexes, the following procedure shall apply to the proposal, adoption and entry into force of additional annexes to this Convention or of annexes to any protocol:

a) Annexes to this Convention or to any protocol shall be proposed and adopted according to the procedure laid down in Article 29;

b) Any Party that is unable to approve an additional annex to this Convention or an annex to any protocol to which it is Party shall so notify the Depositary, in writing, within one year from the date of the communication of the adoption by the Depositary. The Depositary shall without delay notify all Parties of any such notification received. A Party may at any time withdraw a previous declaration of objection and the annexes shall thereupon enter into force for that Party subject to subparagraph (c) below;

c) On the expiry of one year from the date of the communication of the adoption by the Depositary, the annex shall enter into force for all Parties to this Convention or to any protocol concerned which have not submitted a notification in accordance with the provisions of subparagraph (b) above.

3. The proposal, adoption and entry into force of amendments to annexes to this Convention or to any protocol shall be subject to the same procedure as for the proposal, adoption and entry into force of annexes to the Convention or annexes to any protocol.

4. If an additional annex or an amendment to an annex is related to an amendment to this Convention or to any protocol, the additional annex or amendment shall not enter into force until such time as the amendment to the Convention or to the protocol concerned enters into force

Article 31 Right to Vote

1. Except as provided for in paragraph 2 below, each Contracting Party to this Convention or to any protocol shall have one vote.

2. Regional economic integration organizations, in matters within their competence, shall exercise their right to vote with a number of votes equal to the number of their member States which are Contracting Parties to this Convention or the relevant protocol. Such organizations shall not exercise their right to vote if their member States exercise theirs, and vice versa.

Article 32 Relationship between this Convention and Its Protocols

1. A State or a regional economic integration organization may not become a Party to a protocol unless it is, or becomes at the same time, a Contracting Party to this Convention.

2. Decisions under any protocol shall be taken only by the Parties to the protocol concerned. Any Contracting Party that has not ratified, accepted or approved a protocol may participate as an observer in any meeting of the parties to that protocol.

Article 33 Signature

This Convention shall be open for signature at Rio de Janeiro by all States and any regional economic integration organization from 5 June 1992 until 14 June 1992, and at the United Nations Headquarters in New York from 15 June 1992 to 4 June 1993.

Article 34 Ratification, Acceptance or Approval

1. This Convention and any protocol shall be subject to ratification, acceptance or approval by States and by regional economic integration organizations. Instruments of ratification, acceptance or approval shall be deposited with the Depositary.

2. Any organization referred to in paragraph 1 above which becomes a Contracting Party to this Convention or any protocol without any of its member States being a Contracting Party shall be bound by all the obligations under the Convention or the protocol, as the case may be. In the case of such organizations, one or more of whose member States is a Contracting Party to this Convention or relevant protocol, the organization and its member States shall decide on their respective responsibilities for the performance of their obligations under the Convention or protocol, as the case may be. In such cases, the organization and the member States shall not be entitled to exercise rights under the Convention or relevant protocol concurrently.

3. In their instruments of ratification, acceptance or approval, the organizations referred to in paragraph 1 above shall declare the extent of their competence with respect to the matters governed by the Convention or the relevant protocol. These organizations shall also inform the Depositary of any relevant modification in the extent of their competence.

Article 35 Accession

1. This Convention and any protocol shall be open for accession by States and by regional economic integration organizations from the date on which the Convention or the protocol concerned is closed for signature. The instruments of accession shall be deposited with the Depositary.

2. In their instruments of accession, the organizations referred to in paragraph 1 above shall declare the extent of their competence with respect to the matters governed by the Convention or the relevant protocol. These organizations shall also inform the Depositary of any relevant modification in the extent of their competence.

3. The provisions of Article 34, paragraph 2, shall apply to regional economic integration organizations which accede to this Convention or any protocol.

Article 36 Entry into Force

1. This Convention shall enter into force on the ninetieth day after the date of deposit of the thirtieth instrument of ratification, acceptance, approval or accession.

2. Any protocol shall enter into force on the ninetieth day after the date of deposit of the number of instruments of ratification, acceptance, approval or accession, specified in that protocol, has been deposited.

3. For each Contracting Party which ratifies, accepts or approves this Convention or accedes thereto after the deposit of the thirtieth instrument of ratification, acceptance, approval or accession, it shall enter into force on the ninetieth day after the date of deposit by such Contracting Party of its instrument of ratification, acceptance, approval or accession.

4. Any protocol, except as otherwise provided in such protocol, shall enter into force for a Contracting Party that ratifies, accepts or approves that protocol or accedes thereto after its entry into force pursuant to paragraph 2 above, on the ninetieth day after the date on which that Contracting Party deposits its instrument of ratification, acceptance, approval or accession, or on the date on which this Convention enters into force for that Contracting Party, whichever shall be the later.

5. For the purposes of paragraphs 1 and 2 above, any instrument deposited by a regional economic integration organization shall not be counted as additional to those deposited by member States of such organization.

Article 37 Reservations

No reservations may be made to this Convention.

Article 38 Withdrawals

1. At any time after two years from the date on which this Convention has entered into force for a Contracting Party that Contracting Party may withdraw from the Convention by giving written notification to the Depositary.

2. Any such withdrawal shall take place upon expiry of one year after the date of its receipt by the Depositary, or on such later date as may be specified in the notification of the withdrawal.

3. Any Contracting Party which withdraws from this Convention shall be considered as also having withdrawn from any protocol to which it is party.

Article 39 Financial Interim Arrangements

Provided that it has been fully restructured in accordance with the requirements of Article 21, the Global Environment Facility of the United Nations Development Programme, the United Nations Environment Programme and the International Bank for Reconstruction and development shall be the institutional structure referred to in Article 21 on an interim basis, for the period between the entry into force of this Convention and the first meeting of the Conference of the Parties or until the Conference of the Parties decides which institutional structure will be designated in accordance with Article 21.

Article 40 Secretariat Interim Arrangements

The secretariat to be provided by the Executive Director of the United Nations Environment Programme shall be the secretariat referred to in Article 24, paragraph 2, on an interim basis for the period between the entry into force of this Convention and the first meeting of the Conference of the Parties.

Article 41 Depositary

The Secretary-General of the United Nations shall assume the functions of Depositary of this Convention and any protocols.

Article 42 Authentic Texts

It deals with the equally authentic texts (Arabic, Chinese, English, French, Russian and Spanish) and deposited with the Secretary- General of the United Nations. It also states that the convention signed on fifth day of June, one thousand nine hundred and ninety- two at Rio de Janeiro.

Identification/Monitoring, Arbitration and Conciliation (Annexes)

Annex I of the CBD stipulates identification and monitoring of:

i. Ecosystems and habitats containing high diversity, large numbers of endemic or threatened species, or wilderness; required by migratory species; of social, economic, cultural or scientific importance; or which are representative, unique or associated with key evolutionary or other biological processes;

ii. Species and communities which are: threatened; wild relatives of domesticated or cultivated species; of medicinal, agricultural or other economic value; or social, scientific or cultural importance; or importance of research into the conservation and sustainable use of biological diversity, such as indicator species; and

iii. Described genomes and genes of social, scientific or economic importance.

Convention provides for arbitration (Art. 1 through 17 of the part 1 of the Annex-II) and conciliation (Art.1 through 6 of the part 2 of the Annex-II).

In case of disputes between two parties, the arbitral tribunal shall consist of three members. Each of the parties to the dispute shall appoint an arbitrator and the two arbitrators so appointed shall designate by common agreement the third arbitrator who shall be the President of the tribunal. The later shall not be a national of one of the parties to the dispute, nor have his or her usual place of residence in the territory of one of these parties, nor be employed by any of them, nor have dealt with the case in any other capacity.

The arbitral tribunal shall render its decisions in accordance with the provisions of this Convention, any protocols concerned and international law. Unless the parties to the dispute otherwise agree, the arbitral tribunal shall determine its own rules of procedure. Decisions both on procedure and substance of the arbitral tribunal shall be taken by a majority vote of its members. The tribunal shall render its final decision within five months of the date on which it is fully constitute unless it finds it necessary to extend the time-limit for a period which should not exceed five more months. The award shall be binding on the parties to the dispute. It shall be without appeal unless the parties to the dispute have agreed in advance to an appellate procedure.

Implementation of CBD

Thematic Programmes

The Conference of the Parties (COP) has established seven thematic programmes of work (listed below) which correspond to some of the major biomes on the planet. Each programme establishes a vision for, and basic principles to guide future work. They also set out key issues for consideration, identify potential outputs, and suggest a timetable and means for achieving these. Implementation of the work programmes depends on contributions from Parties, the Secretariat, relevant intergovernmental and other organizations. Periodically, the COP and the SBSTTA review the state of implementation of the work programmes.

☆ Agricultural Biodiversity

☆ Dry and Sub-humid Lands Biodiversity

☆ Forest Biodiversity

☆ Inland Waters Biodiversity

☆ Island Biodiversity

☆ Marine and Coastal Biodiversity

☆ Mountain Biodiversity

Cross-cutting Issues

The Conference of Parties has also initiated work on key matters of relevance to all thematic areas. These cross-cutting issues correspond to the issues addressed in the Convention's substantive provisions in Articles 6-20, and provide bridges and links between the thematic programmes. Some cross cutting initiatives directly support work under thematic programmes, for example, the work on indicators provides information on the status and trends of biodiversity for all biomes. Others develop discrete products quite separate from the thematic programmes. The work done for these cross-cutting issues has led to a number of principles, guidelines, and other tools to facilitate the implementation of the Convention and the achievement of the 2010 biodiversity target.

☆ Aichi Biodiversity Targets

☆ Access to Genetic Resources and Benefit-sharing

☆ Biological and Cultural Diversity

☆ Biodiversity for Development

☆ Climate Change and Biodiversity

☆ Communication, Education and Public Awareness

☆ Economics, Trade and Incentive Measures

☆ Ecosystem Approach

☆ Ecosystem Restoration

☆ Gender and Biodiversity

☆ Global Strategy for Plant Conservation

☆ Global Taxonomy Initiative

☆ Impact Assessment

☆ Identification, Monitoring, Indicators and Assessments

☆ Invasive Alien Species

☆ Liability and Redress - Art. 14 (2)

☆ Protected Areas

☆ Sustainable Use of Biodiversity

☆ Tourism and Biodiversity

☆ Traditional Knowledge, Innovations and Practices - Article 8 (j)

☆ Technology Transfer and Cooperation

The Cartagena Protocol on Biosafety, and its Nagoya—Kuala Lumpur Supplementary Protocol on Liability and Redress

The Cartagena Protocol on Biosafety is an additional agreement to the Convention on Biological Diversity. It aims to ensure the safe transport, handling and use of living modified organisms (LMOs) resulting from modern biotechnology that may have adverse effects on biodiversity, also taking into account risks to human health. The Protocol establishes procedures for regulating the import and export of LMOs from one country to another. There are two main sets of procedures, one for LMOs intended for direct introduction into the environment, known as the advance informed agreement (AIA) procedure, and another for LMOs intended for direct use as food or feed, or for processing (LMOs-FFP). Under the AIA procedure, a country intending to export an LMO for intentional release into the environment must notify in writing the Party of import before the first proposed export takes place. The Party of import must acknowledge receipt of the notification within 90 days and must communicate its decision on whether or not to import the LMO within 270 days. Parties are required to ensure that their decisions are based on a risk assessment of the LMO, which must be carried out in a scientifically sound and transparent manner. Once a Party takes a decision on the LMO, it is required to communicate the decision as well as a summary of the risk assessment to a central information system, the Biosafety Clearing-House (BCH). Under the procedure for LMOs-FFP, Parties that decide to approve and place such LMOs on the market are required to make their decision and relevant information, including the risk assessment reports, publicly available through the BCH.

The Protocol also requires Parties to ensure that LMOs being shipped from one country to another are handled, packaged and transported in a safe manner. The shipments must be accompanied by documentation that clearly identifies the

LMOs, specifies any requirements for the safe handling, storage, transport and use and provides contact details for further information. The Cartagena Protocol is reinforced by the Nagoya - Kuala Lumpur Supplementary Protocol on Liability and Redress. The Supplementary Protocol specifies response measures to be taken in the event of damage to biodiversity resulting from LMOs. The competent authority in a Party to the Supplementary Protocol must require the person in control of the LMO (operator) to take the response measures or it may implement such measures itself and recover any costs incurred from the operator.

The Nagoya–Kula Lumpur Supplementary Protocol

☆ Provides flexibility in regulatory approaches by allowing Parties to apply existing or new domestic laws that may be general or specific as regards response measures to damage n Creates an enabling environment and builds further confidence in the safe development and application of modern biotechnology

☆ Contributes to the prevention or mitigation of damage by creating incentives for operators to ensure safety in the development or handling of LMOs

The Nagoya Protocol on Access to Genetic Resources and Benefit-Sharing

The fair and equitable sharing of the benefits arising out of the utilization of genetic resources is one of the three objectives of the Convention on Biological Diversity. At the tenth Conference of the Parties, held in Nagoya, Japan, in October 2010, the Nagoya Protocol on Access to Genetic Resources and the Fair and Equitable Sharing of Benefits Arising from their Utilization was adopted (Anonymous, 2014a). It is a new international agreement which aims at sharing the benefits arising from the utilization of genetic resources in a fair and equitable way, thereby contributing to the conservation and sustainable use of biodiversity. The Nagoya Protocol further builds on the access and benefit-sharing provisions of the CBD by creating greater legal certainty and transparency for both providers and users of genetic resources. It does this by establishing more predictable conditions for access to genetic resources and helping to ensure benefit sharing when genetic resources leave the contracting Party providing the genetic resources. Genetic resources, whether from plant, animal or micro-organisms, are used for a variety of purposes ranging from basic research to the development of products. In some cases, traditional knowledge associated with genetic resources that comes from indigenous and local communities (ILCs), provides valuable information to researchers regarding the particular properties and value of these resources and their potential use for the development of, for example, new medicines or cosmetics. Users of genetic resources include research and academic institutions, and private companies operating in various sectors such as pharmaceuticals, agriculture, horticulture, cosmetics and biotechnology (Dirk and Minou, 2014). When a person or institution seeks access to genetic resources in a foreign country, it should obtain the prior informed consent of the country in which the resource is located; this is one of the fundamental principles of access and benefit-sharing. Moreover, the person or institution must also negotiate and agree on the terms and conditions of access and use of this resource. This includes the sharing of benefits arising from the use of this resource with the provider as a prerequisite for access to the genetic resource and its use. Conversely, countries, when acting as providers of genetic resources, should provide fair and non-arbitrary rules and procedures for access to their genetic resources (Anonymous, 2002).

National Biodiversity Strategies and Action Plans (NBSAPs)

The Convention on Biological Diversity calls for each Party to develop a National Biodiversity Strategy and Action Plan (NBSAP) to guarantee that the objectives of the Convention are fulfilled in each country (Article 6). The national biodiversity strategy reflects the country's vision for biodiversity and the broad policy and institutional measures that the country takes to fulfil the objectives of the Convention, while the action plan comprises the concrete actions to be taken to achieve the strategy. The strategy should include ambitious but realistic and measurable national targets developed in the framework of the Strategic Plan for Biodiversity 2011-2020, and its twenty Aichi Targets adopted at the tenth meeting of the Conference of the Parties. The strategy and action plan are developed by each Party in accordance with national priorities, circumstances and capabilities. It is essential that all sectors whose activities impact on biodiversity, and those societal groups who depend on biodiversity, be brought into the NBSAP process early. This engenders a broad ownership of the NBSAP whereby all stakeholders in biodiversity are engaged in its development and implementation. It also enables 'mainstreaming' which means the integration of biodiversity considerations into relevant legislation, plans, programmes and policy, such as National Development Plans; National Strategies for Sustainable Development; Poverty Reduction Strategy Papers; Strategies to achieve the Millennium Development Goals; National Programmes to Combat Desertification; National Climate Change Adaptation or Mitigation Strategies; and relevant private-sector policies. While the NBSAP can take the form of a single biodiversity planning document, it can also be conceived as comprising a 'basket' of elements on, for example, laws and administrative procedures; scientific research agendas, programmes and projects; communication, education and public awareness activities; forums for inter-ministerial and multi-stakeholder dialogue-which together provide the means to meet the three objectives of the Convention, thereby forming the basis for national implementation. The NBSAP should be a living process by which increasing information and knowledge, gained through the monitoring and evaluation of each phase of implementation, feed an ongoing review and improvement.

Strategic Goals and Targets for 2020

Taking Action for Biodiversity

The world is now on a path to building a future of living in harmony with nature. In October 2010, in Japan, governments agreed to the Strategic Plan for Biodiversity 2011-2020 and the Aichi Targets as the basis for halting and eventually reversing the loss of biodiversity of the planet. This plan provides an overarching framework on biodiversity, not only for the biodiversity-related conventions, but for the entire United Nations system and all other partners engaged in biodiversity management and policy development. To build support and momentum for this urgent task, the United Nations General Assembly at its 65th session declared the period 2011-2020 to be *"the United Nations Decade on Biodiversity, with a view to contributing to the implementation of the Strategic Plan for Biodiversity for the period 2011-2020"* (Resolution 65/161).

The United Nations Decade on Biodiversity will serve to support the implementation of the Strategic Plan for Biodiversity and promote its overall vision of living in harmony with nature. Its goal is to mainstream biodiversity at different levels. Throughout the United Nations Decade on Biodiversity, governments are encouraged to develop, implement and communicate the results of national strategies for implementation of the Strategic Plan for Biodiversity. The actions taken by individuals, stakeholders and governments are important steps, one building on the other, towards protecting the life support systems that not only ensure human well-being, but support the rich variety of life on this planet.

Achieving the Aichi Targets by 2020

There are five strategic goals and 20 ambitious yet achievable targets (Table 1.2). Collectively known as the Aichi Targets, they are part the Strategic Plan for Biodiversity. Their purpose is to inspire broad-based action in support of biodiversity over this decade (2011-2020) by all countries and stakeholders promoting the coherent and effective implementation of the three objectives of the Convention on Biological Diversity: conservation of biodiversity; sustainable use of biodiversity; fair and equitable sharing of the benefits arising from the use of genetic resources.

Table 1.2: Aichi Biodiversity Goals and Targets

Strategic Goal A: Address the underlying causes of biodiversity loss by mainstreaming biodiversity across government and society
Target 1
By 2020, at the latest, people are aware of the values of biodiversity and the steps they can take to conserve and use it sustainably.
Target 2
By 2020, at the latest, biodiversity values have been integrated into national and local development and poverty reduction strategies and planning processes and are being incorporated into national accounting, as appropriate, and reporting systems.
Target 3
By 2020, at the latest, incentives, including subsidies, harmful to biodiversity are eliminated, phased out or reformed in order to minimize or avoid negative impacts, and positive incentives for the conservation and sustainable use of biodiversity are developed and applied, consistent and in harmony with the Convention and other relevant international obligations, taking into account national socio economic conditions.

Strategic Goal B: Reduce the direct pressures on biodiversity and promote sustainable use

Target 5

By 2020, the rate of loss of all natural habitats, including forests, is at least halved and where feasible brought close to zero, and degradation and fragmentation is significantly reduced.

Target 6

By 2020 all fish and invertebrate stocks and aquatic plants are managed and harvested sustainably, legally and applying ecosystem based approaches, so that overfishing is avoided, recovery plans and measures are in place for all depleted species, fisheries have no significant adverse impacts on threatened species and vulnerable ecosystems and the impacts of fisheries on stocks, species and ecosystems are within safe ecological limits

Target 7

By 2020 areas under agriculture, aquaculture and forestry are managed sustainably, ensuring conservation of biodiversity

Target 8

By 2020, pollution, including from excess nutrients, has been brought to levels that are not detrimental to ecosystem function and biodiversity.

Target 9

By 2020, invasive alien species and pathways are identified and prioritized, priority species are controlled or eradicated, and measures are in place to manage pathways to prevent their introduction and establishment.

Target 10

By 2015, the multiple anthropogenic pressures on coral reefs, and other vulnerable ecosystems impacted by climate change or ocean acidification are minimized, so as to maintain their integrity and functioning

Strategic Goal C: To improve the status of biodiversity by safeguarding ecosystems, species and genetic diversity

Target 11

By 2020, at least 17 per cent of terrestrial and inland water, and 10 per cent of coastal and marine areas, especially areas of particular importance for biodiversity and ecosystem services, are conserved through effectively and equitably managed, ecologically representative and well connected systems of protected areas and other effective area-based conservation measures, and integrated into the wider landscapes and seascapes.

Target 12

By 2020 the extinction of known threatened species has been prevented and their conservation status, particularly of those most in decline, has been improved and sustained.

Target 13

By 2020, the genetic diversity of cultivated plants and farmed and domesticated animals and of wild relatives, including other socio-economically as well as culturally valuable species, is maintained, and strategies have been developed and implemented for minimizing genetic erosion and safeguarding their genetic diversity.

Strategic Goal D: Enhance the benefits to all from biodiversity and ecosystem services

Target 14

By 2020, ecosystems that provide essential services, including services related to water, and contribute to health, livelihoods and well-being, are restored and safeguarded, taking into account the needs of women, indigenous and local communities, and the poor and vulnerable.

Target 15
By 2020, ecosystem resilience and the contribution of biodiversity to carbon stocks has been enhanced, through conservation and restoration, including restoration of at least 15 per cent of degraded ecosystems, thereby contributing to climate change mitigation and adaptation and to combating desertification.
Target 16
By 2015, the Nagoya Protocol on Access to Genetic Resources and the Fair and Equitable Sharing of Benefits Arising from their Utilization is in force and operational, consistent with national legislation.
Strategic Goal E: Enhance implementation through participatory planning, knowledge management and capacity building
Target 17
By 2015 each Party has developed, adopted as a policy instrument, and has commenced implementing an effective, participatory and updated national biodiversity strategy and action plan.
Target 18
By 2020, the traditional knowledge, innovations and practices of indigenous and local communities relevant for the conservation and sustainable use of biodiversity, and their customary use of biological resources, are respected, subject to national legislation and relevant international obligations, and fully integrated and reflected in the implementation of the Convention with the full and effective participation of indigenous and local communities, at all relevant levels.
Target 19
By 2020, knowledge, the science base and technologies relating to biodiversity, its values, functioning, status and trends, and the consequences of its loss, are improved, widely shared and transferred, and applied.
Target 20
By 2020, at the latest, the mobilization of financial resources for effectively implementing the Strategic Plan for Biodiversity 2011-2020 from all sources, and in accordance with the consolidated and agreed process in the Strategy for Resource Mobilization, should increase substantially from the current levels. This target will be subject to changes contingent to resource needs assessments to be developed and reported by Parties.

Major Challenges to Implementing the Convention on Biological Diversity

Following are some of the major challenges to implement the CBD and promoting sustainable development.

☆ Meeting the increasing demand for biological resources caused by population growth and increased consumption, while considering the long-term consequences of our actions.

☆ Increasing our capacity to document and understand biodiversity, its value, and threats to it.

☆ Building adequate expertise and experience in biodiversity planning.

☆ Improving policies, legislation, guidelines, and fiscal measures for regulating the use of biodiversity.

☆ Adopting incentives to promote more sustainable forms of biodiversity use.

☆ Promoting trade rules and practices that foster sustainable use of biodiversity.

☆ Strengthening coordination within governments, and between governments and stakeholders.

☆ Securing adequate financial resources for conservation and sustainable use, from both national and international sources.

☆ Making better use of technology.

☆ Building political support for the changes necessary to ensure biodiversity conservation and sustainable use.

☆ Improving education and public awareness about the value of biodiversity.

Biological Diversity Act-2002

Biodiversity in India: Status and Trends

India is one of the recognized mega-diverse countries of the world, harbouring nearly 7-8 per cent of the recorded species of the world, and representing 4 of the 35 globally identified biodiversity hotspots (Himalayas, Indo-Burma, Western Ghats and Sri Lanka, Sundaland). India is also a vast repository of traditional knowledge associated with biological resources. So far, over 91,200 species of animals and 45,500 species of plants have been documented in the ten bio-geographic regions of the country. Inventories of floral and faunal diversities are being progressively updated with several new discoveries through the conduct of continuous surveys and exploration. Along with species richness, India also possesses high rates of endemism. In terms of endemic vertebrate groups, India's global ranking is 10 in birds, with 69 species; fifth in reptiles with 156 species; and seventh in amphibians with 110 species. Endemic-rich Indian fauna is manifested most prominently in Amphibia (61.2 per cent) and Reptilia (47 per cent). India is also recognized as one of the eight Vavilovian centres of origin and diversity of crop plants, having more than 300 wild ancestors and close relatives of cultivated plants, which are still evolving under natural conditions.

The varied edaphic, climatic and topographic conditions and years of geological stability have resulted in a wide range of ecosystems and habitats such as forests, grasslands, wetlands, deserts, and coastal and marine ecosystems (Anonymous, 1994). Arid and semi-arid regions cover 38.8 per cent of India's total geographical area. The cold arid zone located in the Trans-Himalayan region covers 5.62 per cent of the country's area. The region is the stronghold of three cat predators – the lion, leopard and tiger. Of the 140 species of known birds, the Great Indian Bustard is a globally threatened species. The flora of the Indian desert comprises 682 species, with over 6 per cent of the total plant species being endemic. The cold desert is the home of rare endangered fauna, such as the Asiatic Ibex, Tibetan Argali, Wild Yak, Snow Leopard, *etc.*, and the flora is rich in endemism and economically important species.

India has a variety of wetland ecosystems ranging from high altitude cold desert wetlands to hot and humid wetlands in coastal zones with diverse flora and fauna. About 4,445 km2 of the country is under mangroves. India is blessed with rich fish diversity that dwells in the inland waters. The major rivers of India and their tributaries traverse through varied geoclimatic zones, displaying high diversity in their biotic and abiotic characteristics throughout their 28,000 km linear drift. The current distribution of 783 species of freshwater fishes, belonging to 89 genera under 17 families, which includes 223 endemic fishes, is recorded in India. In total, the Indian fish population represents 11.72 per cent of species, 23.96 per cent of genera, 57 per cent of families and 80 per cent of the global fishes. The country is the third largest producer of fish in the world, with 2,411 fish species.

India has a vast coastline of 7,517 km, of which 5,423 km belong to Peninsular India and 2,094 km to the Andaman, Nicobar and Lakshadweep Islands, and an EEZ of 2.02 million km2 with a very wide range of habitats (*e.g.* estuaries, lagoons, mangroves, backwaters, salt marshes, rocky coasts, stretches and coral reefs, all of which are characterized by rich and unique biodiversity components). Another crucial ecosystem for India is its forest, covering 23.39 per cent of the geographical area of the country (of which 75 per cent occurs in the North-Eastern states) and counting over 16 major forest types and 251 sub-types. Against the global trend of deforestation, it is worth underlining the achievement made by India in stabilizing its area under forest cover over the years.

The mountain ecosystems of India are largely described under two global hotspots, *viz.*, the Eastern Himalayas and the Western Ghats and Sri Lanka. They contribute prominently in geographic extent, biophysical and socio-cultural diversity and uniqueness. The extent of species endemism in vascular plants alone ranges from 32 per cent to 40 per cent in the mountain ecosystems. Other groups, such as reptiles, amphibians and fish show more than 50 per cent of species endemism in Western Ghats. Of the 979 bird species recorded from the Himalayan region, four Endemic Bird Areas have been delineated for priority conservation measures and, likewise, identification of Key Biodiversity Areas (KBAs) has been initiated in Western Ghats.

As per the IUCN Red List version 2010.4, 94 species of mammals, 78 species of birds, 66 species of amphibians, 30 species of reptiles, 122 species of fish, 113 species of invertebrates and 255 species of plants in India are listed as Critically Endangered, Endangered and Vulnerable. So far, 758 animal and plant species are listed as globally threatened in India by IUCN, which is about less than 1 per cent (*i.e.* 0.55 per cent) of species documented in India.

For India, conservation of biodiversity is crucial not only because it provides several goods and services necessary for human survival, but also because it is directly linked with providing livelihoods to and improving socio-economic conditions for millions of local people, thereby contributing to sustainable development and poverty alleviation (Venkataraman, 2006). An example of a benefit derived from biodiversity in India is reflected by the forest sector, which is increasingly being looked upon as a major performer in poverty alleviation programmes. India's forests neutralize nearly 11 per cent of India's greenhouse gas emissions. Nearly 200 million people are dependent on forests for livelihood in India.

Pursuant to the CBD, a first major step was the development of the National Policy and Macrolevel Action Strategy (1999) that called for consolidating existing biodiversity conservation programmes and initiating new steps in conformity with the spirit of the Convention. This was followed by implementation of a UNDP/ GEF-sponsored NBSAP Project (2000-2004) that yielded micro-level action plans adequately integrating crosscutting issues and livelihood security concerns. Some of the major programmes that contribute to its implementation include: Protected Areas (PA) network and its steady growth over the years, consolidation of Biosphere Reserves (BRs) (15), establishment of more species-specific reserves, growth in designated Ramsar sites, augmentation of *ex-situ* efforts through the establishment of the network of Lead Gardens and initiatives in the conservation of genetic resources, *etc.*

Subsequent to the approval of the National Environment Policy (NEP) in 2006, preparation of the National Biodiversity Action Plan was taken up by revising the 1999 document in consonance with the NEP, using the NBSAP project report as one of the inputs. The National Biodiversity Action Plan (2008) defines targets, activities and associated agencies for achieving the goals, drawing upon the main principle in the NEP that human beings are at the centre of concerns of sustainable development and they are entitled to a healthy and productive life in harmony with nature. Work is currently in progress to develop national targets within the framework of the Strategic Plan for Biodiversity (2011-2020).

Consequent to the Convention on Biological Diversity (CBD) and also due to recent global developments, biodiversity became the property of a nation on which the sovereign rights exist for a country in contrast to the utopian traditions that, biodiversity is a common heritage of mankind (Singh, 2004). In this context, sharing of genetic resources has become a focal and critical issue at international/national levels. The global developments and their implications in protection of biodiversity wealth at national level have a bearing on all the states as agriculture and forestry are state subjects under the Indian constitution. Following the ratification of CBD and after widespread consultations, India also enacted the Biological Diversity Act in 2002 and notified the Rules in 2004, to give effect to the provisions of the CBD, including those relating to its third objective on Access and Benefit Sharing (ABS). India was one of the first few countries to enact such legislation. The Act is to be implemented through a three-tiered institutional structure: National Biodiversity Authority (NBA), State Biodiversity Boards (SBBs), Biodiversity Management Committees (BMCs) at the local level, in line with the provisions for decentralized governance contained in the Constitution. The Biological Diversity Act is a path-breaking and progressive legislation which has the potential to positively impact biodiversity conservation in the country (Venkataraman, 2009).

The Biodiversity Act - 2002 was introduced primarily to addresses access to genetic resources and associated knowledge by foreign individuals, institutions or companies, to ensure equitable sharing of benefits arising out of the use of these resources and knowledge to the country and the people (Varaprasad and Pandravada, 2005; Varaprasad *et al.*, 2007). The National Biodiversity Authority (NBA) was established in 2003 to implement India's Biological Diversity Act (2002).

The NBA is an Autonomous Body and it performs facilitative, regulatory and advisory function for the Government of India on issues of conservation, sustainable use of biological resources and fair and equitable sharing of benefits arising out of the use of biological resources. It also advises the State Governments in the selection of areas of biodiversity importance to be notified under Sub-Section (1) of Section 37 as heritage sites and measures for the management of such heritage sites;

The NBA with its headquarters in Chennai, Tamil Nadu, delivers its mandate through a structure that comprises of the Authority, Secretariat, SBBs, BMCs and Expert Committees. Since its establishment, NBA has supported creation of SBBs in 25 States and facilitated establishment of around 37,769 BMCs.

Salient Provisions of Biological Diversity Act-2002

The Act has the following main provisions:

   i. Regulation of access to biological resources of the country with the purpose of securing equitable share in benefits arising out of biological resources associated knowledge

  ii. Promoting conservation and sustainable utilization of biological resources

 iii. According respect and protecting knowledge of local communities related to Biodiversity

  iv. Establishing the National Biodiversity Authority (NBA) and its functions and powers

   v. Securing benefits to local people who are conservers of biological resources and holders of related knowledge and information

  vi. Conservation and development of areas of importance from the standpoint of biological diversity by declaring them as biological diversity heritage sites;

 vii. Protection and rehabilitation of threatened species

viii. Involvement of institutions of state governments in the broad scheme of the implementation of the Biological Diversity Act through constitution of committees.

   ☆ Section - 3: All foreign national require approval from NBA for obtaining Biological Resources.

   ☆ Section-4: Indian individuals/entities to seek approval before transferring knowledge/research and material to foreigners.

   ☆ Section - 5: Guidelines for Government sponsored collaborative research projects.

   ☆ Section - 6: Prior approval of NBA before applying for any kind of IPR based on research conducted on biological material and or associated knowledge obtained from India.

   ☆ Section - 7: Indians required to provide prior intimation to State Biodiversity Boards for obtaining biological material for commercial purposes. SBB can regulate such access.

☆ Growers and cultivators of Biological Diversity and *vaids* and *hakims* who are practicing Indian system of medicines and local people exempted.

☆ Section - 8: Establishment of NBA, its composition.

☆ Section - 13: Committees of NBA.

☆ Section - 18: Functions and powers of NBA.

☆ Section - 19: Approval by the NBA.

☆ Section - 21: Determination of equitable benefit sharing by NBA.

☆ Section - 22: Establishment of State Bio-diversity Boards.

☆ Section - 23: Function of the State Bio-diversity Boards.

☆ Section - 24: Powers of State Bio-diversity Boards

☆ Section - 26 : National Biodiversity Fund.

☆ Section - 32: State Biodiversity Fund.

☆ Section - 36: Central Government to develop National strategies plans *etc.* for conservation of biodiversity.

☆ Section - 36: (1A): Central Government to issue direction to State Governments to take corrective measures for conservation of biodiversity.

☆ Section - 36 (3) (i): Impact assessment of developmental projects on biodiversity.

☆ Section - 36 (3) (ii): Regulate release of GMOs.

☆ Section - 36 (4): Measures for protecting the traditional knowledge.

☆ Section - 37: Biodiversity heritage sites.

☆ Section - 38: Notifications of threatened species.

☆ Section - 39: Designation of repositories.

☆ Section-40: Exemption for normally traded commodities from purview of the act.

☆ Section - 41: Establishment of Biodiversity Management Committees by local bodies.

☆ Section - 42: Local Biodiversity Fund.

☆ Section - 52 A: Appeals to High Court on the decision of NBA/SBB.

☆ Section - 53 B: Orders of NBA/SBB at par with civil courts.

☆ Section - 55: Penalties - imprisonment up to 5 years and or a fine of 10 lakhs or to the extent of damage caused.

☆ Section - 59: Act to have effect in addition to other Acts.

☆ Section - 61: Cognizance of offences.

☆ Section - 62: Power of Central Government to make Rules.

☆ Section - 63: Power of State Government to make Rules.

☆ Section - 64: Power to make regulations.

☆ Section - 65: Power to remove difficulties.

India places high priority on establishing a mandatory obligation to disclose source of origin, proof of Prior Informed Consent (PIC) and Access and Benefit Sharing (ABS) relating to genetic resource inventions and traditional knowledge. (The Indian Patent Law requires disclosure of origin, but not the other elements).

The Biological Diversity Act 2002 covers various aspects of genetic resources. The Act mainly deals with access to genetic resources by foreign companies, individuals or organizations. It also deals with requests to transfer the results of any related research out of India. It has provision to decide how benefits of the research are to be shared with local communities. The Act has provision of state-level boards that regulate commercialization of any bio-resources by Indians and provision of Biodiversity Management Committees at local government level that are designed to help conserve and document biodiversity and traditional knowledge.

Section 3 of the Act, under Chapter-I, *i.e.* Regulation of Access to Biological Diversity, states that certain persons cannot undertake biodiversity related activities without approval of National Biodiversity Authority (NBA).

"3. (1) No person referred to in sub-section b (2) shall, without previous approval of the National Biodiversity Authority, obtain any biological resource occurring in India or knowledge associated thereto for research or for commercial utilization or for bio-survey and bio-utilization"

Section 3 implies that no biological resources of Indian origin can be accessed by any person outside India without prior approval of the NBA. The NBA has its head office in Chennai and is linked with the Ministry of Environment and Forests (MoEF). The limitation posed to the ICAR-AU system in the wake of follow-up of the Section 3 of the Biological Diversity Act, 2002, encompasses all GRFA including plants, animals, fish and micro-organisms alike.

Regarding IPRs, the BD Act has specific provisions. Section 6 of the Act makes it explicit that "no person shall apply for any IPR, by whatever name called, in or outside India for any invention based on any research or information on a biological resource obtained from India without obtaining the previous approval of the NBA before making such application". Further, it is stated in Section 19 (2) that "Any person who intends to apply for a patent or any other form of intellectual property protection whether in India or outside India referred to in sub section (1) of Section 6, may take an application in such form and in such manner as may be prescribed to the National Biodiversity Authority". It is, however, to be noted that the IPR related provisions of the BD Act exclude national plant variety protection as stated in the Section 6 (3) *i.e.* "the provisions of this section (Section 6) shall not apply to any person making an application for any right under any law relating to protection of plant varieties enacted by the Parliament". It is obviously referring to the PPV and FR Act, 2001 of India and making further allowance for any of its subsequent revision as passed by the Parliament.

Chapters III, IV and V spanning from Sec. 8 to 21 relate to establishment of the NBA, its powers and functions and approvals to be accorded/taken from the NBA. Section 21 of the Act provides for securing determination of equitable benefit sharing by the NBA.

The Act approve of its commitment made to oppose the grant of IPRs in any country outside India on any Indian bioresource and associated knowledge (Sec. 18 (4). Section 36 (1) of the BDA explicitly provides that, "The Central Government shall develop national strategies, plans, programmes for the conservation and promotion and sustainable use of biological diversity, promotion of *in-situ* and *ex-situ* conservation of biological resources, incentives for research, training and public education to increase awareness with respect to biodiversity".

The BD Act provides for certain exemptions. The major exemptions are:

a. Exemption (from access regulation, giving prior intimation for commercial utilization, Sec. 7) to use biological resources to local people and communities of the area, including growers and cultivators of biodiversity and *'vaids'* and *hakims'*, who have been practicing indigenous medicine (Sec. 7, second paragraph).

b. Exemption through notification of normally traded commodities from the purview of the Act (Sec.40).

c. Exemption for collaborative research through government sponsored or government approved institutions subject to overall policy guidelines and approval of the Central Government (Sec. 5).

Sections related to collaborative research projects as also important operational mechanisms as related to agro-biodiversity that need to be set in place under the BD Act and related rules and other regulatory provisions would be further discussed in this chapter.

Implementation of the Biodiversity Act

Biodiversity Act, 2002 has a three-tiered structured for its implementation.

i. National Biodiversity Authority (NBA) at National Level.

ii. State Biodiversity Boards (SBB) at State Level.

iii. Biodiversity Management Committee (BMC) at local level of Panchayats, Zilla, Taluks and Municipalities.

National Biodiversity Authority (NBA)

As per the subsection (1) of Section 8 of the BD Act, 2002 has been established in Chennai, Tamil Nadu. This Authority is the nodal agency to advise the Government of India on matters pertaining to conservation, sustainable use and equitable sharing, grant permissions where applicable and provide technical assistance to bodies established according to the ACT *etc*. The Authority regulates activities referred to in Section 3, 4 and 6 of the BD Act, 2002 and by regulations issue guidelines for access to biological resources and for fair and equitable benefit sharing. It also has powers grant approval for under taking any activity referred to in Section 3, 4 and 6. The other functions include to advise the Central Government on matters relating to the conservation of biodiversity, sustainable use of its components and equitable sharing of benefits arising out of the utilization of biological resources and to advise the State Governments in the selection of areas of biodiversity importance to be notified under subsection (1) of Section 37 as heritage sites and measures for the management of such heritage sites. It may also may, on behalf of the Central Government, take any measures necessary to oppose the grant of intellectual property rights in any country outside India on any biological resource obtained from India or knowledge associated with such biological resource which is derived from India. The mechanism of its operation is through the members of the Authority with the Chairperson as the Chief Executive. The NBA has established several expert committees to ensure implementation of the Act. The five committees are in charge of developing guidelines on collaborative research; material transfer agreements; patents and benefit sharing; normally traded commodities; rare, threatened, endangered and endemic species; and database on bio-resources and traditional knowledge. Each committee is made up of 10 to 15 experts drawn from government departments, research institutes, private agencies and academia.

State Biodiversity Boards (SBBs)

The State Biodiversity Boards (SBBs) focus on advising the State Governments, subject to any guidelines issued by the Central Government, on matters relating to the conservation of biodiversity, sustainable use of its components and equitable sharing of the benefits arising out of the utilization of biological resources. The SSBs also regulate, by granting of approvals or otherwise requests for commercial utilization or bio-survey and bio-utilization of any biological resource by Indians. The local level Biodiversity Management Committees (BMCs) are responsible for promoting conservation, sustainable use and documentation of biological diversity including preservation of habitats, conservation of land races, folk varieties and cultivars, domesticated stocks and breeds of animals and microorganisms and chronicling of knowledge relating to biological diversity. State Biodiversity Boards (SBBs) which have to be established by each state (Section 22 of the BD Act, 2002). The SBB, whose structure is stipulated by the Act like that of the NBA has functions relating to advising the State Government, subject to any guidelines issued by the Central Government, on matters relating to the conservation of biodiversity, sustainable use of its components and equitable sharing of the benefits arising out of the utilization of the biological resources; to regulate access to biological resources by granting of approvals or otherwise requests for commercial utilization or bio survey and bio utilization of any biological resource by Indians and also to perform such other functions as may be necessary to carry out the provisions of this Act or as may be prescribed by the State Government. Under the provisions of Section 7, the SBB may prohibit or restrict any such activity that is detrimental or contrary to the objectives of conservation and sustainable use of biodiversity or equitable sharing of benefits. The SBBs have to consider the views of the local people through the local institutions for this purpose. This provision substantiates the intent to preserve, sustain the bio-resources and to ensure participation of the local people in decision-making.

Biodiversity Management Committees (BMCs)

The work of the SBB is strengthened by the inputs of the Biodiversity Management Committees (BMCs) established at the level of Panchayats, municipalities or Corporations. Section 41 of BD Act stipulates every local body shall constitute a Biodiversity Management Committee (BMC) within its area for the purpose of promoting conservation, sustainable use and documentation of biological diversity including (i) preservation of habitats, (ii) conservation of landraces, folk varieties and cultivars, domesticated stocks/breeds of animals and microorganisms and (iii) chronicling of knowledge relating to biological diversity.

The main function of BMC is to prepare People's Biodiversity Register (PBR) in consultation with local people. These registers are mandated to contain comprehensive information on availability and knowledge of local biological resources, their medicinal or any other use or any other traditional knowledge associated with them. The Committees shall also maintain a Register giving information about the details of the access to biological resources and traditional knowledge granted, details of the collection fee imposed and details of the benefits derived and the mode of their sharing. The Registers are mandated to contain comprehensive information on availability and knowledge of local biological resources, their medicinal or any other use or any other traditional knowledge associated with them.

The other functions of the BMC are to render advice on any matter referred to it by the State Biodiversity Board or Authority for granting approval, to maintain data about the local vaids and practitioners using the biological resources. The BMC are to be provided guidance and technical support by the NBA and the SBB. In fact, the Authority shall take steps to specify the form of the People's Biodiversity Registers and the particulars it shall contain and the format for electronic database. At present, efforts are being undertaken by NGO's. SBB and NBA to formulate model PBR registers. The formats of methodology and PBR developed by West Bengal, Kerala and Madhya Pradesh are also being examined by expert committees for further validation.

Exemptions for Research Projects under the BD Act, 2002

Collaborative projects with international, regional and national research organizations and other international agricultural research systems have immensely contributed towards enrichment of Indian agro-biodiversity (Kloppenburg, 1988; Smale, 1996; Smale *et al.*, 2002). Any disruption or impediment to germplasm flow may have serious repercussions on the diversity and resultantly on food security of the country.

The Sections 3 to 7 of BD Act, 2002 specially cover the provisions to this effect and have directly a bearing on the functioning provisions to this effect and have directly a bearing on the functioning of ICAR/DARE. Provisions of Section 5 of the Act are the harbinger to counteract the limitations posed by the Section 3 and also 4. Section 5 is reproduced below verbatim.

"The provisions of section 3 and 4 shall not apply to collaborative research projects involving transfer or exchange of biological resources or information relating thereto between institutions, including Government sponsored institutions of India, and such institutions in other countries, if such collaborative research projects satisfy the conditions specified in sub-section (3)".

In relation to Genetic Resources for Food and Agriculture (GRFA), there are no specific provisions in the BD Act, 2002 or the Biological Diversity Rules, 2004. The BD Act, 2002 permits exchange or transfer of biological material or information for such approved collaborative projects which follow GOI guidelines, without making a reference to National Biodiversity Authority (NBA) if, (i) the collaborative projects conform to the policy guidelines issued by the Central Government on this behalf [Section 5 (3) (a)], (ii) be approved by the Central Government [Section 5 (3) (b)].

Biodiversity Heritage Sites (BHS)

"Biodiversity Heritage Sites" are well defined areas that are unique, ecologically fragile ecosystems - terrestrial, coastal and inland waters and, marine having rich biodiversity comprising of any one or more of the following components: richness of wild as well as domesticated species or intra-specific categories, high endemism, presence of rare and threatened species, keystone species, species of evolutionary significance, wild ancestors of domestic/cultivated species or their varieties, past pre-eminence of biological components represented by fossil beds and having significant cultural, ethical or aesthetic values and are important for the maintenance of cultural diversity, with or without a long history of human association with them (Rodgers and Panwar, 1990).

Under Section 37 of Biological Diversity Act, 2002 the State Government in consultation with local bodies may notify in the official gazette, areas of biodiversity importance as Biodiversity Heritage sites (Kannaiyan and Gopalam, 2007a,b). The National Biodiversity Authority issued guidelines for the selection and management of BHS in 2009. Areas having any of the following characteristics may qualify for inclusion as BHS.

Criteria for Biological Heritage Sites

☆ Areas that contain a mosaic of natural, semi-natural, and manmade habitats, which together contain a significant diversity of life forms.

☆ Areas that contain significant domesticated biodiversity component and/ or representative agroecosystems with ongoing agricultural practices that sustain this diversity.

☆ Areas that are significant from a biodiversity point of view as also are important cultural spaces such as sacred groves/trees and sites, or other large community conserved areas.

☆ Areas including very small ones that offer refuge or corridors for threatened and endemic fauna and flora, such as community conserved areas or urban greens and wetlands.

☆ All kinds of legal land uses whether government, community or private land could be considered under the above categories.

☆ As far as possible those sites may be considered which are not covered under Protected Area network under the Wildlife Protection Act 1972 as amended.

☆ Areas that provide habitats, aquatic or terrestrial, for seasonal migrant species for feeding and breeding.

☆ Areas that are maintained as preservation plots by the research wing of Forest department.

☆ Medicinal Plant Conservation Areas.

Designated Repositories (DR) for Biological Resources

National Biodiversity Authority, Chennai has designated several national organizations with a mandate to act as Designated Repository (DR) (Table 1.3) for safe deposit of holotypes/isotypes/paratypes of new taxa discovered in India and samples of biological resources accessed by foreign citizens/entities for research or sent/carried abroad by Indian citizens/institutions for research [vide Sections 39 (1-3) and 19 (1) of the Biological Diversity Act, 2002 read with Rule 14 (6) (viii) of the Biological Diversity Rules, 2004, and clause 4 (6) of the Regulation on ABS Guidelines notified in November, 2014] (www.nbaindia.org).

Table 1.3: Designated Repositories under Biological Diversity Act in India

Sl.No.	Category of Biological Resources	Name of the Designated Repository
1.	Flora	Botanical Survey of India and its Regional centres.
	(angiosperms, gymnosperms, pteridophytes, bryophytes, lichens, macro fungi, macro algae)	Indian Council of Forestry Research (FRI, Dehradun and IFGTB, Coimbatore)
		NBRI, Lucknow
2.	Fauna, Fauna in Protected Areas	Zoological Survey of India and its Regional centres.
		TFRI, Jabalpur for termites, butterflies and moths
		Wildlife Institute of India, Dehradun
3.	Genetic Resources	ICAR-NBPGR, New Delhi- Cultivated Plants and their Wild Relatives
		ICAR -NBAGR, Karnal, Domestic Animals
		ICAR –NBFGR, Lucknow- Fish
		ICAR –NBAIM, Mau- Agriculturally Important Microorganisms
		ICAR –NBAII, Bengaluru- Agriculturally Important Insects, Mites, Spiders
4.	Marine Flora and Fauna	National Institute of Oceanography, Goa
5.	Microorganisms	IMTECH, Chandigarh (actinobacteria, bacteria, fungi and yeasts)
		NCCS, Pune (bacteria, fungi (including yeasts), recombinant DNA materials (in the form of clones in bacterial host and bacteriophages),
		ICAR-IARI, New Delhi (fungi/blue-green algae)
6.	Viruses	NIV, Pune

Source: NBA website www.nbaindia.org.

Access and Benefit Sharing elements

Access and benefit-sharing refers to the way genetic resources—whether plant, animal or microorganism—are accessed in countries of origin, and how the benefits that result from their use by various research institutes, universities or private companies are shared with the people or countries that provide them. Ensuring the fair and equitable sharing of benefits from the use of genetic resources is one of the three objectives of the Convention on Biological Diversity (CBD) (Pushpangadan, 2002; Tiwari, 2006). Biological Diversity Act of India, 2002 and Biodiversity Rules, 2004 have included several Access and benefit sharing elements (Tables 1.4 and 1.5). When the biological resources are accessed for commercial utilization or the bio-survey and bio-utilization leads to commercial utilization, the applicant shall have the option to pay the benefit sharing ranging from 0.1 to 0.5 per cent at the following graded percentages of the annual gross ex-factory sale of the product which shall be worked out based on the annual gross ex-factory sale minus government taxes as detailed as follows - Annual Gross ex-factory sale of product Rupees 1,00,00,000 (0.1 per cent), between Rupees 1,00,00,001 and 3,00,00,000 (0.2 per cent) and above Rupees 3,00,00,000 (0.5 per cent) (Anonymous, 2014b).

Table 1.4: ABS Elements in the Biological Diversity Act, 2002, India

ABS Element	Relevant Articles
Competent National Authority	
Biodiversity Management Committees	Art. 41 (2)
Functions and powers (NBA)	Art. 18
Functions and powers (SBB)	Art. 23-24
National Biodiversity Authority (NBA)	Art. 8
State Biodiversity Board (SBB)	Art. 22
Prior Informed Consent (PIC) Procedures	
Approval of NBA for access	Art. 3, 5, 7, 19 (1)
Mutually Agreed Terms (MATs)	
Approval of NBA for transfer of research results (collaborative research projects)	Art. 4-5, 20 (1-2)
Biodiversity Management Committees: collection of fees	Art. 41 (3)
Determination of equitable benefit-sharing by NBA	Art. 21
Fees	Art. 19 (3), 20 (3)
Compliance Mechanisms	
Penalties (foreigners and locals)	Art. 55-57
Traditional Knowledge Associated to Genetic Resources	
Duties of Central and State Governments	Art. 36 (5)
Exemption for local people and communities (access)	Art. 7
Other	
Approval from NBA for application for intellectual property rights	Art. 6, 19 (2)
Constitution of Trust Funds	Art. 27, 32, 43

Source: NBA, Chennai (www.nbaindia.org).

Table 1.5: ABS Elements in the Biological Diversity Rules (2004), India

ABS Element	Relevant Articles and Sections
Competent National Authority	
General function of the Authority	Rule. 12
Prior Informed Consent (PIC) Procedures	
Procedure for access, traditional knowledge, intellectual property rights application, material transfer to third party	Rule. 14, 17-19
Mutually Agreed Terms (MATs)	
Criteria for equitable benefit sharing	Rule. 20
Compliance Mechanisms	
Dispute settlement	Rule. 23
Revocation of access or approval	Rule. 15
Traditional Knowledge Associated to Genetic Resources	
Same as PIC	Rule. 14, 17-19
Other	
Procedure for seeking prior approval before applying for IPRs	Rule. 18

Source: NBA, Chennai (www.nbaindia.org).

Fair and Equitable Benefit Sharing Options

The following options, either one or more, may be applied in accordance with mutually agreed terms between the applicant and the NBA, on a case by case basis, in accordance with the provisions of sub-rule (3) of rule 20 of the Biological Diversity Rules, 2004. These options are indicative in nature and other options, as approved by the NBA in consultation with the Central Government, may also be adopted:

(a) Monetary Benefits Options

(i). Up-front payment; (ii). One-time payment; (iii). Milestone payments; (iv). Share of the royalties and benefits accrued; (v). Share of the license fees; (vi). Contribution to National, State or Local Biodiversity Funds; (vii). Funding for research and development in India; (viii). Joint ventures with Indian institutions and companies; (ix). Joint ownership of relevant intellectual property rights.

(b) Non-monetary Benefits Options

(i). Providing institutional capacity building, including training on sustainable use practices, creating infrastructure and undertaking development of work related to conservation and sustainable use of biological resources; (ii). Transfer of technology or sharing of research and development results with Indian institutions/individuals/entities; (iii). Strengthening of capacities for developing technologies and transfer of technology to India and/or collaborative research and development programmes with Indian institutions/individuals/entities; (iv). Contribution/collaboration related to education and training in India on conservation and sustainable use of biological resources; (v). Location of production, research, and development units and measures for conservation and protection of species in the area from where biological resource has been accessed, contributions to the local economy and income generation for the local communities; (vi). Sharing of scientific information relevant to conservation and sustainable use of biological diversity including biological inventories and taxonomic studies; (vii). Conducting research directed towards priority needs in India including food, health and livelihood security focusing on biological resources; (viii). Providing scholarships, bursaries and financial aid to Indian institutions/individuals preferably to regions, tribes/sects contributing to the delivery of biological resources and subsequent profitability if any; (ix). Setting up of venture capital fund for aiding the cause of benefit claimers; (x). Payment of monetary compensation and other non-monetary benefits to the benefit claimers as the NBA may deem fit.

Biodiversity Fund

National, State and Local level biodiversity funds are established under Sections 27, 32 and 43 of the Act respectively. The funds are intended for channeling benefits to bonafide claimers and for conservation and promotion of biological resources and/or the associated knowledge and socio-economic development of these areas (Pandravada *et al.*, 2008; 2012).

National Biodiversity Fund (NBF)

As per the Section of 27 (1) of the Act any grants and loans, all charges, royalties received by the NBA may be called National Biodiversity Fund. The fund shall be applied for channelling benefits to the benefit claimers, conservation and promotion of biological resources and development of areas from where such biological resources or knowledge associated thereto has been accessed and for socio-economic development of areas in consultation with the local bodies.

State Biodiversity Fund (SBF)

The State Biodiversity Fund created as per Section 32 (1) of the Act receives any grants and loans made to the SBB/any grants or loans made by the NBA/all sums received by the SBB from other sources may be decided upon by State Government as State Biodiversity Fund. The SBF shall be used for the management and conservation of heritage sites, compensating or rehabilitating any section of the people economically affected by notification, conservation and promotion of biological resources and socioeconomic development of areas from where such biological resources or knowledge associated thereto has been accessed.

Local Biodiversity Fund (LBF)

The Local Biodiversity Fund is created at every area notified by the State Governments where any institution of self-Government is functioning and there shall be credited thereto, any grants or loan made by NBA, fees/levy charges by way of collection fees from any person for accessing or collecting any biological resources for commercial purposes. The LBF shall be used for conservation and promotion of biodiversity in the areas falling within the jurisdiction of the concerned local body and for the benefit of the community.

Biological Diversity Rules, 2004

To implement the Biological Diversity Act 2002, Rules have been notified in 2004 (Anonymous, 2004). The salient features of the BD Rules 2004 are:

☆ Procedures for appointment of Chairperson and Members of the Authority, conduct of authority meetings, and general functions of the authority are described in the Rules 3-8, 10 and 12 respectively.

☆ The process to regulate activities for access to biological resources and associated traditional knowledge in accordance with the Sections 3 (Access to Biological Resources), 4 (Transfer of Research Results) and 6 (Seeking 'No objection Certificate' for obtaining patent) under the Biological Diversity Act, 2002, are given in Rule 14, 17 and 18 respectively.

☆ The Procedure to revoke written agreements, action in prohibiting access and recovery of damages (Rule 15)

☆ Restricting access of endangered, endemic and rare species, restricting access in case of adverse environmental impact, genetic erosion, ecosystem function and purposes contrary to national interest as well as international agreements (Rule 16)

☆ Imposing terms and conditions for ensuring equitable sharing of benefits on access, transfer of results of research, application for patent/IPR claims (Rule 20)

☆ Constitution of Biodiversity Management Committee (BMC) and Preparation, maintenance and validation of People's Biodiversity Register (PBR) in consultation with the local people should be done as per Rules 22 (2) and 22 (6) respectively.

☆ Appeal for settlement of disputes between NBA and SBB or between SBBs is dealt under Rules 23.

The Rules also have prescribed formats *viz.,* Form-I (Access to Biological resources and associated traditional knowledge), Form-II (Transfer of Research results), Form-III (Seeking 'No objection Certificate' for obtaining patent) and Form-IV (Seeking approval for Third Party Transfer).

Conclusions

Biodiversity is the basis for ecosystem services, which are essential for human well-being and economic development. In addition to its intrinsic value, biodiversity and ecosystems are of tremendous economic value as well. However, many ecosystem services are not traded on markets and their value is not properly reflected in existing market prices for goods and services. Consequently, there is concern in economic decision-making in their conservation and sustainable use. The economic quantification under the Convention on Biological Diversity seeks to elucidate this 'hidden' economic value of ecosystem services and the underlying biodiversity and incorporate it into market prices through the use of incentive measures that favour the conservation and sustainable use of biodiversity by the native people. Zoos, aquariums, botanical gardens and nature shows are popular.

Yet, beyond the knowledge of a few charismatic species, the general public is not aware of the critical role biodiversity plays in providing the essentials for our survival and well-being. This means less public support for actions and policies towards a more sustainable relationship between humanity and the biodiversity of the planet. The International Day for Biological Diversity, held every 22 May and organised around special themes, provides an excellent opportunity for states and individuals to celebrate and commemorate the importance of biodiversity.

Levels of greenhouse gases in the atmosphere are rapidly increasing, warming the Earth's surface and lower atmosphere. Higher temperatures lead to climate change that includes effects such as rising sea levels, changes in precipitation patterns that can produce floods and droughts and the spread of vector-borne diseases. Biodiversity can also help reduce the effects of climate change. The conservation of habitats, for example, can reduce the amount of greenhouse gases released into the atmosphere. The conservation and sustainable use of biological diversity is critically important for meeting the food, fodder, fibre, health, water and other needs of growing world population for which purpose, access to and sharing of both genetic resources and technologies are essential. In this regard, the CBD and BDA are determined to facilitate conservation and sustainable utilization of biological diversity for the benefit of present and future generations.

References

Anonymous. 1994. *Conservation of Biological Diversity in India*: An Approach. Ministry of Environment and Forests, Government of India

Anonymous. 2000. *Sustaining life on earth: How the Convention on Biological Diversity promotes nature and human well-being*. Secretariat of the CBD, Canada 21p.

Anonymous, 2002. Bonn Guidelines on Access to Genetic Resources and Fair and Equitable Sharing of the Benefits Arising out of their Utilization (2002), by Secretariat of the Convention on Biological Diversity, Canada.

Anonymous, 2004. *The Biological Diversity Act, 2002 and Biological Diversity Rules, 2004*. National Biodiversity Authority, Chennai, P.57

Anonymous, 2014a. Nagoya Protocol on Access to Genetic Resources and the Fair and Equitable Sharing of Benefits Arising from their Utilization (ABS), entered into force on 12 October 2014, A supplementary agreement to the Convention on Biological Diversity. CBD Secretariat, Canada

Anonymous, 2014b. Guidelines on Access to Biological Resources and Associated Knowledge and Benefits Sharing Regulations (GABRAKBSR, 2014), enacted by Ministry of Environment, Forests and Climate Change, through National Biodiversity Authority.

Cropper, A. 1993. Convention on Biological Diversity. Environmental Conservation, 20, pp 364-364.

Dirk Lanzerath and Minou Friele (Eds.) 2014. Concepts and values in biodiversity. Routledge studies in biodiversity politics management. Taylor and Francis Group, London 363 p.

https: //www.cbd.int/convention accessed on 20.04.206.

http: //www.nbaindia.org accessed on 21.04.2016

Kannaiyan, S. and Gopalam, A (Eds). 2007. *Agro-biodiversity* volume II. Associated Publishing Company, New Delhi p.372.

Kannaiyan, S. and.Gopalam, A (Eds). 2007. *Biodiversity in India – Issues and Concerns.* Associated Publishing Company, New Delhi, p.430.

Kloppenburg, J. 1988. Seeds and Sovereignty. The use and control of plant genetic resources. Durham, NC. Duke University Press.

National Biodiversity Authority 2003 available at http: //nbaindia.org 22/4/2016.

Pandravada, S.R., Sivaraj, N., Kamala, V., Sunil, N., Sarath Babu, B and Varaprasad, K.S. 2008. Agri-biodiversity of Eastern Ghats - Exploration, Collection and Conservation of Crop Genetic Resources. Proceedings of the National Seminar on Conservation of Eastern Ghats. Environment Protection Training and Research Institute, Hyderabad. P. 19 – 27.

Pandravada, S.R., Sivaraj, N., Kamala, V., Sunil, N., Sarath Babu, B and Chakrabarty, S.K. 2012. *Rytaangaaniki medhoparamina hakkulu pondadaaniki vupayogapade chattaalu.* National Bureau of Plant Genetic Resources, Regional Station, Rajendranagar, Hyderabad-500 030, Andhra Pradesh. 4 pp.

Pushpangadan, P. 2002. Biodiversity and emerging benefit sharing arrangements-challenge and opportunities for India. *Proc. Indian Natl. Acad B* 68: 297-314.

Putterman, D.M. 1994. Trade and biodiversity convention. *Nature,* 371: 553-554.

Rodgers, W.A. and Panwar, H.S. 1990. *Planning a Protected Area Network in India.* Vol. 1 and 2. Wildlife Institute of India, Dehra Dun.

Sarat Babu G.V. and Sujata Arora, 2004. Convention on biological diversity and Indian biodiversity legislation. (In) Dhillon B.S., Tyagi, R.K., Arjun Lal and Saxena, S (Eds.). *Plant genetic resource management.* Narosa Publishing House, New Delhi pp.375- 383.

Smale, M. 1996. Understanding global trends in the use of wheat diversity and international flows of wheat genetic resources. *Economics working paper* No.96-02., "CIMMY" Mexico.

Smale, M., Reynolds, M.P., Warburton, M., Skovmand, B., Trethowan, R., Singh, R.P., Ortiz-Monasterio, I and Crossa, J. 2002. Dimensions of diversity modern spring bread wheat in developing countries since 1965. *Crop Science* 42: 1766-1779.

Singh, B.K. 2004. *Biodiversity conservation and management.* Mangaldeep publishers, Jaipur

Tiwari, S.P. 2006. *Regulatory and operational mechanisms as related to agro-biodiversity, Second and Revised Edition.* NAARM (ICAR) Publication, pp.1-216.

UNEP, 1992. Convention on Biological Diversity. United Nations Environment Programme. Nairobi, Kenya.

Varaprasad,K.S and Pandravada,S.R. 2005. Pantala Janyusampada Parirakshanalo Todpade Sanketika Paddatulu Mariyu Chattalu. *Annadata*. 37 (3): 22-23.

Varaprasad, K.S. and Sivaraj, N. 2010. Plant genetic resources conservation and use in light of recent policy developments. *Electronic journal of Plant Breeding* 1 (4): 1276-1293.

Varaprasad,K.S., Pandravada,S.R., Sivaraj,N and Sharma,S.K. 2007. Agrobiodiversity hotspots in Eastern Ghats - Issues and Challenges. Workshop on Agrobiodiversity hotspots and access and benefit sharing. NBA, Chennai, PPVFRA, New Delhi and Annamalai University, Annamalai Nagar. Abstracts. p. 59.

Venkataraman, K. 2006. Biodiversity legislations in like minded mega diversity countries. (In) D.D. Verma, S. Arora and R.K. Rai (Eds.) Perspectives in Biodiversity. Ministry of Environment and Forests, Govt. of India, New Delhi pp.79-92

Venkataraman, K. 2009. India's biodiversity act 2002 and its role in conservation. *Tropical Ecology* 50 (1): 23-30

Chapter 2

Assessment on Endemism and Biodiversity among 34 Hotspots of the World

Aparna Banerjee and Rajib Bandopadhyay*

UGC-Centre of Advanced Study, Department of Botany,
The University of Burdwan, Golabag, Burdwan – 713 104, West Bengal
*E-mail: rajibindia@gmail.com

Global scenario of biodiversity is changing as the sensitive factors; such as land use, vegetation, climate, atmospheric green house gas concentration *etc.* are changing fast (Sala, 2000). The biodiversity is considered by diversity of life on Earth, and is calculated at different levels of biological set up together with genes, species and ecosystems along with its interactions. In 1992, Rio de Janeiro world summit, importance of biodiversity was the core point, which resulted in the Convention on Biological Diversity (CBD). The aims of CBD were 'conservation of biological diversity, sustainable use of its components, and fair and equitable sharing of the benefits from the use of genetic resources'. But anthropogenic impacts resulted in gigantic biodiversity loss, including ecosystem disturbance, global warming, sea-level increase, habitat destruction, and loss of species genetic diversity. Now a days, species are heading towards extinction at the highest rate from the time when the mass extinction of dinosaurs (Banerjee, 2016).

Biodiversity Hotspots

A 'biodiversity hotspot' is a biogeographic stretch which should be both an important pool of biodiversity and also in danger of extinction. Species richness, endemism and threat to extinction- are the 3 factors that usually determine a hotspot. A biodiversity hotspot should accomplish two important criteria: it must include at least 1500 endemic vascular plants and also have 30 per cent or less of its natural flora, *i.e.* it should be threatened. Around the world, total 34 biodiversity hotspots

(Figure 2.1) correspond to only 2.3 per cent of the Earth's land area, but they still consist of more than half of the world's endemic plant species and almost 43 per cent of endemic mammals, birds, reptiles and amphibian species (Banerjee, 2016).

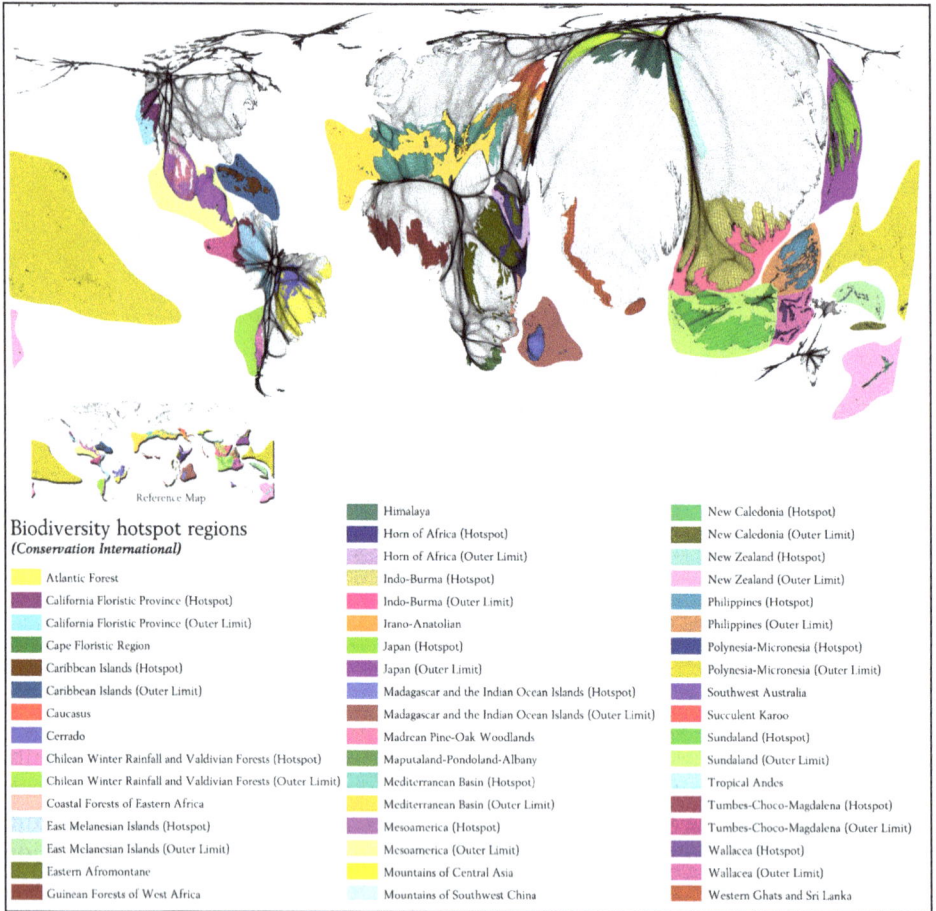

Figure 2.1: Different Biodiversity Hotspots of the World and its Outer Limit Region.
(http://www.viewsoftheworld.net/?p=2330)

There are a total of 34 biodiversity hotspots distributed in 10 different zones around the world (Table 2.1).

North and Central America

California Floristic Province

This hotspot presents a higher level of endemism in plants than animals. Of the 7,031 vascular plant varieties found over here; 2,153 taxa of 25 genera are endemic. Today, only about 80,000 km² or 24.7 per cent of the original vegetation

Table 2.1: Different Biodiversity Hotspots and their Unique Features

Different Zones	Biodiversity Hotspots	Unique Features
North and Central America	California Floristic Province	Home to giant sequoia, world's largest living organism
	Madrean pine-oak woodlands	Important center of pine and oak diversity
	Mesoamerica	4th in terms of total biodiversity
		2nd highest mammal and bird diversity
The Caribbean	Caribbean Islands	Center of globally threatened species
South America	Atlantic Forest	Centers of endemism
	Cerrado	largest savanna region in South America and biologically richest savanna in world
	Chilean Winter Rainfall-Valdivian Forests	Home to endemic amphibian family Rhinodermatidae
	Tumbes-Chocó-Magdalena	Most threatened tropical forest
	Tropical Andes	1st in terms of total biodiversity
		Highest bird diversity with both resident and migrant bird species
Europe	Mediterranean Basin	3rd in terms of total biodiversity
Africa	Cape Floristic Region	Temperate Mediterranean-type region with entire floral kingdom
		Highest non-tropical concentration of higher plant species
	Coastal Forests of Eastern Africa	Contains 11 species of wild coffee, with 8 endemic; none are exploited commercially
	Eastern Afromontane	Center of endemic chameleons
	Guinean Forests of West Africa	1st in terms of mammal diversity
	Horn of Africa	Entirely arid hotspot
	Madagascar and the Indian Ocean Islands	10th largest hotspot
		8th in terms of remaining intact habitat
	Maputaland-Pondoland-Albany	Highest tree diversity among temperate forests
		2nd richest African floristic region after Cape Floristic Region
	Succulent Karoo	Entirely arid hotspot

Different Zones	Biodiversity Hotspots	Unique Features
Central Asia	Mountains of Central Asia	Consists 2 Asia's major mountain ranges, the Pamir and the Tien Shan; arid but consist of about 20,000 glaciers
South Asia	Eastern Himalaya	It is all biodiversity hotspot, megadiversity country (India), Crisis ecoregion and among global ecoregion
	Indo-Burma	8 Endemic Bird Area are here
	Western Ghats and Srilanka	Only 20per cent of original forest cover remains in virgin state
South East Asia and Asia-Pacific	East Melanesian Islands	Richest diversity of mammals among bats
	New Caledonia	One of the smallest hotspots in the world
	New Zealand	Biogeographically similar with New Caledonia by undersea Norfolk Rise; split away from Gondwanaland together 40 million years ago.
	Philippines	Entirely both hotspot and megadiversity country
	Polynesia-Micronesia	Wide about 12 principal vegetation biomes are here
	Southwest Australia	One of five Mediterranean-type ecosystems in the world
	Sundaland	2nd in terms of total biodiversity
	Wallacea	3rd highest in terms of mammal endemism after Madagascar and Sundaland
East Asia	Japan	Japanese giant salamander can grow to more than one meter length and is one of the world's largest amphibians
	Mountains of Southwest China	One of the 25 richest and most threatened hub of plant and animal life on Earth
West Asia	Caucasus	Highest biological diversity among temperate forests of the world
	Irano-Anatolian	Known for threatened and endemic viper reptiles

remains intact. It has giant sequoia forests, California oak woodlands, redwood forests, sagebrush steppes, prickly pear shrublands, coastal sage scrub, chaparral, juniper-pine woodland, upper montane-subalpine forests, alpine forests, riparian forests, cypress forests, mixed evergreen forests, douglas fir forests, coastal dunes, mudflats and salt marshes. A few examples of plants that are both endemic to the province and endangered are Baker's larkspur, Gowen cypress, Hickman's potentilla, Point Reyes bird's beak, Santa Cruz Tarplant and Santa Rosa Island Manzanit *etc.* (Hickman, 1993).

Madrean Pine-oak Woodlands

With a unique area of 461,265 km², this pine-oak woodland is composed *Quercus* sp., *Pinus* sp., *Pseudotsuga* sp. and *Abies* sp. This region is habitat to 25 per cent of Mexico's plant species and Mexico alone is home to 44 of the 110 species of pine and more than 135 species of oak, which is over 30 per cent of the world's total oak species. Two endemic oak species found only in the Sierra Madre Occidental are *Quercus carmenensis* and *Q. deliquescens*. Forests in the Baja California Peninsula have an important diversity of pine trees, including *Pinus lambertiana*, which can grow up to 70 meters and produces pine cones of 70 cm in length. Around 525 bird species are found in the hotspot, of which more than 20 are endemic in the middle Sierra Madre Occidental, the Trans-Mexican Volcanic Belt and the southern Sierra Madre Oriental. Bird Life International defined 4 Endemic Bird Areas overlapping with the hotspot and together support more than 15 restricted-range species, most of which became threatened. Nearly 330 mammals occur in the hotspot, six are endemic. More than 380 species of reptiles happen, nearly 40 of which are endemic. Amphibian diversity in the hotspot is remarkable with nearly 200 species, about 25 per cent of which are endemic. There are more than 80 fish species present here, around 25 per cent of which are endemic. The best-known invertebrates in the Madrean Pine-Oak Woodlands are butterfly; approximately 160-200 species inhabit, of which about 45 are endemic (Lot, 1993).

Mesoamerica

This hotspot has an approximate 24,000 species of vascular plants, of which 21 per cent are endemic. This region is the fourth highest in terms of total diversity by the Tropical Andes, Sundaland, and Mediterranean hotspots, while the figure for endemism ranks 10th on the global list. Vertebrate diversity is even more impressive; as mammal diversity is the second highest on the hotspot list, with 521 mammal species where 195 species are rodents alone, exceeded only by the Guinean Forests of West Africa. Of these, 210 *i.e.* remarkably 40 per cent are endemic to the hotspot. Resident bird species number 1,052 and migrant species 141, for a total of 1,193, second only to the Tropical Andes. Of these, 251 *i.e.* 21 per cent are endemic, again exceeded only by the Tropical Andes. BirdLife International recognizes 17 Endemic Bird Areas within the hotspot, exceeded only by the Tropical Andes. The region is a critical flyway for at least 225 migratory species; three of the Western Hemisphere's four migratory bird routes cover Mesoamerica (http://www.eoearth.org/view/article/150625/).

The Caribbean

Caribbean Islands

Caribbean Islands Hotspot comprises 30 nations and territories, each characterized by unique and wide-ranging biodiversity. It is one of world's greatest centers of endemic biodiversity as a result of the region's geography and climate: an archipelago of tropical and semi-tropical islands. Caribbean's biodiversity is at solemn risk of species extinctions. Above 700 species are globally threatened, making Caribbean one of the top hotspots assessed by CEPF for globally threatened species. All of the 189 amphibian native species of Caribbean Islands are endemic. Caribbean Islands support 92 terrestrial mammal species, of which 23 are now considered extinct. Of 69 existing species, 51 are endemic and 27 are globally threatened, amounting to 39 per cent of known mammal species; as assessed by Global Mammal Assessment (IUCN and Conservation International) in 2008 with support of CEPF. More than 560 bird species have been recorded in the Caribbean Islands Hotspot (Raffaele et al., 1998); of which 148 are endemic and 9 per cent are globally threatened. Bird endemism is high at species level; remarkable 36 genera of birds are endemic, as well as 2 endemic families. With more than 520 native species, Caribbean islands are rich in reptiles, 95 per cent of which are endemic. The Caribbean Islands is home to 1,447 native genera and about 11,000 native species of seed plants (Cycadopsida, Coniferopsida, Magnoliopsida and Liliopsida). Generic endemism is especially significant; about 13.2 per cent comprising 191 genera are endemic. The hotspot supports 167 species of freshwater fish, about 65 of which are endemic.

South America

Atlantic Forest

The high biodiversity in the Atlantic Forest is a function of extreme environmental variation in this biome. One of the most important factors for this variation is the 38° latitudinal span of the hotspot. The geographic distribution of lizards in the Atlantic Forest, for example, is significantly affected by latitude, with only one wide-ranging species in this area. The second major source of variation is elevation, as forests extend from sea level up to 1,800 meters, with corresponding gradients of biodiversity. Finally, inland forests differ noticeably from coastal ones. These factors combine to generate a unique diversity of landscapes supporting extraordinary biodiversity. The Atlantic Forest contains an estimated 250 species of mammals (55 endemic), 340 amphibians (90 endemic), 1,023 birds (188 endemic), and approximately 20,000 trees, 50 per cent are endemic. More than two-thirds of the primates' species are endemic. Centers of endemism have been recognized in the Atlantic Forest (http://wwf.panda.org/about_our_earth/ecoregions/atlantic_forests.cfm).

Cerrado

The Cerrado region of Brazil comprises 21 per cent of the country's total area and is the most extensive woodland-savanna of South America. With a distinct dry season, it supports a unique array of drought and fire -adapted plant species and surprising numbers of endemic bird species. The Cerrado has a high diversity of vertebrates; 150 amphibian species, 120 reptile species, 837 bird species and 161 mammal species (Myers, 2000). One recent study found 57 lizard species in one cerrado area with the high diversity driven by the availability of open habitat. *Ameiva ameiva* is the largest lizard found in the Cerrado and is the most important lizard predator where it is found (Noguiera, 2009). There is relatively high diversity of snakes in the Cerrado (22-61 species, depending on site) with Colubridae being the richest family. The open nature of the *cerrado* vegetation most likely contributes to the high snake diversity (Franc, 2008).

Chilean Winter Rainfall-Valdivian Forests

Continental island bounded by Pacific Ocean on the west, Andes Mountains on the east and Atacama Desert in north, Chilean Winter Rainfall-Valdivian Forests harbours richly endemic flora and fauna. The hotspot covers 397,142 km² of central-northern part Chile and western edge of Argentina, stretching from Pacific coast to Andean mountains. It encompasses about 40 per cent of Chile's total land area. Of nearly 4,000 vascular plants found over here, 50 per cent are endemic. It represents about three-fourth of all Chilean plant species and 40 per cent is endemic nearly. Plant diversity in winter-rainfall area is around 3,539 species compared to Valdivian Forests which support only about 1,284 species; 1,769 *i.e.* 50 per cent of which is endemic. Typically, birds are not very well represented over here and consist of around 225 species. Mammal endemism is relatively low, with roughly 70 species and 15 endemics. Whereas endemism levels for both reptiles and amphibians are high. Regarding two-thirds of hotspot's more than 40 reptile species, are endemic. Three-quarters of more than 40 amphibian species of this hotspot, is endemic. Although the hotspot has a relatively small fish fauna, with only over 40 native species, it has remarkable two endemic families: mountain catfishes (Nematogenyidae) and perch-like fishes of genus *Percilia* (Perciliidae).

Tumbes-Chocó-Magdalena

Tumbes-Chocó-Magdalena biodiversity hotspot includes tropical moist forests and tropical dry forests of Pacific coast of South America and Galapagos Islands. It is 1,500 km long and encircles 274,597 km² area. As the hotspot is near to Pacific Ocean, factors that threaten its biodiversity are farming interference, deforestation and population growth. Whereas the Panamanian and Colombian portion of hotspot are relatively intact, approximately 98 per cent of native forest in coastal Ecuador is cleared; portraying it the most threatened tropical forest of world.[1] The hotspot includes ecoregions of Chocó-Darién moist forests, Ecuadorian dry forests, Guayaquil flooded grasslands, Gulf of Guayaquil-Tumbes mangroves, Galápagos Islands xeric scrub, Magdalena Valley montane forests, Magdalena-Urabá moist forests, Manabí mangroves, Tumbes-Piura dry forests, Piura mangroves and Western Ecuador moist forests. Galápagos Islands are home to nearly 700 species of vascular plants, 25 per cent of which are endemic, including six endemic flowering plant genera. Tumbes-Chocó-Magdalena forests are globally important for bird endemism, holding nearly 900 bird species and 110 of which are endemic;

including 14 endemic bird genera, 10 of which have only single species. There are more than 285 mammal species over here; 11 of these are endemic. It is estimated that more than 320 reptile species inhabit over here, of which nearly 100 are endemic. The region's most diverse reptile family is Colubridae, world's largest snake family, with 122 species (16 endemic). With more than 200 amphibian species, around 30 are endemic. The hotspot has about 250 fish species; about 50 per cent of which are endemic, with 7 endemic genera.

Tropical Andes

The richest and most diverse region on Earth, Tropical Andes spans 1,542,644 km², from western Venezuela to northern Chile and Argentina, including Colombia, Ecuador, Peru, and Bolivia. It is home to an estimated 30,000-35,000 vascular plant species, accounting 10 per cent of world's total species. It is also world's leader in plant endemism, with an estimated 60 per cent of these species found nowhere else on Earth; meaning nearly 7 per cent of the world's vascular plants are endemic to 0.8 per cent of Earth's land area. There are also roughly 330 endemic genera and a single endemic family Columelliaceae. Tropical Andes harbour more than 1,700 bird species, nearly 600 of which are endemic, highest in world. It also has 66 endemic bird genera and 21 Endemic Bird Areas defined by BirdLife International. There are almost 570 mammal species over here; about 75 of these are endemic and virtually 70 are threatened. The hotspot also has 6 endemic genera represented by only one species: Garlepp's mouse (*Galenomys garleppi*), Andean rat (*Lenoxus apicalis*), little or mountain coati (*Nasuella olivacea*), puna mouse (*Punomys lemminus*), fish-eating rat (*Anotomys leander*) and yellow-tailed woolly monkey (*Oreonax flavicauda*). There are more than 600 reptile species identified in Tropical Andes hotspot, more than 270 of which are endemic and three endemic genera. It is the most important region in the world for amphibians, with around 980 species and more than 670 endemics with 8 endemic genera. Best-known amphibians from Tropical Andes are brightly colored poison dart frogs of Dendrobatidae family. Around 450 amphibian species are listed on the 2004 IUCN Red List as threatened. There are more than 375 species of freshwater fishes in the hotspot, of which more than 130 are endemic.

Europe

Mediterranean Basin

The Mediterranean Basin Hotspot is one of the most extraordinary places on Earth and is remarkable for both its high level of biological diversity and spectacular nature. Its location at the intersection of two major landmasses Eurasia and Africa and the huge topographical variety and altitudinal differences- from sea level to 4,165 meters in the west Morocco and 3,756 meters in the east Turkey; are major contributing factors to its biodiversity. The mammal fauna of the Mediterranean Basin includes more than 330 species. Of these, 87 terrestrial mammals are endemic to the hotspot, with rodents, shrews, moles and hedgehogs. The avifauna of the hotspot consists of 600 species, including 16 endemics. There are a significant number of species that migrate from Europe to Africa crossing the Mediterranean Basin at various points. There are 357 species of reptiles including two species of marine turtle; of which 170 species are endemic. A total of 115 amphibian species occur in the basin, including 71 endemics. The freshwater fish in the region are derived from the rich faunas of Eurasia and Africa. Of the 400 species of freshwater fish in the hotspot, 253 are endemic.

Africa

Cape Floristic Region

78,555 km^2 of Cape Floristic Region is located entirely within borders of South Africa. It is one of the 5 temperate Mediterranean-type regions on Conservation International hotspots list and is one of only two hotspots that encompass an entire floral kingdom like New Caledonia. It is home to greatest non-tropical concentration of higher plant species in world, with 9,000 species; of which more than 69 per cent are endemic. 5 of South Africa's 12 endemic plant families and 160 endemic genera are found only over here. Species richness and local endemism is greatest in Cape Peninsula (471 km^2) which alone supports 2,256 species including 90 endemics. Of 320 land bird species, only 6 are endemic and are considered as an Endemic Bird Area by BirdLife International. Happily none of the endemic birds are considered threatened. There are about 90 species of mammals here, 4 of which are endemic. Reptile diversity in Cape Floristic Region is relatively high, with about 100 species, 25 per cent endemic. 5 tortoise species are found almost exclusively within this Region. It includes the angulate tortoise (*Chersina angulata*); parrot-beaked tortoise (*Homopus areolatus*); geometric tortoise (*Psammobates geometricus*) and leopard tortoise (*Geochelone pardalis*). There are more than 40 species of amphibians in here with 16 endemics, including Table Mountain ghost frog *Heleophryne rosei*, which is found only on the slopes of Table Mountain between 240-1,060 meters above sea level. Of nearly 35 native freshwater fish species in here, about 12 are endemic. Of more than 230 species of butterflies, about 30 per cent are endemic.

Coastal Forests of Eastern Africa

Coastal Forests of Eastern Africa stretches along eastern edge of Africa, from small patches of coastal forest along Jubba and Shabelle Rivers in southern Somalia, south through Kenya; stretching further south into Tanzania along the coast of Mozambique, ending at Limpopo River with a total area of 291,250 km^2; south of which is Maputaland-Pondoland-Albany Hotspot. There are about 4,050 vascular plant species and approximately 1,750 *i.e.* 43 per cent are endemic with 28 monotypic endemic genera. The best-known plants here are the species of African violets (*Saintpaulia* spp.). It also contains 11 species of wild coffee, 8 of which are endemic and none of these are commercially exploited. Nearly 200 mammals are found over here with 11 endemics. This relatively tiny hotspot boasts 3 endemic monkey species. The hotspot supports considerable populations of threatened large African herbivores, including black rhinoceros *Diceros bicornis* and savannah elephants *Loxodonta africana*. The hotspot has over 85 amphibian species, with 6 endemics; one species Loveridge's snouted toad *Mertensophryne micranotis*, the only member of its genus; is remarkable as it breed by internal fertilization. Nearly 220 fish species live in fresh waterways of Coastal Forests of Eastern Africa and more than 30 are endemic. Some species of fish have remarkable adaptations to survive in hotspot's temporary coastal swamps and floodplains. For example, air-breathing lungfishes *Protopterus amphibious* and *P. annectens* can survive in a dormant state for over a year in cocoons underneath dried mud. Levels of endemism within some invertebrate groups are significantly higher than among vertebrates. About 80 per cent of millipedes and 68 per cent of molluscs are endemic.

Eastern Afromontane

This hotspot has widely scattered but biogeographically similar mountain ranges in eastern Africa, from Saudi Arabia and Yemen in north to Zimbabwe in south with around one million km^2 areas. The hotspot holds nearly 7,600 species of plants, of which more than 2,350 are endemic. Grasslands of Southern Rift are particularly rich in orchids including more than 500 species; Nyika Plateau supports nearly 215 orchid species with 4 species and two endemic subspecies. High plateau of Ethiopian Highland has giant *Lobelia rhynchopetalum* growing about 2-3 meters height before producing a single dark bluish-purple inflorescence of around 9 meters height. About 1,300 bird species occur here and about 110 are endemic. Eastern Arc and Southern Rift Mountains has an Endemic Bird Area defined by BirdLife International. Nyika National Park on Nyika Plateau in Southern Rift supports world's largest breeding population of blue swallow *Hirundo atrocaerulea*. Two Endemic Bird Areas are also included in Albertine Rift; Southern Highlands and Central Ethiopian Highlands. It is home to nearly 500 mammal species and bout 100 endemics. Africa's larger flagship mammals; elephant *Loxodonta Africana* and leopard *Panthera pardus* are here. Three primate species are endemic to Eastern Arc Mountains and Southern Rift; Sanje mangabey *Cercocebus sanjei*, Udzungwa red colobus *Procolobus gordonorum* and mountain dwarf galago *Galagoides orinus*. The well-known mountain gorilla *Gorilla beringei beringei* is restricted to about 380 individuals in Virungas. Ethiopian wolf *Canis simensis* is an endemic species found in Afroalpine ecosystem of Ethiopian Highlands with less than 450 individuals in 7 small and isolated populations. Nearly 350 reptile species are found here with about 90 endemic. It is also home to about 230 amphibian species, nearly 70 of which are endemic. Albertine Rift alone is home to about 19 per cent of Africa's amphibian species, including more than 30 endemic species and 3 monotypic endemic genera: Parkers tree toad *Laurentophryne parkeri*, Itombwe golden frog *Chrysobatrachus cupreonitens* and African painted frog *Callixalus pictus*. Great Rift Lake in this area is a vital region for freshwater fish diversity with about 890 fish species and about 620 are endemic. Lake Malawi is home to more than 380 fish species, nearly 90 per cent are endemic. Up to 1,300 butterfly species occur in Albertine Rift alone, including around 120 endemic species and one endemic genus *Kumothales*.

Guinean Forests of West Africa

Guinean Forests of West Africa hotspot encompasses all the lowland forests of West Africa. It is home to an estimated 9,000 vascular plant species, about 20 per cent are endemic. Nearly 2,500 plant species have been recorded on Mount Cameroon. Because of their relative isolation from rest of the hotspot, this region supports around 185 highly endemic floras. Among many economically important plant species; oil palm *Elaeis guineensis* is native here, valuable timber species including African ebony *Diospyros gracilis*, African mahogany (*Entandophragma and Khaya*) and iroko *Milicia excelsa* are widely exploited. It supports about 785 bird species, of which almost 75 species and 7 genera are endemic. Dozens of region's bird species are threatened by extensive deforestation occurring all over here. BirdLife International has recognized 6 Endemic Bird Areas lying partly or entirely here. Conservation of Mount Oku forest is last remaining hope for 2 species- Bannerman's turaco *Tauraco bannermani* and banded wattle-eye *Platysteira laticincta*. Mammal diversity here is very high. West Africa's estimated 320 species represent more than 25 per cent of total mammal species found on entire Africa. More than 60 of these are endemics with 18 primate species. Guinean Forests are renowned for primate diversity, with about 30 species. So, these forests have been identified as Africa's most critical primate conservation area. It encompasses populations of Africa's great apes; chimpanzees *Pan troglodytes* and western lowland gorillas *Gorilla gorilla gorilla*. While reptiles are not well documented in West Africa, it is suggested with more than 200 species and 25 per cent endemic. Among distinctive endemics; Los Archipelago worm lizard (*Cynisca leonina*), Benson's mabuya (*Mabuya bensonii*) and Liberia worm snake (*Typhlops leucostictus*) are some of leading reptiles. Amphibians are also poorly documented here, about 225 amphibians inhabit here with 80 endemics. Huge and threatened Goliath frog *Conraua goliath* is here growing up to 30 cm long and weighs about 3.3 kg. Fish diversity is quite remarkable here, with more than 510 freshwater fishes, 35 per cent of which are endemic. About 25 per cent of the world's 350 species of killifish inhabit here with 50 per cent endemics. Cichlid fishes are also major, with more than 50 per cent of over 60 species are endemic to the hotspot. Four of the five endemic genera of cichlids are found only in Lake Barombi Mbo of Cameroon.

Horn of Africa

This entire hotspot covers more than 1.5 million km² areas. The leading vegetation types are *Acacia-Commiphora* bushland, evergreen bush land, succulent shrub land, dry evergreen forest and woodland, semi-desert grassland and low-growing dune and rock vegetation. Small areas of mangrove are found on both African and Arabian sides of the hotspot, as well as riverine vegetation along Wabe Shabelle and Awash rivers. Horn of Africa is one of the only 2 hotspots that are entirely arid; other is the Succulent Karoo of south western Africa. It is supposed that these two arid regions were joint by an arid corridor in the Pleistocene and earlier Tertiary. Several flowering plant genera are totally restricted to just these two regions; *Kissenia* with one species in arid Horn and one in Succulent Karoo and *Wellstedia* with six species in arid Horn and one in Succulent Karoo. Of about 5,000 vascular plant species of the region; about 2,750 are endemics. Additionally, Horn of Africa is home to two endemic plant families: Barbeyaceae and Dirachmaceae; represented by two threatened species *Dirachma socotrana* and *D. somalensis*. Several native tree species provided raw materials for some of Horn of Africa's most important commodities, including frankincense, myrrh and dragon's blood or cinnabar; as resin producing trees. Dragon's blood is also used as medicine and dye. Of 697 bird species recorded, 24 are endemic. Bird Life International defined 4 Endemic Bird Areas entirely within the hotspot. Nearly 220 mammal species are found in Horn of Africa, with about 20 endemics. Hamadryas or sacred baboon *Papio hamadryas* that was treated as sacred in ancient Egypt and mummified is now endemic to arid Horn, living on hillsides and adjacent southern Red Sea and Gulf of Aden. Having over 90 endemics of 285 species, Horn of Africa's highest levels of endemism occurs in reptiles. Unlike reptiles, amphibians are rather poorly documented; nearly 30 species recorded and 6 endemics. About 100 freshwater fish species inhabit in Horn of Africa, 10 of which are endemic. Populations of *Aphanius dispar* is introduced to some waters as part of an anti-malaria program.

Madagascar and the Indian Ocean Islands

With 587,045 km² area, *i.e.* 0.4 per cent of earth's surface, Madagascar is largest oceanic island and 4th largest island on Earth. In terms of original extent of its native habitat, Madagascar and the Indian Ocean Islands is 10th largest of the 34 biodiversity hotspots identified by Conservation International. It ranks 8th among the hotspots in terms of remaining intact habitat which is roughly 18 per cent of the original extent, according to estimates of tropical forest cover. Madagascar and its neighbouring islands have astonishing 8 plant families, 4 bird families and 5 primate families endemic in total. Madagascar's more than 50 lemur species are the island's fascinating worldwide ambassadors for conservation, though unfortunately more than 15 of which is now extinct. Seychelles, Comoros and Mascarene islands in Indian Ocean support a number of Critically Endangered bird species. Seychelles are also home to the only endemic family of amphibians: Sooglossidae; and Aldabra giant tortoise, region's most heralded endemic reptile. Together these smaller Indian Ocean islands add 1 endemic plant family, 50 endemic genera and 904 endemic species to Madagascar, raising total 11 endemic plant families, about 310 endemic genera and 8,904- 10,504 endemic species. No other hotspot has this much endemism; only Tropical Andes, Sundaland and Mediterranean exceed the number. Madagascar's overall bird diversity has reported 250 species with 115 endemic. BirdLife International highlights 11 Endemic Bird Areas here, 6 of which are at critical level. List of critically endangered birds in this hotspot is among the highest in the 34 hotspots identified by Conservation International. Of Madagascar's 300 reptile species, 274 are endemic with 36 out of 64 genera. Mascarenes add 5 endemic reptile species, Comoros 22 with 7 endemic and Seychelles 15 with 14 endemic. Island of Aldabra is home to endemic giant tortoise *Geochelone gigantea*. Unitedly this hotspot is home to about 327 reptile species with 301 endemic. Entire hotspot includes 190 amphibian species, of which 187 are endemic. Madagascar and Indian Ocean Islands Hotspot is 7th among hotspots in terms of amphibian diversity and 5th in endemism. Mammalian diversity in Madagascar is also unique, 78 of 117 species are endemic; excluding bats and introduced species approaching 100 per cent.

Maputaland-Pondoland-Albany

This hotspot is along east coast of southern Africa; from southern Mozambique and South African Mpumalanga in north, through eastern Swaziland to Eastern Cape of South Africa in south. It is geologically complex, as 3 high diversity areas are merged: Maputa land in north, Pondo land in south and Albany in southwest. About 80 per cent of South Africa's remaining forests fall here. These warm temperate forests are home to nearly 600 tree species having highest tree diversity of any of world's temperate forests. This hotspot is important plant endemism center and 2nd richest African floristic region after Cape Floristic Region. Total about 8,100 plant species from 243 families occur here and about 25 per cent are endemic; including 39 endemic genera and one endemic family Rhynchocalycaceae, represented singly by *Rhynchocalyx lawsonioides* in Pondoland. It is part of BirdLife International's Southeast African Coast Endemic Bird Area, with 4 restricted-range species: Rudd's apalis *Apalis ruddi*, pink-throated twinspot *Hypargos margaritatus*, Neergaard's sunbird *Nectarinia neergaardi* and lemon-breasted seedeater *Serinus citrinipectus*. Of nearly 200 mammal species found here, 4 are endemic; red bush squirrel *Paraxerus palliates*, four-toed elephant shrew *Petrodromus tetradactylus*, Marley's golden mole *Amblysomus marleyi* and giant golden mole *Chrysospalax trevelyani*. Above 200 reptile species are here and 30 are endemic. In restricted habitat, it has about 7 dwarf chameleon *Bradypodion* spp. and endemic Natal black snake *Macrelaps microlepidotus*. Hotspot's All 72 amphibian species are frogs with 11 endemic and 8 threatened. About 70 native freshwater fish species, 20 are endemic; including 4 barb species of *Barbus* spp. It has an exceptionally rich and diverse invertebrate fauna; including number of butterfly and moths, velvet worms, gigantic earthworms, dung beetle *etc.*

Succulent Karoo

From southwestern South Africa to southern Namibia, Succulent Karoo covers 102,691 km^2 deserts. It is mainly consist of winter rainfall desert and is divided into 2 zones; Namaqualand extending along west coast of South Africa and southern Namibia. The mild climate has contributed to evolution of a rich array of endemic species. 2nd zone Southern Karoo experiences peaks of rainfall in spring and autumn and has more extreme climate variations than the Namaqualand desert. Dwarf shrub land dominated by leaf succulents is found throughout the hotspot. These drought-adapted plants have thick, fleshy leaves or stems for water storage. In Succulent Karoo, there are about 1700 species of leaf succulents and is dominant is among the world's deserts. For an arid region, this hotspot has extremely high plant endemism. In total, there are more than 6350 vascular plant species in this hotspot, nearly 40 per cent of which is endemic. Avifauna of the Succulent Karoo includes more than 225 species, one of which is endemic: Barlow's lark *Certhilauda barlowi*. Major concentrations of large mammalian vertebrates include elephant *Loxodonta africana*, black rhinoceros *Diceros bicornis* and African buffalo *Syncerus caffer*. Reptile diversity is relatively high in Succulent Karoo, with more than 90 species, about 15 of which are endemic. All of the endemics are geckos and lizards. As rest of South Africa, tortoise diversity is very high in Succulent Karoo, with 7 taxa and 2 endemic: Namaqualand tent tortoise *Psammobates tentorius trimeni* and Namaqualand speckled padloper *Homopus signatus signatus*. Amphibians are poorly represented here with just over 20 species. All of these are frogs, including endemic desert rain frog *Breviceps macrops*. There are around 26 species of freshwater fish here, none of which are endemic. Of 70 species of scorpions found in here, nearly

20 are endemic. Monkey beetle that is exclusively found in southern Africa, are concentrated in this hotspot.

Central Asia

Mountains of Central Asia

Mountains of Central Asia consists 2 of Asia's major mountain ranges, Pamir and Tien Shan. Hotspot's 860,000 km^2 area include southern Kazakhstan, most of Kyrgyzstan and Tajikistan, eastern Uzbekistan, western China, northeastern Afghanistan and part of Turkmenistan. Flora over here is a mix of Boreal, Siberian, Mongolian, Indo-Himalayan and Iranian. There are about 5,500 vascular plant species here with nearly 1,500 endemic; 64 endemic genera, including 21 of Umbelliferae family and 12 of Compositae family. Endemic flora includes several tree species, grasses (*Atraphaxis muschketovii* and *Stipa karatavica*) and wild onion. Although nearly 500 bird species occur regularly, none are endemic. Six of the 140 mammals found here are endemic: Menzibier's marmot *Marmota menzbieri*, found only in western Tien Shan above 2,000 meters; Ili pika *Ochotona iliensis*, a small lagomorph species found only in Chinese portion of Tien Shan; 2 susliks or ground squirrels *Spermophilus ralli* and *S. relictus*; Pamir shrew *Sorex bucharensis* and Alai mole vole *Ellobius alaicus* only from Alai Mountains in southern Kyrgyzstan. Nearly 60 reptiles are here, though only one is endemic; skink *Asymblepharus alaicus*. Diversity is highest in lower altitudes, desert and semi-desert areas. Only 7 amphibian species are recorded, 4 are endemic; including salamander *Ranodon sibiricus* in Dzhungarian Alatau Range at northern Tien Shan end; frog *Rana terentievi* from southern Tajikistan. This arid region has about 30 freshwater fish species, 5 are endemic. Endemism is centered in Lake Issyk-Kul Basin of Kyrgyzstan, which has no connection with other water bodies. Additionally, Kugitang blind cave fish *Troglocobitis starostini* is only in Kugitang Mountains. Rich insect diversity is observed in alpine meadows. Eleven of 26 apollo butterfly species are endemic. There are also 87 endemic molluscs, including Kokand freshwater clam *Colletopterum kokandicum*, restricted to a single lake in Fergana Valley.

South Asia

Eastern Himalaya

Among 34 biodiversity hotspots in the world, Eastern Himalaya range (EH) is one of the richest with almost 750,000 km^2 region covering Nepal, Bhutan, Indian states West Bengal, Sikkim, Assam and Arunachal Pradesh, southeast Tibet and northern Myanmar. EH region is in attention as a part of crisis ecoregions, biodiversity hotspots, endemic bird areas, megadiversity countries and Global 200 Ecoregions. Also, 15 per cent area of EH has 99 protected areas. From a total of 60 in HinduKush-Himalayan region, the EH region has 25 ecoregions. This area is geologically young and shows high altitudinal difference from tallest alluvial grasslands among world and subtropical broadleaf forest in foothills to temperate broadleaf forest in midhills, conifer forest in higher hills and alpine meadows above tree line. More than 7000 plants species, 175 mammal species and above

Figure 2.2: Landscapes of different Eastern Himalayan Region.
A: Upland forest at Chele la, Bhutan; B: Alpine region at Dochula, Bhutan; C: Alpine zone at Gnathang valley, Sikkim and D: Lowland forest at Pedong, Sikkim.

500 bird species are recorded from EH, which alone consists of many endemic and endangered flora and fauna. Thus the EH biodiversity hotspot is itself unique; but is also threatened due to deforestation and habitat fragmentation, animal poaching, mining, construction of roads and dams, pollution due to agrochemicals *etc.* (Banerjee, 2016).

Indo-Burma

Indo-Burma encompasses 2,373,000 km² of tropical Asia, east to Ganges-Brahmaputra lowlands. Indo-Burma hotspot itself is home to 7000 endemic plants and has 1.9 per cent of world's total endemic vertebrates. There are over 1,260 bird species found in Indo-Burma with more than 60 endemic. There are about 430 mammal species here with about 70 species and seven endemic genera. Nearly 520 reptile species are found over here with 12 genera and over 200 endemic species. Nine of endemic genera are represented by a single species, including pit viper from Vietnam *Triceratolepidophis sieversorum*. Endemism is specially high among Colubidae family snakes. There are more than 280 amphibian species in Indo-Burma hotspot with over 150 endemic. High endemism is not at genus; only 3 of 46 genera are restricted here: *Ophryophryne*, *Bufoides* and *Glyphoglossus*. *Bufoides* and *Glyphoglossus* comprise single species: Khasi Hills toad *B. meghalayanu* from northeastern India, while *G.*

molossus is localized but widespread in hotspot. Indo-Burma has a remarkable freshwater fish fauna, with more than 1,260 species or about 10 per cent of world's freshwater fishes. More than 560 of these are endemic with 30 genera and one family; Indostomidae or armored sticklebacks; a family remotely related to marine seamoths.

Western Ghats and Sri Lanka

Western Ghats extending along west coast of India covers an area of 180,000 km^2; comprises major portion of Western Ghats and Sri Lanka Hotspot. Because it is a largely montane area that receives between 2,000-8,000 mm annual rainfall within a short span of 3-4 months, it performs important hydrological and watershed functions. Only one-third of Western Ghats region is under natural vegetation. Also existing forests are highly fragmented and facing rising degradation. It is amazingly rich in biodiversity. Although total area is less than 6 per cent of Indian land area, Western Ghats contains more than 30 per cent of all plant, fish, herpetofauna, bird and mammal species of India. Minimum 6,000 vascular plant species are here with more than 52 per cent endemic. Western Ghats harbours approximately 5,000 vascular plants species belonging to nearly 2,200 genera with about 1,700 endemic species. More than 450 known bird species are here with about 35 endemic. More than 20 species are endemic to Sri Lanka, mostly from lowland rainforests and montane forests of island's southwestern region. Both Western Ghats and Sri Lanka are considered as Endemic Bird Areas by BirdLife International. It is home to about 140 mammal species, although less than 20 are endemic. While mammal diversity is lower here than in some other tropical hotspots, it does support a significant diversity of bats with about 50 species and one endemic genus represented by *Latidens salimalii* in High Wavy Mountains of Western Ghats. It has important Asian elephant population; *Elephas maximus*. Highest level of vertebrate endemism is in reptiles and amphibians. Of region's more than 260 reptile species, about 66 per cent are endemic. Nearly 140 of about 190 strictly freshwater fish species are endemic here. There are also 9 endemic genera, including Malpulutta; found only in Sri Lanka. More than 100 of nearly 140 tiger beetle species are endemic. Srilanka hosts more than 50 known species of endemic freshwater crabs (Parathelphusidae family) with all endemic.

South East Asia and Asia-Pacific

East Melanesian Islands

East Melanesian Islands is northeast and east of New Guinea and includes Bismarck and Admiralty Islands, Solomon Islands and Vanuatu islands. In total, this hotspot includes about 1,600 islands with land area of nearly 100,000 km^2 more than double of Polynesia-Micronesia Hotspot. Because most of the islands of this hotspot were never in land contact with New Guinea, their fauna and flora are a mix of recent long-distance immigrants and indigenous lineages derived from ancient Pacific-Gondwanaland species. There are an estimated 8,000 vascular plant species, about half of which are endemic to the region. Overall, it is home to more than 360 regularly occurring bird species with around 40 per cent endemic. The region has 6 Endemic Bird Areas defined by BirdLife International; counting

Solomon group with more than 60 endemic species. This hotspot is known by skinks and geckos and majority of hotspot's more than 110 reptiles (50 per cent endemic) are of Gekkonidae and Scincidae familiy. The region also has six endemic reptile genera; where 5 are represented by a single species, including large prehensile-tailed skink *Corucia zebrata*, a lizard that lives in trees and feeds primarily on the leaves of epiphytes. More than 40 amphibian species are recorded from this hotspot, with over 90 per cent endemic. There are also 4 endemic genera of amphibians, 2 of which have single species: *Palmatorappia solomonis* from Solomon Islands and *Ceratobatrachus guentheri* on Solomon Islands and Bougainville and Buka islands. It is estimated that there are more than 50 freshwater fish species in the hotspot, although only a few are endemic, among them *Stenogobius alleni* is found on New Britain Island of Papua New Guinea, and *Stiphodon astilbos* in Vanuatu. Butterflies are best-known invertebrates; Ornithoptera (birdwing) butterfly, particularly *O. allotae* and *O. Victoriae*; large and spectacular species and blue emperor swallowtail (*Papilio ulysses*) in Bismarcks and Solomons.

New Caledonia

New Caledonia is one of the smallest hotspots in the world. This group of islands is located in South Pacific of Melanesian region. The region's 18,972 km² area consist of main island of Grande Terre and smaller Loyalty Islands to east, Belep and Surprise Islands to north and Isle of Pines to south. Endemism is specially high among vascular plants. About 3270 plant species are recorded with 74 per cent endemic. There are also 108 endemic genera and 5 endemic plant families: Amborellaceae, Paracryphiaceae, Strasburgeriaceae, Oncothecaceae and Phellinaceae. Out of around 100 birds found in New Caledonia, more than 20 are endemic. There are 3 endemic genera, 2 of which are monotypic, including kagu *Rhynochetos jubatus*, only living member of the endemic family Rhynochetidae. All of New Caledonia's 9 land mammal species are bats; 5 Microchiropters and 4 Megachiropters or flying fox. Six of them are endemic, including a newly described long-eared bat *Nyctophilus nebulosus* discovered in 1991 around Nouméa and 2ⁿᵈ time from same area in 2000. There is an extremely high level of reptile endemism in this hotspot. More than 60 of about 70 terrestrial reptiles are endemic, with 11 of 23 genera. Nearly all of these species are lizards in two families of geckos and one family of skinks. Best known are giant gecko of endemic genus *Rhacodactylus*. Hotspot's two snake species, pacific boa *Candoia bibroni* and a small burrowing blind snake *Ramphotyphlops willeyi* of Typhlopidae family restricted to Loyalty Islands. There are no native amphibians reported on New Caledonia. Aquatic diversity is high in terms of hotspot's size with about 85 freshwater fish species, although less than 10 are endemic. Most notable species is endemic galaxiid *Galaxias neocaledonicus*, northernmost representative of a group mostly restricted to southern tips of New Zealand, Australia, South America, and Africa. A single genus *Protogobius* is endemic. Among invertebrates, rich endemic diversity of land snails are there, although only 200 species are described out of 400-600 species. Largest of these snails *Placostylus fibratus* grow more than 15 cms long and weigh up to 100 grams. The hotspot has an estimated 37 species of macro-crustaceans with 40 per cent endemic.

New Zealand

An archipelago 2000 km² southeast of Australia in southern Pacific Ocean, New Zealand hotspot covers 270,197 km² on 3 main islands (North Island, South Island and Stewart Island) and several smaller surrounding islands. Plant endemism is very high in New Zealand. Nearly 1900 of about 3400 vascular plant species are endemic. Endemism also extends to genus level; 35 plant genera are endemic. Endemic monotypic genus *Desmoschoenus spiralis*, a coastal plant, is used by Maori people in traditional building construction. Fern *Loxoma cunninghamii* is one of the hotspot's "living fossils". Nearly 200 bird species regularly occur in New Zealand, almost 90 of which are endemic. 5 Endemic Bird Areas are identified by BirdLife International. The hotspot also has 17 endemic bird genera and 3 endemic bird families: Acanthisittidae, Apterygidae and Callaeidae. Also, it is only hotspot to have an endemic bird Order, represented by flightless kiwis (*Apterygiformes*); national bird of New Zealand. Land mammal species native to hotspot are endemic bats, one of which is the only living representative of Mystacinidae family: New Zealand lesser short-tailed bat (*Mysticina tuberculata*) and 2nd land mammal species found in New Zealand is long-tailed wattled bat (*Chalinolobus tuberculatus*). Nearly 40 species of reptiles are found in New Zealand and all are unique to hotspot; fauna comprises geckos and skinks only and there are no native snake species. In addition, remarkably five of six reptile genera are endemic. It also boasts entire endemic order tuataras (Rhynchocephalia), which resemble iguanas and primitively existed since dinosaur age and have well-developed third eye. Amphibians are represented by just 4 primitive frog species of endemic family Leiopelmatidae and all 4 are threatened, among them Archey's frog (*Leiopelma archeyi*) occurring on North Island is severely affected by chytrid fungus. Of nearly 40 freshwater fish species native to New Zealand, about 25 are endemic. Fish fauna is dominated by Galaxiidae family. It is home to 28 gastropod species of for which IUCN Red List is available, 6 of which are considered to be Data Deficient by IUCN Red List. Two species of land snail are threatened, *Rhytida clarki* and *R. oconnori*.

Philippines

The world's 2nd largest archipelago country after Indonesia, Philippines includes more than 7,100 islands covering 297,179 km² in westernmost Pacific Ocean. Philippines is north to Indonesia and east of Vietnam. It is one of the few nations that is entirety both a hotspot and a megadiversity country. One-third of about 9,250 vascular plant species native to Philippines are endemic. Plant endemism is mainly at species level; there are no endemic plant families and 26 endemic genera; Gingers, begonias, gesneriads, orchids, pandans, palms and dipterocarps particularly. There are more than 150 species of palms in the hotspot and around two-third is endemic. Of about 1,000 orchid species found in Philippines, 70 per cent is endemic. There are over 530 bird species in Philippines with 35 per cent endemic and over 60 threatened. BirdLife International identified 7 Endemic Bird Areas here: Mindoro, Luzon, Negros and Panay, Cebu, Mindanao and Eastern Visayas, Sulu archipelago, and Palawan. At least 165 mammal species are found in Philippine and over 100 are endemic; one of the highest levels of mammal endemism in any hotspot. Endemism is high at generic level as well, with 23 of 83 genera. Largest and most impressive mammal is tamaraw *Bubalus mindorensis*, a dwarf water buffalo living on Mindoro Island. Reptiles are about 235 species with 68 per cent endemic and 6 endemic genera, counting snake genus *Myersophis*, represented singly by *Myersophis alpestris*. Philippine flying lizards of *Draco* genus are well represented with about 10 species. Endemic freshwater crocodile *Crocodylus mindorensis* is most threatened crocodile in world. Of about 90 amphibian species, 85 per cent are endemic; unique panther flying frog *Rhacophorus pardalis* has special adaptations for gliding, including extra skin flaps and webbing between fingers and toes to generate lift during glides. Frog genus *Platymantis* is well represented with 26 endemic species; with 22 threatened. Philippines has more than 280 inland fish, including 9 endemic genera and over 65 endemic species. *Sardinella tawilis*, a freshwater sardine is found only in Taal Lake. About 70 per cent of Philippines' nearly 21,000 recorded insect species endemic. About one-third of 915 butterflies found are endemic and over 110 of more than 130 species of tiger beetle are found nowhere else.

Polynesia-Micronesia

This hotspot includes all the islands of Micronesia and Polynesia, plus Fiji; scattering across 40 million km² Pacific Ocean. This vast span has about 4,500 islands of 11 countries, 8 territories and one U.S. state Hawaii. It is one of the smallest hotspots in terms of land area and covers only 46,488 km² area. Polynesia-Micronesia is native to about 5,330 vascular plant species with more than 3,070 endemic. One family Degeneriaceae is endemic to Fiji and includes single tree species *Degeneria vitiensis*. Birds are dominant terrestrial vertebrates in Pacific and are home to about 290 regularly occurring species, with about 160 endemic. Avian endemism increases with isolation and topographic diversity of the islands and most endemic species are found in larger and higher islands. BirdLife International identified 15 Endemic Bird Areas here, having 8 Critically Endangered bird species, 6 endangered and 4 vulnerable. Only terrestrial mammal native here is 15 bat species with 11 endemic. Fijian monkey-faced flying fox *Pteralopex acrodonta*, one of the most primitive species of fruit bats in world, is only mammal endemic to Fiji. There are more than 60 native terrestrial reptile species in Polynesia-Micronesia, including 7 snakes, saltwater crocodile *Crocodylus porosus* and more than 50 lizards. Over 30 reptile species are endemic. Reptile fauna is almost entirely Indo-Pacific in origin, with exception of two iguana species, Fiji banded iguana *Brachylophus fasciatus* and Fiji crested iguana *Brachylophus vitiensis*, which are endemic to Fiji-Tonga area. Although there are no truly freshwater fishes in Polynesia-Micronesia other than introduced species, nearly 100 native species are found in freshwater as adults (having pelagic marine larval stages); about 20 are endemic. Invertebrate diversity in Polynesia-Micronesia is high for particularly land snails. Of the 13 major indigenous pulmonate land snail families on Pacific Islands, 4 are endemic to the central Pacific.

Southwest Australia

Southwest Australia occupies about 356,717 km² area of Western Australia. Of around 5,570 vascular plant species found here, nearly 2,950 are endemic. A major number of genera; 87 of 697 are found nowhere else in world. Additionally, 4 families are endemic: Ecdeiocoleaceae, Emblingiaceae, Eremosynaceae and monotypic Cephalotaceae, represented by carnivorous Albany pitcher plant *Cephalotus follicularis*. Over 280 native bird species occur here, 12 are endemic. Level of endemism is slightly higher than in other Mediterranean-type hotspots and is considered an Endemic Bird Area by BirdLife International. It is home to about 22 parrot species with 3 endemics. This hotspot has about 60 native mammal species with 12 endemic; including mouse-sized, nectarivorous honey possum *Tarsipes rostratus*, only representative of Tarsipedidae family. Of about 175 reptile species found here, 15 per cent are endemic. Western swamp turtle *Pseudemydura umbrina*; an endemic, monotypic genus member; is the most threatened reptile in Australia and is one of 25 most threatened freshwater turtles worldwide. Wild population stands less than 100 and is now only found in one or two swamps near Perth. There are more than 30 amphibian species recorded for this hotspot, nearly two-thirds are endemic, including four species representing endemic genera: turtle frog, Nicholl's toadlet, sandhill frog and harlequin frog. This hotspot has very little freshwater habitat leading to low fish diversity. Thus, only about 20 native freshwater fish species is recorded over here. Yet endemism is comparatively high among species, about half of endemic species are included in 3 whole genera. Most remarkable one is salamander fish *Lepidogalaxias salamandroides*; only member of the hotspot's single endemic family Lepidogalaxiidae.

Sundaland

Sundaland's biological diversity is exceptionally high covering western half of Indo-Malayan archipelago with about 17,000 equatorial islands. It has 2 biggest islands in world: Borneo (725,000 km^2) and Sumatra (427,300 km^2). Sundaland is one of the biologically richest hotspots on Earth, having about 25,000 species of vascular plants with 60 per cent endemics. Plant family Scyphostegiaceae is confined here and is represented singly by *Scyphostegia borneensis*. At least 117 endemic plant genera are here; 59 are in Borneo, 17 in Sumatra and 41 on Malay Peninsula. Sumatran forests contain more than 100 dipterocarp species, nearly a dozen of which are endemic and Java has more than 270 endemic orchids. Prominent plants include genus *Rafflesia* with 16 species. *Rafflesia arnoldii* flowers largest in world. Of approximately 770 regularly occurring bird species in Sundaland, nearly 150 are endemic and about 40 are threatened. With 20 confined species, Bornean Mountains are one of five Endemic Bird Areas recognized by BirdLife International, in addition to Sumatra and Peninsular Malaysia, Enggano, Java and Bali Forests, and Javan Coastal Zone. Of Sundaland's more than 380 mammal species, over 170 are endemic. Moreover 17 of 136 genera are also endemic. Borneo boasts most endemic mammal species; about 25. Slowly maturing orang-utans with low reproductive rate are threatened by habitat loss due to logging, fires and agricultural activity. Other flagships are Proboscis monkey *Nasalis larvatus*, found only on Borneo and 2 threatened rhinoceros species, least known of the 5 surviving rhino species on Earth. There are over 450 species of reptiles, roughly 250 of which are endemic, including 24 genera. One of the exclusive reptiles here is endemic false gharial *Tomistoma schlegelii*, freshwater crocodilian species growing up to 4.7 meters found in Sumatra and Borneo. It is also home to several Endangered and Vulnerable tortoise and freshwater turtle species. Sundaland is home to about 240 amphibian species with nearly 200 endemic. There are about 1000 known freshwater fish species here with more than 25 per cent endemic. Borneo tops with about 430 species and about 160 endemics. One of best known fish here is dramatic Asian bony tongue or golden arowana *Scleropages formosus*, highly priced aquarium fish.

Wallacea

Wallacea hotspot covers central islands of Indonesia, east of Java, Bali and Borneo and west of New Guinea and whole Timor Leste; occupying total land area of 338,494 km². Flora of this region is not well known and about 10,000 vascular plants are there, with roughly 15 per cent endemic and about 12 endemic genera. Major trade value trees are tall kauri *Agathis* spp., magnificent yellow-flowered legume *Pterocarpus indicus* and gum tree *Eucalyptus deglupta*. There are about 650 frequently occurring bird species in Wallacea and almost 265 are endemic in 29 endemic genera. More than 125 of about 220 mammal species are endemic. One of the most unusual mammals in Wallacea is *Babyrousa babyrussa*. Babirusas are pig-like animals known by male's long recurved tusks penetrating upper lip. 2 species of anoas or dwarf buffaloes are endemic to Sulawesi forests: lowland anoa *Bubalus depressicornis* and mountain anoa *Bubalus quarlesi*. There are a number of endemic primates here. The island of Sulawesi is home to at least 7 species of endemic macaques and at least 5 endemic tarsier. Among around 220 reptile species, nearly 100 are confined. There are also 3 endemic genera, all snakes: *Calamorhabdium* with 3 species and both monotypic *Rabdion* and *Cyclotyphlops*. Best-known reptile species is Komodo dragon *Varanus komodoensis*, largest lizard in world, with males growing up to 2.8 meters length and weigh about 50 kgs. It is found only on Komodo, Padar, Rinca islands and Flores of Lesser Sundas, driest portion of Indonesia. About 50 amphibian species are native with more than 30 endemic, including Sulawesi toad *Bufo celebensis* and *Oreophryne monticola*. More than 300 freshwater fish species found here are tolerant of both fresh and saltwater with 75 endemics. Most of invertebrate fauna of Wallacea remained poorly known; except for bird-wing butterflies of swallowtail family. There are also 109 tiger beetles here, with 79 endemic. Northern Moluccas is habitat of world's largest bee *Chalocodoma pluto* with females growing up to 4 cms in length.

East Asia

Japan

The entire archipelago of Japan was declared a Biodiversity Hot Spot in 2005 because it is rich in unique animal and plant life and because this unique animal floral and faunal range is threatened by the encroachment of people. The Japanese Archipelago is composed of about 3,000 small islands covering approximately 370,000 km², of which about 62,000 km² is protected. The T-junction of 3 tectonic plates generates numerous hot springs and volcanoes creating a basis for unique environments and therefore propagating unique life. About 25 per cent of the vertebrate species occurring in this hotspot are endemic, including the Critically Endangered Okinawa woodpecker and the Japanese macaque, the famous snow monkeys. Japan has high amphibian diversity of as well, with 75 per cent endemics. Japan is home to roughly 5,600 species of vascular plants, about a third of which is endemic. Nearly 370 bird species occur regularly in Japan, although only 13 endemic. Japan is home to around 90 species of mammals and 50 per cent of these are endemic to the hotspot. Over 65 reptile species, almost 30 are endemic. The reptile fauna includes a number of important threatened species. Endemism is particularly high among amphibians, with 44 of 50 species found only in Japan. The hotspot has a relatively small freshwater fish fauna, with almost 215 native species, more than 50 of which are endemic. Some invertebrate groups are very well documented in Japan; for example, nearly 240 butterfly species are native to Japan, and nearly 25 species of tiger beetles are recorded from the hotspot with about 25 per cent endemic.

Mountains of Southwest China

An area of extreme topography, the Mountains of Southwest China biodiversity hotspot is home to several of the world's best-known mammals, including the giant panda, the red panda, the golden monkey and the snow leopard. It is one of the 25 richest and most threatened reservoirs of plant and animal life on Earth. The hotspot, which stretches from southeast Tibet through western Sichuan and extends into central and northern Yunnan, is also home to more than 12,000 species of higher plants, of which 29 per cent are unique to this hotspot. Of the more than 600 bird species occurring in the hotspot, only a single bird species is endemic – the white-speckled laughing thrush *Garrulax bieti*. Given its size and temperate climate, the Mountains of Southwest China hotspot is also home to a surprisingly wide diversity of reptiles and amphibians. There are more than 90 reptile species in the hotspot, comprising a little over 20 lizard species, and nearly 70 species of snakes. About 15 species are endemic, including the Szechwan pit viper *Trimeresurus xiangchengensis* and Kingdonward's bloodsucker *Calotes kingdonwardi*. Amphibians are represented by around 90 species in the hotspot, with the genera *Scutiger, Oreolalax* and *Amolops* being particularly well represented. The hotspot has more than 90 freshwater fish species, almost a quarter of which are endemic, including two endemic genera. The majority of fish in the hotspot are from two families, Cyprinidae and Balitoridae, while most of the endemic fish are from two genera, *Schizothorax* and *Triplophysa*.

West Asia

Caucasus

The Caucasus hotspot has the greatest biological diversity of any temperate forest region in the world. It shelters 6,500 species of vascular plants, at least 1,600 of which (25 percent) are unique to the region, and a number of endemic animals, including the East and West Caucasian turs, Armenian mouflon and Caucasian salamander. The 2010 IUCN Red List identifies 50 species of globally threatened animals in the Caucasus. The Caucasus Ecoregion is located at the border of Europe and Asia between the Black Sea and Caspian Sea. The ecoregion is considered on the of world's biodiversity hotspots with over 6,400 plant species, 348 birds, 131 mammals, 86 reptiles, 17 amphibians and 127 freshwater fishes.

Irano-Anatolian

Covering 899,773 km^2 areas of major parts of central and eastern Turkey, part of southern Georgia, Nahçevan Province of Azerbaijan, part of Armenia, northeastern Iraq, northern and western Iran and Northern Kopet Dagh Range in Turkmenistan; climate is continental with hot summers and very cold winters; with annual rainfall from 100-1,000 mm in winter and spring. Principal habitat is oak-dominant mountainous forest steppe, deciduous forests in Anatolia and Zagros mountains and juniper forest in southern slopes of Elburz mountains and Kopet Dagh. Wide zone of subalpine and alpine vegetation covers mountain peaks above timberline and thorn-cushions are found in subalpine zone. There are permanent glaciers in alpine zone of Turkey's Cilo and Hakkâri mountains. It is home to about 6,000 plant species, with 2,500 endemic. Of around 360 regularly occurring bird species; none are endemic. Yet, several globally threatened birds have important breeding populations in here. More than 140 mammals are found with roughly 10 endemic. Most of these are rodents, including Dahl's jird *Meriones dahli* and *Microtus quzvinensis*; a vole of northern Iran. There are more than 115 reptile species with about 12 endemic; including 4 endemic and threatened vipers of very restricted ranges: Darevsky's viper, mountain viper, Wagner's viper and Latifi's viper. Roughly 20 amphibian species occur over here, including two endemic and threatened salamanders of *Neurergus* genus: *N. Microspilotus* and *N. kaiseri*, found only in Iran. Both species are declining in number due to habitat destruction, pollution and drought. About a third of 90 freshwater fish species are endemic; mainly in closed-basin lakes and rivers. Many of these are globally threatened, including *Salmo platycephalus*. Endemic fish *Chalcalburnus tarichi* is effectively used model species by local conservation NGOs in Van Province to start conservation interest in Eastern Turkey. Invertebrate fauna here is not well studied; it is mostly rich in butterflies, with at least 350 species. About 240 of these are found in Turkey and nearly 20 are endemic. Several globally threatened species are here, with single-site endemic *Polyommatus dama*. This region the richest part of Palearctic region for scorpions with more than 40 described species; about 50 per cent of which is endemic (http://www.cepf.net/Pages/default.aspx).

All the 34 biodiversity hotspots around the world though contain only 2.3 per cent of world's land area, but are the hub of endemic plants and animals with several threatened species as described by IUCN red list. Additionally as a result of global warming and increasing pollution when rate of positive carbon emission is increasing; Eastern Himalaya biodiversity hotspot contains one such country that tops in negative carbon emission- Bhutan. This is a remarkable pollution control for a complete mountain locked country. Still Bhutan is not safe from the affect of nthropogenic impact because of its vulnerable position in EH region. Because of glacier melting and landslide Bhutan is facing severe species loss.

Threat to Biodiversity and Anthropogenic Impact

The scenario of biodiversity is changing globally and the magnitude of the change is so large that it is creating negative impact on overall ecosystem. Animal poaching, illegal farming, increase in concentration of atmospheric CO_2 due to industrialization, deforestation and faster urbanization, jhum cultivation *etc.* are some of the reason which is creating serious threat to biodiversity. Anthropogenic impact is increasing the ecological footprint day by day resulting into ecological outshoot. Specifically the equatorial region of earth is rich in biodiversity but it is also in vulnerable condition and most prone to biodiversity loss because of global warming, sea level increase, loss of mangroves, pollution and most importantly human encroachment. Though the equatorial region that is inhabitant of the poorest country of the world are deprived of even basic needs and does not contribute to increasing pollution that much; but to its position in equator, air pollutants flow towards the equatorial troposphere slowly. Also marshy environment of tropical region contribute to increasing methane concentration. Combining effect of all these; the most biodiversity rich tropical region is facing devastating key stone species loss.

To Overcome the Impossibilities and Biodiversity Conservation

To overcome the anthropogenic impact and major biodiversity change, more and more conservation effort is needed. Tumbes-Choco-Magdalena is currently the most threatened tropical forest and it needs immediate conservation. Caribbean island is in limelight for being the center of globally threatened species. Atlantic forest is declared as center of endemism. In terms of total biodiversity richness,

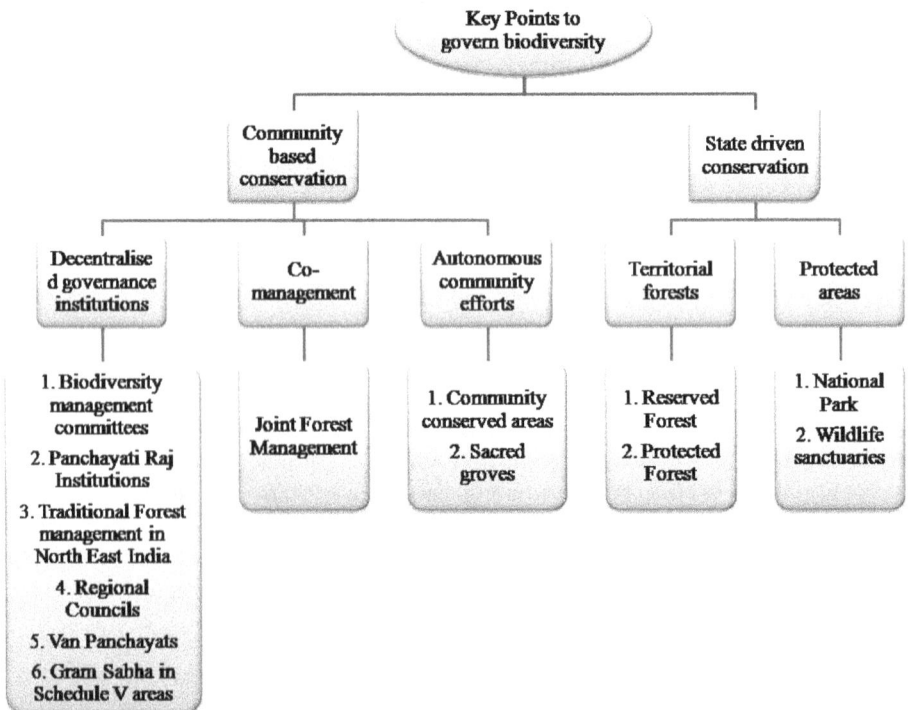

Figure 2.3: Key Factors of Biodiversity Conservation in India.

(http://www.indiawaterportal.org/articles/conservation-across-landscapes-study-various-mechanisms-biodiversity-conservation-india)

Tropical Andes, followed by Sundaland, Mediterranean and Mesoamerica are in the top list. Biodiversity conservation has normally followed one of the two approaches- setting aside protected areas and community-based conservation (Figure 2.3). Setting aside protected areas is important is government pay attention to country's protected areas and forests. But community-based conservation is particularly of importance because different committees and communities locally observe the biodiversity zones and here direct surveillance and quick attention is present. Also detection of new biodiversity hotspots across the earth will enable better conservation approach of endemics and threatened species.

Acknowledgement

Authors are thankful to UGC-Centre of Advanced Study, Department of Botany, The University of Burdwan for pursuing research activities. Aparna Banerjee is also thankful to UGC for the financial assistance of JRF (State Funded) [Fc (Sc.)/RS/SF/BOT./2014-15/103 (3)].

References

Banerjee, A. and Bandopadhyay R. 2016. Biodiversity hotspot of Bhutan and its sustainability. *Current Science*, 110 (4): 521-527.

Escalante, P. P., Navarro S.A.G., and Peterson, A.T. 1993. A geographic, ecological, and historical analysis of land bird diversity in Mexico. In T. P. Ramamoorthy, R. Bye, A. Lot and J. Fa. (Eds.),*Biological Diversity of Mexico: Origins and Distribution*. pp. 281-307. Oxford: Oxford University Press.

Franc, F.G.R., Mesquita, D.O., Nogueira, C.C., and Araujo, A.F. 2008. Phylogeny and ecology determine morphological structure in a snake assemblage in the central Brazilian Cerrado. *Copeia* 1: 23-38.

Goncalves, G., Marques, P.A., Granadeiro, C.M., Nogueira, H.I., Singh, M.K., and Gracio, J. 2009. Surface modification of graphene nanosheets with gold nanoparticles: the role of oxygen moieties at graphene surface on gold nucleation and growth. *Chemistry of Materials*, 21 (20), 4796-4802.

Hickman, J.C. 1993 (Ed). The Jepson Manual: Higher Plants of California. University of California Press, Berkeley, California

Myers, N., Mittermeier, R.A., Mittermeier, C.G., Da Fonseca, G.A., and Kent, J. 2000. Biodiversity hotspots for conservation priorities. *Nature*, 403 (6772): 853-858.

Raffaele, H., Wiley, J.0., Garrido, A.K., and Raffaele. J. 1998. A guide to the birds of the West Indies. Princeton University Press.

Sala, O.E., Chapin, F. S., Armesto, J. J., Berlow, E., Bloomfield, J., Dirzo, R., and Leemans, R. 2000. Global biodiversity scenarios for the year 2100. *Science*, 287 (5459): 1770-1774.

Chapter 3

Underutilized Fruits of Hot Arid Region

R.S. Singh, A.K. Singh, Sanjay Singh and Vikas Yadav

ICAR-Central Institute for Arid Horticulture,
Beechwal, Bikaner – 334 006, Rajasthan
E-mail: aksbicar@gmail.com

The Indian arid zone occupies 38.70 million hectares in the states of Rajasthan, Gujarat, Haryana, Punjab, Karnataka, Andhra Pradesh, is characterized by low and erratic rainfall (between 100 and 500 mm) with a coefficient of variation varying from 40 to 70 per cent and extremes of temperature (1–48°C), high wind velocity and sandy soils. Inhospitable climate, too deep or too shallow soils with low moisture, and poor water holding capacity and fertility, deep underground water, which is often brackish or saline, coupled with intense biotic pressure permits only specialised fruit plants, which are well adapted to these climatic, edaphic and biotic adversities and fluctuations. The soils of the north-western arid region described as 'desert soils' and 'grey brown soils' of the Order Aridisols having light texture. The environmental condition of arid region is harsh with very low rainfall and varies from 100 mm in north-western sector of Jaisalmer to 450 mm in the eastern boundary or arid zone of Rajasthan. Due to low and erratic rainfall pattern in arid region, appropriate technology is needed to increase productivity. The water holding capacity of soil is also poor. With the increasing biotic and abiotic pressure, most of the arid and semi-arid regions are confronted with the challenges of producing more per unit land with uncertain and dwindling supplies of water. Arid zone is characterized by high and low temperature and erratic rainfall, which limit the scope for high crops productivity. However, these conditions greatly favour the development of high quality in a number of fruits such as date palm, mulberry, phalsa, ber, aonla, bael, kinnow, lasoda *etc.* It is now realized that there is a limited scope for quantum jump in fruit production in the traditional production areas. The amelioration of the extreme conditions is also considered vital for life support to the inhabitants of this area. The area expansion and yield potential of arid horticultural crops has increased many folds, because of development of new varieties and advancement in agro-techniques in arid regions.

The major problem in development of horticulture scenario in arid parts of the country is lack of sufficient quality seed and planting material. Planting materials is a precious input to increase quality production of fruit crops. Thus, production of quality planting material and their supply to farmers will boost up the arid horticultural crops production.The post harvest management is essential to overcome the losses at different stages of grading, packing, storage, transport and finally marketing of both fresh and processed products. The weak processing infrastructure, as it exists today, has been one of the contributing factors for ineffective utilization of the raw materials resulting in huge post harvest losses.Lack of sufficient processing units for production of quality output is a major bottleneck for the extension of arid fruit crops. Marketing of horticultural produce is a major constraint in the production and disposal system and has a major role to play in making the industry viable. There is vast scope of agribusiness in horticulture sector. High cost of inputs and lack of enough incentives for production of quality varieties/ species, product diversification, value addition, *etc.* also hinder crops development. Lack of proper storage facility and knowledge and equipment for grading and packaging of fruits are also the constraints for the growers of hot arid region.

Constraints in Arid Region

The soil health of arid region is very poor in availability of nutrients, water holding capacity, texture *etc.* Most of arid areas (about 64.6 per cent) are duny where the soils often contain only about 3.2-4.0 per cent clay and 1.4-1.8 per cent silt. Besides this, about 5.9 per cent area is covered by soils having hard pan, 5.6 per cent is under hills and pediments, 6.8 per cent area is alluvial dunes and 1.6 per cent is sierozems extending from the soils of Haryana and the Punjab. In the peninsular India, a considerable part of arid region has red sandy soil and some parts have mixed black soils. The soils are poor in organic matter having organic carbon of 0.03 per cent in bare sand dunes to 0.1 per cent in the stabilized dunes. The water holding capacity of soil is also poor. Soils are generally rich in total potassium and boron but are low in nitrogen, phosphorus and micronutrients such as copper, zinc and iron. The soils often have high salinity. The ground water resource is not only limited owing to poor surface and sub-surface drainage but is also saline in quality. The irrigation water resources in the region are seasonal rivers and rivulets, surface wells and some runoff water storage devices (*e.g., nadi, tanka, khadins*) and canal irrigation in arid region. Thus, the water resources in arid region are limited and can irrigate hardly 4 per cent of the area.

The annual average rainfall in the arid regions is very low, erratic and varies from 100 mm in north-western sector of Jaisalmer to 450 mm in the eastern boundary or arid zone of Rajasthan. Most of the precipitation in north western arid region occurs during July-September in about 19-21 rain spells. Due to low and erratic rainfall pattern in arid region, appropriate technology is needed to increase productivity. Water is precious input in hot arid region of the country therefore, adoption of micro-irrigation system is desirable to save water and enhance productivity. For arid climatic conditions, the variety is needed which is resistant to biotic and abiotic stresses for sustainable production. In some parts of arid region, occurrence of frost is also common feature during winter season which affects vegetative growth of plants as well as productivity, quality of fruits especially in ber, lasoda, mulberry, phalsa and aonla. There is no heat tolerance variety of arid horticultural crops which should be developed to achieve higher production.

The high capital cost involved in establishing orchards, or rejuvenation of existing old unproductive plantation poses serious constraint in area expansion. The situation becomes all the more difficult in view of the large number of small holdings devoted to these crops which are essentially owned by weaker section, who have no means to invest, nor can afford to stand the burden of credit even if available. High cost of inputs and lack of enough incentives for production of quality varieties/species, product diversification, value addition, *etc.* also hinder crops development. Lack of proper storage facility and knowledge and equipment for grading and packaging of fruits and vegetables is also constraint for the growers of hot arid region. Another problem faced by the farmers is proper marketing of produce to get better returns. The development of storage facilities and agro- based processing units in the region is also needed for development of horticulture (More *et al.*, 2012).

Prospects of Arid Horticulture

There is a tremendous scope in expansion of horticultural crops in arid region and it has vast potential for changing scenario of horticulture of the country. Vast land resource, surplus family labours, increasing canal irrigated area, developing infrastructural facilities, plenty of solar and wind energy, *etc.* are the strength in arid region for research and development of arid horticulture. Further, low pressure of diseases/insects in the region is good scope for production of seed and planting material of horticultural crops. Ber is commercially grown in more than 80,000 ha with production of 0.9 million tones under semi arid and arid regions of the country. It requires more attention for value addition. Pomegranate area and production (1.14 million tones) is increasing very fast in dry parts of the country. Since, it has vast scope of export of this crop from semi arid and arid regions of the country. In Maharashtra alone the area under pomegranate cultivation is about 93,500 ha with production of 6,01,500 tones. Presently, the export value from pomegranate is Rs.92 million. This is being exported to Middle East, U.K., Germany, UAE, The Netherlands, Bahrain, Kuwait and Egypt. The crops like fig, mulberry, custard apple, lasoda, ker, aonla, bael and tamarind are also coming very well under dry land conditions. At present, fig is cultivated in more than 3,000 ha area in Maharashtra, Karnataka. Likewise, custard apple is grown in the state of Maharashtra, A.P., Karnataka, Rajasthan and Tamil Nadu. In foot hills of Arawali, custard apple is grown naturally and its potential should be exploited. Aonla is a medicinal fruit plant and cultivated in more than 55,000 ha area and produced 150 tones fruits/year.

Date palm is most suitable fruit tree of dry hot arid region and it is grown in Rajasthan, Gujarat, Punjab and Haryana. During the year 2009-10, the area of date palm cultivation in Gujarat has increased from 12,493 (2004-05) to 16,688 ha and with production of 12, 3490 tones fresh fruit. However, date is imported to India from Gulf countries due to its meager production. In Kutch region of Gujarat, mainly tissue culture plants of cultivar Barhee has been planted in about 1000 acre (Muralidharan *et al.,* 2008). There is considerable increase in area and production of date palm in Kutch region and estimated income from dates is around 17 crores.In recent years, imported tissue culture plants of cvs. Barhee, Khalas, Khunezi, Medjool, Khadrawy, Zamli and Saggai have been planted in districts of western Rajasthan. The planting of male palms are also essential for pollination and male cvs Ghanami and Al-in- city have been planted for this purpose. Looking to potential of date palm in hot arid region, area is being increased. In Bikaner district alone, tissue culture plantation of these date palm cultivars have been made in about 500 ha. Out of these, maximum plantation is under vegetative growth stage. In Tamil Nadu state also, date palm plantations have been done by the farmers in about 100 ha. Bael is also an important fruit crop of semi arid and arid region. Earlier there were no systematic orchards of bael. Now, looking to its nutritional and medicinal value, attention is being given on its commercial production in Rajasthan, Madhya Pradesh, Gujarat, Punjab and Uttar Pradesh. A survey report of revenue department for the year 2009-10, about 32 thousand ha area under plantation of 86 lakh 90 thousand fruit plants in the State of Rajasthan. The area under aonla cultivation has been increased substantially in the state in comparison to pomegranate.

Plant Genetic Resources of Arid Zone Fruits

The genetic resource conservation of major arid horticultural crops is being maintained in field repository at ICAR-CIAH, Bikaner. Among 318 ber (*Ziziphus mauritiana*) genotypes, the ber varieties Gola, Seb, Umran, Kaithali and Banarasi Kadaka are performing well under hot arid climate. Tikadi is found tolerant to frost/low temperature during winter season in arid region. Out of 154 genotypes of pomegranate (*Punica granatum*), Jalore Seedless, Ganesh, G-137, Mridula, Phule Arakta and Bhagawa are the better genotypes for fruit yield and quality. Besides this, anardana types pomegranate were also evaluated and Amlidana and Collection -12 from HP were found promising (Singh *et al.,* 2011). The varietal evaluation of aonla (*Emblica officinalis*) revealed that the NA-7 (Neelam) is a prolific bearer followed by Chakaiya and NA-6 (Amrit). Among 17 bael (*Aegle marmelos*) genotypes, NB-5 and NB-9 are performed well under irrigated hot arid conditions (Anon., 2011). New bael NB-16 and NB-17 have been recommended for cultivation from NDUAT, Faizabad. A five-year old budded plant of NB-5 yields about 40 fruits/tree while NB-9 yields about 29 fruits/ree. The fruit size of NB-5 is smaller (1.0 kg/fruit) than NB-9 (1.4 kg/fruit). The fruit quality is excellent in Goma Yashi, earlyness in Thar Divya and high yield in Thar Neelkanth bael varieties developed by CHES (ICAR-CIAH) are becoming popular among farmers due to deired traits under semi arid conditions. In National repository of date palm (*Phoenix dactylifera*), sixty one indigenous and exotic cultivars/genotypes are maintained and evaluated for different horticultural traits. In addition to this, a number of minor underutilized crops like ber, lasoda, ker, mulberry, bael, phalsa, pilu are also being conserved and evaluated for different traits (More and Singh, 2008). Some exotic fruits like Marula nut, Argan, Carob and Cactus pear have been introduced and evaluated for its utilization. Marula nut can be easily multiplied through stem cuttings treated with 1000 ppm IBA (Singh and Bhargava, 2008).

Table 3.1: Germplasm Conservation at National Field Repository of ICAR-CIAH, Bikaner and ICAR-CHES Godhra

Fruit Crops					
Crops	*Scientific Name*	*No.*	*Crops*	*Scientific Name*	*No.*
Ber	*Ziziphus mauritiana*	318	Marula nut	*Sclerocarya birrea*	01
Bordi	*Ziziphus rotundifolia*	22	Karonda	*Carissa congesta*	08
Date palm	*Phoenix dactylifera*	61	Wood apple	*Feronia limonia*	01
Bael	*Aegle marmelos*	17	Ker	*Capparis decidua*	06
Cactus pear	*Opuntia ficus indica*	20	Manila Tamarind	*Pithecelobium dulce*	02
Phalsa	*Grewia subinaequalis*	06	Fig	*Ficus carica*	03
Mulberry	*Morus alba*	15	Lasoda	*Cordia myxa*	65

Table 3.2: Germplasm Conservation at ICAR-CHES, Godhra, Gujarat

Fruit Crops					
Crops	*Scientific Name*	*No.*	*Crops*	*Scientific Name*	*No.*
Ber	*Ziziphus mauritiana*	55	Phalsa	*Grewia subineaqualis*	30
Custard apple	*Annona squamosa*	38	Fig	*Ficus indica*	05
Karonda	*Carissa congesta*	40	Wood apple	*Feronia limonia*	43
Bael	*Aegle marmelos*	181	Mahua	*Madhuca latifolia*	50
Jamun	*Syzigium cuminii*	66	Chironji	*Buchanania lanzan*	30
Tamarind	*Tamarindus indica*	25	Khirni	*Manilkara hexandra*	30

Suitable Fruit Crops and Varieties

Among suitable fruit crops such as ber, date palm, kinnow, bael, pomegranate, aonla, phalsa and lasoda are potential crops for arid region. Some important underutilized arid horticultural crops like khejri, ker, pilu, mulberry and lasoda, which can be successfully grown with low input cost. The fruit crop and their varieties should be specific for cultivation in hot arid environment. The Institute (ICAR-CIAH) has released 19 varieties of arid fruits crops which include, Thar Bhubharaj, Thar Malti, Thar Sevika, Goma Kirti of ber, Thar Shobha of khejri and Goma Aishwarya of aonla, Thar Gold in lasoda, Thar Lohit andThar Harit in mulberry. Besides the some promising lines in ber, mulberry, lasoda and bael, has been identified for evaluation and release (More *et al.*, 2008). At Institute level, Goma Khatta, a pomegranate genotype for anardana purpose has released. Goma Pratik of Tamarind and Goma Yash, Thar Divya and Thar Neelkanth of bael and Goma Priyanka of Jamun are released by the Institute and its regional Station ICAR-CHES, Godhra for cultivation in arid and semi-arid parts of the country.

Varietal variation in endurance to drought has also been observed in horticultural crops. Early ripening cultivars seem to escape stress conditions caused by the receding soil moisture sotred in the soil profile during the monsoon. Ber cultivars Gola, Seb and Mundia for extremely dry areas, Banarasi Kadaka, Kaithli, Umran and Thar Bhubhraj, for dry regions, and Sanaur-2, Umran and Mehrun for comparatively humid regions have been recommended. Apart from morphological parameters, plants should also have physiological parameters for endurance to drought for commercial cultivation in this region. Some physiological parameters identified in ber are no mid day depression in photosynthetic rate, low rate of transpiration, maintenance of leaf water balance, growth, canopy development, dry matter allocation, high water use efficiency, *etc*. It has been demonstrated that plant having capacity for drought endurance are able to maintain turgour, dry matter allocation, leaf and fruit growth even under low soil moisture level.

Table 3.3: Suitable Varieties of Arid Fruits

Fruit Crops	Improved Cultivars	Propagation Method	Spacing	Yield (kg plant)
Bael	Goma Yashi, Thar Divya, Thar Neelkanth NB-5, NB-9, Pant Urvashi, Pant Aparna, Pant Shivani and Pant Sujata, CISH Bael-1	Patch budding (May–July) Softwood grafting (June)	6 x 6 m	60-100
Aonla	Chakiya, Kanchan, NA-6, NA -7, NA-10, Goma Aishwarya, Laxmi-52, Krishna, Anand -2	Patch budding (May-August)	8mx8m	40–150
Indian jujube	Gola, Mundia (early), Seb, Banarasi, Kaithli, Goma Kirti (medium), Umran, Illaichi, Tikdi (late), Thar Sevika, Thar Bhubhraj and Thar Malti	I-budding (July–August)	6mx6m	40–100
Sour lime	Kagzi lime, Sai Sarbati, Vikram, Pramalini	Air layering and budding		20–30
Karonda	Thar Kamal Pant Manohar, Pant Suvarna, Pant Sudarshan	Seeds and cutting (August)	4 mx 4 m	10–20
Datepalm	Halawy, Barhee, Medjool, Khuneizi, Khalas, Chip chap	Suckers and Tissue culture	10m x 10 m	50-80
Tamarind	Goma Prateek, Pratisthan', 'Yogeshwar', 'PKM-1, 'Urigam	Softwood Grafting	8 m x 8m	270-300 kg pods
Mulberry	Thar Lohit and Thar Harit	Cutting	8mx6m	10-20 kg
Lasoda	Thar Gold	Cutting and seed	8mx5m	40-60 kg
Phalsa	Thar Pragati	Cutting, seed	2.5 m x 2.5m	

Biodiversity in Important Underutilized Arid Fruit Crops

Phalsa

Phalsa (*Grewia subinaequalis* D.C.) a member of family Tiliaceae, is one of the oldest fruits known to Indian. Phalsa has been mentioned in Vedic literature as having certain medicinal properties. It is capable of growing under neglected and water scarcity conditions where only a few other crops could survive. Besides, it is an important catch crop in commercial orcharding. The ripe phalsa fruits are purple in colour at maturity and may turn black at ripening on the bushes. The mildly acidic fruits are rich in vitamin A, C and minerals. Because of short shelf life, its fruits are suitable for local market or need to be processed immediately after harvesting. According to *Ayurveda*, the ancient Indian treatise on medicine, the fruits are a cooling tonic and aphrodisiac. Mucilaginous extract of the bark is useful in clarifying sugar and fiber of phalsa is used for making ropes. Being fine grained, strong and flexible, its wood is used for archer's bows, shingles, spear handles and poles. Its stems are used both for garden poles as well as basket making. Fruits can be processed into juices and candies. Phalsa is a tall shrub with rough bark on the stem, and have numerous long, slender, drooping branches where the young branchlets are densely covered with a coating of hairs. The widely spaced, thick, alternate, deciduous and large leaves are broadly heart shaped or ovate, pointed at the apex, oblique at the base measure up to 20 cm in length and 15 cm in width and coarsely toothed, with a light, whitish blush on the underside. In late spring, bright yellow-orange flowers of 10 to 19 mm are borne in dense cymes in the leaf axils. The small round fruits, purple, crimson or cherry red in colour when ripe, borne on a 2 to 3 cm long peduncle, produced in huge numbers in open, branched clusters. Fruits ripe gradually on bushes during summer months, while ripening, the fruit skin turns from light green to cherry red or purplish red finally becoming dark purple or black. Phalsa fruits followed a double sigmoid type of growth curve and the respiration curve is of non-climacteric type. Thus, well drained and loamy soils are best for its cultivation. Phalsa could be grown in wasteland without much care and in un-amended sodic soils having up to 30 per cent exchangeable Na. The phalsa can grow successfully all over India except at higher elevations. It is a subtropical crop and does best in regions where there is distinct winter and summer. It has been found to thrive well even in the arid and warmer parts of South India. The freshly extracted seeds should be used for raising seedlings. Therefore, few local types of phalsa have been reported in some parts of country they are Local and Sharbati. Two other distinct types, tall and dwarf has been recognized. Recently at ICAR-CHES, Godhra a new high yielding cv.Thar Pragati has been developed through selection. Two phalsa types *i.e.* dwarf and tall were recognized (Pareek and Panwar, 1981). Dhawan *et al.* (1983) reported differences in physical parameters, chemical composition and electrophoretic pattern of seed proteins in two distinct phalsa types, *i.e.* tall and dwarf. Pruning is a very important operation in phalsa cultivation as phalsa bears fruit on current season's growth, there is need for annual severe pruning before spring for better quality fruits. Late December or early January has been found optimum time of pruning under north Indian conditions and annual winter pruning to a height of about 100 cm from ground level encourages new shoots and higher yield of marketable fruit than more drastic pruning. After 40-45 days of flowering, the fruits start ripening. Phalsa fruits begin to ripe in hot summer during March–April in south and May-end in north India. Fruits are small-sized and ripen over a period of about a month. Gradual but steady ripening of few fruits in a cluster during summer necessitates frequent harvesting. Pre-harvest application of ethephon or etherel (1000 ppm) when few fruits start to change their colour reduces the number of pickings. Fruits should neither be under-ripe nor over-ripe. It should be firm at the time of ripening. Fruits for storage and transport should be harvested at colour turning stage, whereas for local market they should be harvested at red ripe stage. Fruits are individually picked by hand and collected in bamboo/pigeon pea plant baskets cushioned with polythene sheet or newspaper cuttings. On an average, a mature plant provides 7-9 kg of fruits under normal conditions. Storage life of fruits depends upon the stage of harvest. Fruits harvested at colour turning stage can be stored for 2–3 days at room temperature and about 7 days in cold store at 7°C. Fruits harvested at red ripe stage can be stored only for a day; hence, they are marketed immediately in local markets. The phalsa fruits can be processed in to various products like juice, nectar, squash and syrup.

Boradi (*Ziziphus mauritiana* var. *rotundifolia*)

Boradi is drought hardy shrub or tree found growing along water courses in depressions and around croplands in Rajasthan. They are vigorous and upright, bearing small round or oval shaped fruits, which are larger than those of *Jherber*. The fruits are used fresh as well as after sun drying. Young seedlings are commercially used as rootstock for *ber*. The foliage of *boradi* is used as a fodder. Its wood has marginal timber value and is used for making farm implements. A full grown *boradi* tree yields 25-35 kg fruits per tree. Variable genotypes exist throughout the region, especially in river belts and natural depressions. Efforts need to be made to identify and conserve the elite types with good qualities and resistance to biotic and abiotic stresses.

Jherber (*Ziziphus nummularia* Burm. F.) Wight and Am. syn. *Z. rotundifolia* Lam.)

It is a bushy plant having purplish branches and stem. It is found in extremely arid locations on shallow, over grazed, rocky, semi-rocky, saline and sandy soils growing at sand dunes, degraded pasture, along sides of the roads and railway tracks. The plant flowers during July-August and fruits ripen during November-December. Fruits are globose-ovoid, shaped drupes of small size (1 cm diameter) and orange brown to dark red colour having little edible pulp of sub acidic taste. Dried fruit powder is preserved for use. Medicinally the fruit is cooling and astringent. The young and tender leaves having 14.7 per cent crude protein are good feed for sheep and goats. The seedlings are also used as dwarfing root stock for *ber*. The range of variability in Jharber is very narrow. There is need to collect superior types having bigger size fruits, high pulp-stone ratio and good quality.

Karonda (*Carissa carandas* L.)

Karonda (*C. congesta* Wt. Syn *C carandus* L.) is native to India. However, *C. grandiflora* (Natal palm) came to India from South Africa, and *C. edulis* Vahl has been introduced from the USA (Singh *et al.*, 1982 and Pareek and Sharma, 1993). Its occurrence is also common in Burma, Malacca dry area of Sri Lanka. It also often grown in the Philippines, Thailand, Combodia, Sout Vietnam and in East Africa. In India, it is said to be grown in Bihar, Chattishgarh, West Bengal, Rajasthan, Gujarat, Punjab and Uttar Pradesh. Karonda belongs to family Apocynaceae, and it is a very hard multipurpose bush of dry areas and can be grown in any type of soil in both tropical and subtropical climate. The *Carrisa* genus has over 30 species originating in South Africa, Australia, Asia and Malaysia. The important species of *Carissa* are *C. grandiflora* D.C., *C. bispinosa* Desf, *C. edulis* Vahl, *C.ovata*, *C. enermis* Vahl. *C. paucinervia* D. *C.* and *C. spinarum* L (Pareek and Sharma, 1993, Singh and Arora, 1978). Karonda grows as small shrub or tree attain in height of 3-5meter. Botanically, karonda is a large dichotomously branched evergreen shrub with short stem and strong thorns in pairs simpe or forked up to 5 cm long in pair in the axil of the leaves. The leaves are evergreen, simple opposite, oval or elliptic 1-3 long dark green leathery, glossy on the upper surface, lighter green and dull on the under surface. Karonda flowers are small (3-5 cm diameter), found in corymbose cymes, hypogynus and pentamerus. Flowering in *Carissa carandus* takes place in spring (Mishra *et al.*, 1968). A wide range of genetic diversity is available in existing population in the form of morphological characters, qualitative and quantitative fruit characters and bearing potential, because of its endemic nature. Fruits of two types of karonda (red green and whitish yellow) were studied by Singh and Singh (1998). Fruits of red green type were significantly larger than those of white yellow type particularly at mature stage. Naturally growing 212 genotypes of karonda were selected from six Tahsil of Kolhapur district by Sawant *et al*, 2003 at College of Agriculture, Kolhapur, and some genotypes are being maintained at field gene bank of MPKV, Rahuri, ICAR-CAZRI, Jodhpur, ICAR-CIAH, Bikaner and its regional Station ICAR- CHES, Godhra (40) and are being maintained. Thar Kamal, Pant Manohar, Pant Suvarna, Konkan Bold, Pant Sudarshan varieties are available for commercial cultivation. The cultivars are classified according to fruit colour *i.e.*, green, pink and white. The fruits are nutritious containing 1.1 per cent protein, 2.9 per cent fat, 0.6 per cent mineral, 1.5 per cent fibre, 2.9 per cent carbohydrate and 42 K cal energy in 100 g edible portion. The mature fruits are processed into quality pickle and candies. Due to seed propagation, considerable heterozygosity and large variation is observed in the fruit characters such as shape, size, pulp quality and yield in existing germplasm in Mount Abu and Western Ghats and Khandala in Maharashtra. This offers a great deal of scope for improvement in karonda by seedling selections.

Ker (*Capparis decidua* (Forsk.) Edgew.)

Thae genus capparis comprises approximately 250 species including shrub, timber and woody climbers.Reprtedly, 26 species of this genus found in India. Ker is one of the important multipurpose plants of desert and arid regions of Indian sub continent, Africa and Saudi Arabia. It is locally referred as *kair* in Rajasthan, *karil* in Uttar Pradesh, ker in Gujarat, *teent* in Haryana *della* in Delhi and Punjab and nepti in western Maharashtra. It is extremely hardy species and provides vegetative cover in hot sandy desert areas and it is distributed over 3540 km^2 in Bikaner and Jodhpur districts of Rajasthan. Ker performs better under water stress condition when temperature is very high. Phenological observations indicate that there are high intra population variations reflecting high degree of homeostatis in the species (Singh and Singh, 2011).

Under natural population rich genetic diversity exists with regard to bearing habit, fruit size, colour of fruits and petals pulp content, *etc.* some afforts were made by ICAR-NBPGR, Regional Station of ICAR-CAZRI, Jodhpur and ICAR-CIAH, Bikaner to identify promising germplasm of ker from existing population. It is a much branched, struggling, glabrous shrub or a small tree having green, zigzag, thorny stem. It is distributed in drier parts of north western India. *Ker* bushes are found on rocky, semi-rocky lands, sand dunes, gullies, ravines and brackish water areas. The plant is tolerant to high saline conditions (10.0-25.0 ECdsm-1. Flowering takes place twice a year, *i.e.* in March-April and September-October and fruits mature in May-June and November-December, respectively. The summer flowering is profuse giving a fruit yield of 10-20 kg per bush or 20-80 quintals per hectare. Fruit is a glabrous ovoid to sub-globose shaped beaked berry, 0.7-1.5 cm in diameter which becomes scarlet red when ripe and contains 2-5 seeds of 2-4 x 2 mm size. The ripe as well as unripe fruits are rich source of carbohydrate (71 and 59.4 per cent), protein (17 and 14.8 per cent), fat (5 and 7.4 per cent), minerals (calcium, phosphorous and iron) and vitamins (A and C). In view of rapid genetic erosion due to population pressure and urbanization, conservation of its valuable gene pool is of immense importance. Owing to lack of vegetative propagation technique, concerted efforts have not been made to collect its variability but some elite types have been identified at ICAR- CIAH, Bikaner, and efforts are continued to evolve an amicable vegetative propagation technique for this valuable species.

Kair is a multipurpose, perennial, small tree, much branched, leafless bushy and thrives well in the most adverse climatic and in the soils of poor fertility. It is highly suitable for stabilizing sand dunes and controlling soil erosion by wind and water. Its berry-shaped unripe fruits are rich in carbohydrates, proteins and minerals used as fresh vegetables and in the preparation of pickles. Dehydrated fruits are used in the off season as vegetable either alone or in combination with other dried vegetables.

Pilu or Mustard or Salt Bush or Tooth Brush Tree (*Salvadora persica* L.)

Salvadora (named after *Salvadora* Spanish botanist) belong to family Salvadoraceae and genus salvadora. Four to five species are found in worm region of Afric and Asia of which two species *Salvdora oleoides* Decene and *Salvadora persica* L. are commonly found in India. The somatic chromosome number in both the species is confirmed as 2n=24. Two basic chromosome number n=12 and 13 are contemplated for the genus (Kumar *et al.*, 2003)

It is a much-branched evergreen shrub or tree having dull grey or grey white bark with deep cracks. The tree is distributed in the states of Gujarat, Rajasthan, Haryana, Punjab and to some extent in Andhra Pradesh, Tamil Nadu and Karnataka, and suitable for afforestration of ravines, saline and alkaline lands, as windbreak and in shelterbelts. It grows luxuriantly amidst the vast stretches of highly saline soils of over 80dSm⁻¹. Pilu is found to be superior to *S. oleoides* in its adaptability, early flowering, salt tolerance and yield. The fruit is sweet in taste and yields a fermented drink. The leaves are used as vegetable and in sauces. Tender shoots are also eaten as salad. The seed is rich in fat similar to that of *S. oleoides* and yield non-edible oil rich in C12 and C14 acids, which are used in soap and detergent industries. A high degree of variability has been observed in natural population of pilu in semi-arid and arid tract of Rajasthan and Gujarat. Variability with regard to fruit size, quality and oil content of seed have been collected and further improvement programme is in progress at different research stations.

Phenological diversity exists in *Salvadora* and tree to tree variations are common in flowering and in maturing fruits, which may reflect genetic variation or genotype x environmental interaction and or both. Jindal *et al.* (1999) reported considerable variability in fruit colour, seed oil content and seed germination showing potential for improvement of fruit production. Parthenocary has been reported by Balasubramanian and Bole (1993) in *S. spinosa.*

Popularly known as *'Mitha Jal'* or *'Pilu,* a large evergreen tree flowers during January to March with fruiting in the month of May. Flowers are pink with red veined petals in small cluster along with leafless shoots borne in the axil of spines in many flowered corymb from the old branches are from short lateral shoots. Pedicel slender and about 12 mm long. The yield of fresh fruits per tree from mature trees is 10 - 15 kg or 2 - 3 kg dried fruit. Fruits have sweet agreeable aromatic, slightly pungent and peppery taste. They can be eaten raw when ripe either by children or may be cooked or preferably dried and stored by nomadic tribes.

Mulberry

Mulberry belongs to to family Moraceae and Asia is considered to be centre of origin of mulberry.The most important use of mulberry globally is in production of silkworm, which feeds exclusively on its leaves. The country with the largest area of mulberry is China followed by India. The mulberry is cultivated in China, Brazil, Cuba, Mexico, Honduras and Panama, Saint Vincent, Dominican Rep. and El Salvador. In republics of Ex-USSR, the most common species are *M. multicaulis, M. tartarica* and *M. nigra* (Datta, 2000) In Indonesia, there are seven species: *M. alba* (varieties *tartarica* and *macrophyla*)., *M. nigra,* M. *multicaulis, M. cathyana* and *M. mierovra* (Kastumata, 1972). In Vietnam they have over 100 varieties mainly *M. alba, M. nigra* (Kastumata, 1972). The red mulberry is native of most bottom land from Canada to South Texas and is supposed to be latest species in the entire genus. However, the trees of *M. alba* var.*tomentosa* are busy and spreading in nature. In India, *M. alba, M. indica, M serreta* and *M. laevigata* are grown naturally in northern parts of the country (Ravindran *et al.*,1997). For conservation of mulberry outside of its natural habitats, a field gene bank has been developed Central Sericultural Germplasm Resource Centre, Hosur, Karnataka which maintains *M. australis, M. cathayana, M. multicaulis* Poir, *M. rotundifolia* Koidz, *M. alba* L., *M. indica* L., *M. teliaefolia* Makino, *M. nigra* L., *M. serrata* Roxb., *M. laevigata* Wall, *M. rubra* L., *M. sinensi* Hort. and *M.bomycis* Koidz.ICAR-CIAH, Bikaner developed two varieties of Mulberry: Thar Lohit and Thar Harit for semi-arid and arid region.

Mulberry is a fast growing deciduous woody perennial plant. It has a deep root system. The leaves are simple, alternate, petiolate, entire or lobed. The number of lobes varies from one to five. Inflorescence is always auxiliary. Most of the species of the genus *Morus* and cultivated varieties are diploid, with 28 chromosomes. However, triploids (2n= (3x)=42) are also extensively cultivated for their adaptability, vigorous growth and quality of leaves. The species vary greatly in longevity. Red mulberry trees live more than 75 years, while black mulberries have been known to bear fruit for hundreds of years. The mulberry makes an attractive tree which will bear fruit while still small and young. Mulberry is known in India as "Kalpa Vruksha" as all the parts of the plant have many uses. It is quite tolerant of drought, pollution and poor soil. If mulberry is used for silkworm rearing it is possible to obtain 30-35 tones/ha of leaf every year. Many farmers feed their animals with surplus foliage but always mix it with straw. Farmers also use the mulberry branches for fuel after pruning. One hectare of mulberry garden yields about 12.1 tones of mulberry sticks. The energy generated/ha (50 percent moisture loss) is 27 830 Kcal (@ 4 600 calories/kg of mulberry wood). Mulberry could be exploited as an "energy crop" in cultivable, wasteland, low-lying areas, and canal bunds.

Blackberry

Blackberry is a species of flowering plant in the family Moraceae, native to south western Asia, where it has been cultivated for so long that its precise natural range is unknown. It is known for its large number of chromosomes, as it has 154 pairs (308 individuals). Species are sometimes confused with black mulberry, particularly black-fruited individuals of the white mulberry, but black mulberry can be distinguished by the uniformly hairy lower leaf surface. The edible fruit is dark purple, almost black, when ripe, 2–3 cm long, a compound cluster of several small drupes; it is richly flavoured, similar to the red mulberry (*Morus rubra*) but unlike the more insipid fruit of the white mulberry (*Morus alba*).

Ber

The genus *Ziziphus* having more than 600 species (Baiely, 1947) of which 18-20 are native to India (Watt, 1893). Singh and Arora (1978) listed eight species having edible fruits found in different part of India. *Z apetala* H. P.F, *Z. funiculosa* Buch- Ham and *Z. incurva* Roxb occur in north eastern hills. All over the dry tracts, particularly north-west India and U. P.; *Z. oenoplia* Mill and *Z. rugosa* Lamk., particularly central and eastern India, and *Z. vulgaris* Lam. (Sinhguli Unab) grows naturally in north western himalayas. Besides, *Z. rupicola* T. Andern is found in central and eastern India and *Z. xylocarpus* wild in M. P. and peninsular region. Maheshwari and Singh (1965) recognised six economically important species in India. These are *Z. nummularia* (Burn F.) Weight and Arn. (jharber-can be used as rootstock), *Z mauritiana* Lamk (Indian ber), *Z. oenoplia* Mill (Makoia, Makoh), *Z. rugosa* (Suran ber), *Z. satva* Gaertn (Z. *vulgaris* Lam kandiari) and *Z. xylocarpus* willd (kathber). *Z. mauriatana*, and *Z. jujuba* are the most important fruit species, the farmer being cultivated in tropical and subtropical regions of India, whereas the letter in the temperate regions of China. Fruits of *Z. nummularia* and *Z. rotundifolia* are edible and letter is used as rootstocks. Although *Z. nummularia* and *Z. rotudifolia* are reported synonyms horticulturally, both are different (Vashishtha, 1982); *Z. rotundifolia* form a tree even if it is headed back, whereas *Z. numularia* form a bush. In the Middle East *Z. spinachristi* is common and found growing wild and fruit known as dom or nabak are appreciated by rural people (Abbas-Niami and Al-Ani, 1988). Flowering, fruit set and identification key have been reported by Bashishtha and Pareek (1979 and 1989).

Several ber germplasm including species, cultivars and other types have been collected at different ICAR Institutes/Regional Station and University in the country and are being mainatained in field gene bank. Among various gene bank centres,ICAR- CIAH, Bikaner; ICAR-NBPGR, Jodhur; MPKV, Rahuri; CCSHAU, Hisar;ICAR- CAZRI, Jodhpur; and GAU, S.K. Nagar are important centres.ICAR-CIAH, Bikaner is maintaining the largest collection in the field repository (338).

'King of arid zone fruits' has commercial importance with its several new high yielding varieties. It is the only fruit crop which can give good returns even under arid and semi-arid conditions and can be grown in a variety of soils and climatic conditions ranging from sub-tropical to tropical. Varieties like *Goma Kirti, Thar Sevika, Thar Bhubhraj* are early-maturing, suitable for staggered harvesting and practically free from powdery mildew. Different varieties can fulfil different purposes like for desired adaptation (Gola, Umran, Thar Bhubhraj, Thar Sevika), for diverse quality traits (Banarasi Karaka, Illaichi), for high and stable yield (Seb, Ponda), tolerance to biotic and abiotic stress (Tikadi, Katha, Bawal-Sel-1, Sanaur-2) have been identified.

Indian Cherry (*Cordia myxa* L.)

Lasoda (*Cordia dicotoma*) belongs to family Ehretiaceae. The other important species are *C. ghraf, C. macleodii, C. rothii, C. wallichiana, C. vestita, C. crenala, C. rothii, C. sebestina.* Cordia is a large genus of about 250 species. Species that have flowers in helicoids cymes and often have herbage that is coarsely hairy. Great variation exists in natural population with respect to morphological characters particularly plant height, spread, leaf size, fruit size fruiting behaviour; quality parameters like fruit colour, pulp content, sweetness, pickling quality seed and pulp ratio *etc.* There is a small fruited type locally called as *gundi* having very small size fruit (1cm) orange to light pink in colour at maturity, but liked by rural people. Indian cherry, locally known as lasoda is another important fruit plant where green unripe fruits are important as fresh vegetable and pickles during April–May when availability of conventional vegetables is scarce. The species provide vegetative cover as tree component of arid farming system, preventing soil erosion and promoting biodiversity and offers least competition with rainy season crops, since its fruiting season is summer when main crops are already harvested. This plant also offers scope in using harvested rain water for fruit production since it requires irrigation only for 2–3 months period during summer season (April–June). The plants with large sized leaves certainly give bigger sized fruits. Lasoda Collection No. 1 has fruit weight up to 13.1g, with high TSS (18.5° Brix) and pulp content (92.13 per cent). Saini *et al,* (2002) reported variability in plant height and spread at CCS University, Hissar. Kaushik and Dwivedi (2004) reported wide range of biodiversity in morphological and quality characters from 50 accessions of lasoda from Haryana. Flowering and fruiting in lasoda germpasm have been reported by Bashishtha *et al.* (1985).Thar Gold variety has been developed by ICAR-CIAH, Bikaner which is precocious in bearing and possessing better qualitative and qualitative characters.

Indian Mesquite (*Prosopis cineraria* (L.) Druce)

There are no definite records about origin of khejri, but Indian thar desert is considered as native place of khejri. It is distributed from Afghanistan, Arabia,

Iran, and Pakistan to India. It is found growing in Rajasthan, Gujarat, Maharashtra, Madhya Pradesh and north Karnataka. Khejri belong to family Leguminoceae and sub family Mimoceae having chromosome number 2n= 28, is an evergreen tree. Its Sanskrit name is *Sami* and botanically described as *Prosopis cineraria* Macbride. The leaves are 2.5- 8.0 cm long, leaves are alternate, bipinnately compound with rachis 3-5 cm long, glabrous, pinnae mostly two pairs, opposite: 1-3 pairs of pinnae and 7-14 pairs of leaflets 4-15 cm long and 2-4 mm broad leaflets 8-21 pairs, sessile, oblique rounded and mucronate at the apex. Flowering takes place during February- March and seeds mature during April-June. The immature pods are rich in crude protein (18 per cent crude protein), carbohydrates (56 per cent) and minerals (0.4 per cent each of phosphorus and calcium and 0.02 per cent iron. The immature pods, are used as vegetable both fresh as well as after dehydration, while ripe dried pods having 9–14 per cent crude protein and 6–16 per cent sugar can be powdered and used in the preparation of bakery items such as biscuits and cookies. The natural regeneration of kheji is through seed; hence wide range of variability exists under natural population. In general there are two ecotypes *i.e.* one having dropping branches with smaller leaf size and other having erected branches with bigger leaf size.

The variability in khejri with respect to branching habit, thorniness and foliage colour have been repoted by Memio (1976) in Sekhawati region of Thar desert. Based on the harvesting stage of tender pods, the ecotypes can be grouped as early, mid and late. The different ecotypes of khejri can be distinguished when they are grown in similar environment (Saroj *et al.*, 2002). Wide genetic variability in terms of pod length, seed weight, number of leaves, leaflets and plant height. ICAR-CAZRI, Jodhpur has selected about 20 plus tree and elite phenotypes for clonal multiplication and related investigation (Kaul *et al.*, 1991). Systematic work on kheji has been initiated at ICAR-CIAH, Bikaner for their *ex-situ* conservation and *"Thar Shobha"* is the improved variety of this crop.

The *'Kalpavriksh' (khejri)* of the Indian desert, improves the growth of companion annual crops and plays significant role in the economy of Indian desert. This tree grows well in all sorts of climatic constraints evidenced by its new foliar growth, flowering and fruiting during extreme dry months (March–June) when most other trees of the desert remain leafless or dormant.

Date Palm

Date palm is one of the oldest domesticated fruit crops. It is believed to be indigenous to countries around Persian Gulf. Date cultivation has historical records dating back 6000 B.C. along the Tigris and the Euphrates River in Iraq and to 2000 BC at Mohanjodaro, along the river Indus. Several thousand years before its cultivation, pre historic man used the fruit from wild palm and carried the seeds over the wide area from India to Middle East. In India, it is believed that the date palm has been introduced by the soldiers of Alexender in 4th century B. C. in the Indus valley. In Rajasthan, date cultivation came in existence by the then ruler Ganga Singh of the erstwhile Bikaner through the plantation of few suckers of Halawy, Khadarawy and Zahidi at Ganganagar, Rajasthan, a part of Thar Desert after development of Ganga canal in the State. Grooves of date are found on the coastal belt from Azmer to Mandvi in kutch District of Gujarat. In north western Rajasthan, few scattered plantation are confined to mainly in the districts of Ganganagar, Bikaner, Jodhpur and Jaisalmer.

Date palm (*Phoenix ductylifera*) is related to the order Palme and family Palmaceae. The datepalm is monocotyledonous plant with strong straight unbranched stem growing to the height more than 35 ft. The commercially grown date palm species produce superior quality fruits. The species has characteristic features of producing axillary off shoots/suckers which may arise near the base of the stem unlike two other species *viz., Phoenix sylvestris* (L.) Roxb., and *Phoenix canariensis* Chaubaud (Canary island palm). It is an ornamental plant. *P. sylvestris* is found growing wild in almost all parts of India and produces fruits of inferior quality. This palm is used for production of crude sugar and neera (tadi). Date palm is dioecious in nature, bearing male and female flowers in different plants, has chromosomes 2n =36. Intraspecific crosses (*P. ductylifera x P. sylvestris*) showed metaxenic effect. Wide range of variability has been observed in Kutch region of Gujarat. Performance of datepalm cultivars under high hyper partially irrigated from western Rajasthan has been reported by Pundir and Porwal (1999).

Date palm can take 4-8 years to bear fruit after planting. In order to get fruit of marketable quality, the bunches must be thinned and bagged or covered before ripening so that remaining fruits grow larger and are protected from weather and pests like birds. Date palm cv. Halawy is the most suitable with respect to growth, early maturing, regular bearing and fruit quality for cultivation, as this cultivar possesses high rate of photosynthesis, water use efficiency and carboxylation efficiency. Khadrawy, Halawy, Khalas, Khuneji, Sewi, Zagloul, Barhee, Shamran, Medjool Zahidi, Khuneizi, Dayari and Sabiah are better with respect to fruiting. The application of ethephon (1000 ppm) produced more bunch weight and yield per tree.

Future Thrust

1. Survey, collection, evaluation and conservation of variability among hot arid fruit species and selection of high table and processing value variables and their popularization.
2. Development of new varieties with high productivity table value
3. Standardization of propagation techniques.
4. Development of package of practices for improved cultivars.
5. Awareness campaign for their health promoting attributes.
6. Development of post harvest products.
7. Target oriented varieties (rich in nutrients, antioxidants and vitamins, abiotic stress tolerant).
8. Screening of germplasm against abiotic and biotic stresses.

References

Abbas, M. F., Al-Niami, J. H.and Al-Ani, R. F.1988 Some physiological characters of fruits of jujubi (*Ziziphus spinachristi* L wild) at different stage of maturity. *J. Horti. Sci.*, 63 (2): 337-339.

Awasthi, O.P., Saroj, P. L., Singh, I. S. and More, T. A. 2007. Fruit Based Diversified Cropping System for Arid Regions, ICAR-CIAH Tech. Bull. No. 25, CIAH, Bikaner, 18p.

Bailey, L.H. (1947). The standard Cyclopaedia of Horticulture, Macmillon and Co., New York

Singh, Bal, J. S. 2003. Genetic resources of underutilized fruits in Punjab subtropics. *Acta Horticulturae*, 623: 325-331.

Balasubramanian, P. and Bole, P. V. 1993. Fruiting, phenology and seasonality in tropical dry evergreen forest in Pt.Calimere Willdlife Sanctury. *Journal of Bombay Natural History Society*, 90 (2): 163-177.

Datta, R. K. 2000. Mulberry cultivation and utilization in India. In: Proceedings of Electronic Conference.

Dhandar, D. G., Saroj, P. L., Awasthi, O.P. and Sharma, B.D. 2004. Crop diversification for sustainable production in irrigated hot arid eco-system of Rajasthan. *Journal of Arid Land Studies*, 148: 37-40.

Dhawan, K., Malhotra, S., Dhawan, S. S., Singh, D. and Dhindsa, K. S. 1993. Nutrient composition electrophoretic pattern of protein in two distinct type of phalsa (*Grwia sunenaequalis* DC). *Plant Food for Human Nutrition*, 44 (3): 255-260.

H. B.Singh and Arora, R. K.1978 Wild edible plants of India, ICAR, New Delhi.

Jindal, S. K. Prakash, J. and Chawan, D. D. 2003. Variability for seed oil, seedling characters, their relationship and breeding systems in *Salvadora oleoides* (eds. Joshi, N. K., Kathji, S. and Kar, A.) (In) Recent Advances in Management of Arid Ecosystem. Proceeding of Symposium held in India March 1997. pp 405-408.

Katsumata 1972. Mulberry species in west Java and their peculiarities. *J. sericultural Sci. Japan*, 42 (3): 113-223.

Kaushik, R. A. and Dwivedi, N. K. (2004).Genetic diversity in lasoda. Indian Horticulture, 1 and 27.

Koul, K. K., Ajay, Parida, Bisht, M.S. and Raina, S. N. 1991.Variability in *Prosopis cineraria* from India's Thar desert. *Agroforestry Today*, 3 (4): 14

Kumar, A., Rao, S. R. and Rathor T. S. 2002. Cytological investigation in some important tree species of Rajasthan IV. Male meiosis studies in the genus *Salvadora* L. *Cytologia*, 67 (2): 105-115.

Maheswari, P. and Singh, U.1965. Dictionary of economic plants of India, ICAR, New Delhi

Memio 1976.Annual progress repot CAZRI, Jodhpur

More, T.A., Samadia, D. K., Awasthi, O.P. and Hiwale, S. S. 2008. Varieties and Hybrids of CIAH, Tech. Bull. No. 30, CIAH, Bikaner, 11p.

Pareek, O. P. and Panwar, H. S.1981. Vegetative, floral and fruit characteristics of two phalsa (*Grwia sunenaequalis* DC) types. *Annals of Arid Zone*, 20: 281-290.

Pareek, O. P. and Sharma, S. 1993.Genetic resources of underutilized fruits. (In) Advanceces in Horticulture, Vol. I (K. L. Chadha, and O. P. Pareek; Eds.), Pub. Malhotra Publishing House, New Delhi Pp: 189-225.

Pareek, O.P. and Sharma, S. 1991. Fruit trees for arid and semi-arid lands. *Indian Farming*, 41: 25-30.

Pundir, J. P. S. and Powal, R. 1999. Performance of different cultivars of date palm under hyper partially irrigated western Palms of Rajasthan (India). Proc. Int. Conf. date palm held at UAE University, Al-Ain (UAE) on 8-10 March, 329-336.

Ravindarn, S., Ananda Rao, A., Girish Naik V., Tendulkar, A., Mukharjee, P. and Thangavelu, K. 1997. Distribution and variation in mulberry germplasm. *Indian J. Plant Genetic Resources*, 10 (2): 233-242.

Saini, R.S., Kaushik, R. A. and Singh, S. 2002. Research note on evaluation of *C. myxa* Roxb germplasm for vegetative growth character under semi-arid conditions. *Haryana J. Hort. Sci.*, 31 (31-2): 62-63

Saroj, P. L., Dubey, K. C. and Tewari, R. K. 1994. Utilization of degraded lands for fruit production, *Indian J. Soil Conservation*, 22: 162-176.

Singh, A. K. and Singh, P. 1998. Power of significance of difference among fruits and seed size parameters of karonda (*Carissa carandus* L.). *Annuls of Agriculture Research*, 19 (1): 66-77.

Singh, D. and Singh, R. K. 2011. A potenial ethnobotenical weather predictor and livelihood security shrub of arid zone of Rajasthan and Gujarat. *Indian journal of Traditional Knowledge*, 10: 146-155.

Singh, H. B. and Arora, R. K. 1978.Wild edible plants of India, ICAR, New Delhi.

Singh, R. Chopra, D. P. and Gupta, A. K. 1982. Some Carissa for western Rajasthan, Indian Horticulture, 27: 10-11

Singh, R.S. and Bhargava, R. 2014. Propagation of Horticultural Plants: Arid and Semi Arid Regions, NIPA, New Delhi, Pp 552

Vashishtha, B. B. and Pareek, O. P. 1979. Flower morphology, fruit set and fruit drop in some ber (*Z.mauritiana*). *Annals of Arid Zone*, 18: 165-169.

Vashishtha, B. B. and Pareek, O. P. 1989. A key to the identification of ber (*Z. mauritiana*) cultivars. *Indian Jouranl of Agricultural Science*, 46: 183-188.

Vashishtha, B. B. 1982.Horticultural qualities of *Ziziphus nummularia*-Bordi ICAR-CAZRI, Monograph, Jodhpur.

Vashishtha, B. B., Raja Ram, and Singh, Ranbir 1985.Studies on flowering and fruiting in lasoda (*Cordya myxa*). *Haryana J. Hort. Sci.*, 14 (3-4): 11-15.

Vishal Nath, Saroj, P.L., Singh, R. S., Bhargava, R. and Pareek, O. P. 2000. *In-situ* establishment of ber orchards under hot arid eco-system in Rajasthan. *Indian Journal of Horticulture*, 57 (1): 21-26.

Watt, G. 1983. A dictionary of economic plants of India, Vol.I Part IV, Pub. Cosmos Publication, Delhi

Chapter 4

Acid Lime

Mahantesh Kamatyanatti[1] and Murlimanohar Baghel[2]

[1]Department of Horticulture (Fruit Science),
CCS Haryana Agricultural University, Hisar – 125 004, Haryana,
E-mail: mahanteshkk23@gmail.com
[2]Division of Fruits and Horticultural Technology,
IARI, New Delhi – 110 012

The citrus (*Citrus species*) is the 3rd major fruit crops of India next to mango and banana, which comprised of mandarin, lime, lemon, grapefruit, pummelo, citron *etc.* It also includes *Poncirus*, *Fortunella* as well as interspecific and intergeneric hybrids. Among diverse groups, mandarins are the leading cultivated and commercial citrus fruit in India with 43 per cent share, followed by sweet orange (25 per cent share), acid lime and lemons with 25 per cent and others with 7 per cent share (Murkute and Singh, 2015).There are four species of lime which have commercial value *viz.*, Acid lime (*Citrus aurantifolia*), Tahiti lime (*C. latifolia*), Sweet lime (*C. limettioides*) and Rangpur lime (*C. limonia*). The fruits of *Citrus aurantifolia* Swingle are small and seeded, which is commercially important in India, Pakistan and Bangladesh besides Mexico (Chadha, 2009). Acid lime fruits are highly acidic with titratable acidity up to 8 per cent and juice content 50-55 per cent. It is mostly used fresh for garnishing salads, curries and vegetable preparations in south Asian and SE Asian countries. Processed products like cordials and squash are very common as refreshing drinks in summer. Pickle is another very widely produced household product from acid lime in India. Beverages such as ready-to-serve (RTS) drinks and carbonated beverages from lime and lemon juice are becoming very common in hot and dry climates of Indian sub-continent (Chadha, 2009).*Citrus aurantifolia* (Christm) Swingleis a polyembryonic plant cultivated in many countries all over the world and grows in hot subtropical or tropical regions such as Southern Florida, India, Mexico, Egypt, and the West Indies (Enejoh *et al.*, 2015).

Among the citrus group, acid lime ranks 4th in area and production in India. The total area under citrus fruit, including, lime, lemon, sweet orange, mandarin, cultivation in India is 1.078 million hectares (14.90 per cent of total fruit crop area) with the total production of 11.148 million tones (12.50 per cent of total fruit production), whereas the average productivity is 10.3 tones/ha (Anonymous, 2015).Out of which, lime/lemon shares about 4.0 per cent (2.864 lakh ha) and 3.20 per cent (28.35 lakh tones) of the total area and production over other citrus crops, respectively, whereas the total productivity is 9.90 tones/ha. Andhra Pradesh tops in the total production (5.827 lakh tones), Maharashtra leads in total area (45,000 ha), whereas Karnataka have highest productivity (23.40 tones/ha).

Table 4.1: The Leading Lime/Lemon Producing States in the Country (Anonymous, 2015)

Sl.No.	States	Area ('000 ha)	Production ('000 tones)	Productivity (tones/ha)
1	Andhra Pradesh	38.9	582.7	15
2	Gujarat	41.1	449.2	10.9
3	Telangana	22.1	331.9	15.0
4	Maharashtra	45.0	306.0	6.8
5	Karnataka	11.5	268.2	23.4
6	Madhya Pradesh	10.8	237.4	22.0
7	Bihar	18.0	128.9	7.2
8	Assam	13.0	103.5	8.0
9	Jharkhand	8.8	87.7	9.9
10	Chhattisgarh	11.5	78.8	6.9

World Scenario of Acid Lime

According to the FAO database, the total world production of lime and lemons was 16.245 million tones and total area under cultivation was 1.056 million hectares, with India leading production of 2.835 million tones in the year 2014 (FAO, 2014). India is the leading producer of limes and lemons with about 16 per cent of the world's overall production, followed by Mexico (~14.5 per cent), Argentina (~10 per cent), Brazil (~8 per cent), and Spain (~7 per cent) [Al-Abadi *et al.*, 2016]. The leading lime and lemon producing countries are listed in the Table 4.2.

History, Origin and Distribution

The genus *Citrus* L., which includes limes and lemons, has been grown in tropical and temperate regions of the globe over 2000 years ago. The lime and lemon citrus horticultural group is genetically highly complex, involving four ancestral species with diploid, triploid and tetraploid compartments. It is believed that sour lime (*Citrus aurantifolia* Swingle) is originated from Malaysian region of south-western Asia (Khan, 2007). 'Mexican' lime was proposed to be a direct hybrid between citron and *C. micrantha* (Scora, 1975; Nicolosi *et al.*, 2000). Probably the crusaders were responsible to carry the sour lime from the east firstly. Later on it was carried by the Arabs across North Africa into Europe (Spain and Portugal). After that Spanish and Portuguese explorers brought to the Americas in the early part of the sixteenth century (Ziegler and Wolfe, 1961). It was naturalized throughout the Caribbean, Eastern Mexico, tropical South America, Central America and the Florida keys. Commercial production in Florida in Orange and Lake Counties was evident by 1883. Later, small commercial plantings occurred in the Florida Keys (~1913 to 1926) and Miami-Dade County (1970s to early 2000s).

Table 4.2: Global Scenario of Limes and Lemons in the Year 2014 (FAOSTAT 2014)

Sl.No.	Countries	Area (ha)	Production (tones)	Productivity (tones/ha)
1	India	2,86,410	28,35,020	98,985
2	Mexico	1,56,429	22,05,079	1,40,964
3	China, mainland	96,000	21,30,500	2,21,927
4	Argentina	47,582	14,02,011	2,94,651
5	Brazil	43,399	2,53,877	11,01,799
6	Spain	37,743	10,90,709	2,88,983
7	USA	21,974	7,47,520	3,40,184
8	Iran	25,734	4,27,715	1,66,206
9	Turkey	27,665	7,25,230	2,62,147
10	Italy	25,996	3,70,458	1,42,506
11	World	10,56,771	1,62,54,214	-

Taxonomy

Acid lime, botanically known as, *C. aurantifolia* is commonly called Lime (Nigeria), Key lime, Mexican lime, Sour lime, Dayap, bilolo, Indian lime or Kagzi lime, Egyptian lime (Enejoh *et al.*, 2015). Limes comprises, both acid and sweet varieties – which are so different from one another in tree and fruit characteristics that they have been given separate species status. There are two kinds of acid limes, the small-fruited Mexican (West Indian, Key) type, *C. aurantifolia* Swingle and the large-fruited Tahiti (Persian, Bearss) lime, *C. latifolia* Tan., which is triploid and therefore seedless. *Citrus aurantifolia* is native to the Malaysian region of south-western Asia, while *C. latifolia* probably originated in the East and then spread to Persia, and then to Tahiti, possibly via Brazil and Australia, and finally to California. *Citrus limettioides* Tan., the sweet lime, commonly referred to as the Indian or Palestine sweet lime, is native to north-eastern India where it is known as *Mitha nimboo*, while in Egypt it goes by the name of *Limunhelou* (Khan, 2007).

The genus *Citrus* belongs to the subtribe Citrinae, tribe Citreaea, subfamily Aurantioideae of the family Rutaceae. Owing to the wide range of genus, species, varieties/cultivar and their hybrids, polyploids, mutations (Chimeras and sports) as well as polyembryony nature, the classification of genus Citrus is quite confused and complicated (Ray, 2002). Even due to such confusion, several scientists tried to classify the genus Citrus. The earlier classification was given by a number of scholars from USA and Japan. Out of which the works of Swingle and Tanaka's classification got major attention. W.T. Swingle (1948) and Swingle and Reece (1967),

Acid Lime–Thorny Fruiting Branch

Acid Lime–Branch with Fruit

Acid Lime–Fruit Developmental Stages

Acid Lime–Leaves with Small Wing

of the United States Department of Agriculture, grouped citrus and their various forms into major 16 species (Genus- Citrus, 2 sub genera *i.e.* Eucitrus (10 species) and Papeda (6 spicies), total 16 species, 8 botanical varieties).Due to the precise grouping by rejecting most of the forms, his classification was recognized as conservative type and considered as a "Lumper". On the other side, a Japanese scientist called Tanaka (1954) separated the genus Citrus into two subgenera *i.e.* Archicitrus and Metacitrus, comprised total 145 species, later 159 (Tanaka, 1969) and finally 162 (Tanaka and Kashio, 1977). The Tanaka's citrus classification was huge elaborative as he included most of the hybrids into separate groups and considered as "Splitter". In the classification of Swingle and Reece (1967) lemons and limes are classified as two separate species, *Citrus limon* (L.) Burm. F. and *C. aurantifolia* (Christm.) Swingle. In the classification of Tanaka (1954), limes and lemons are classified as 37 species. Another classification was given by Hodgson (1961), who critically studied these two classifications and made a compromised classification. This genus is further divided into two subgenera (Citrus and Papeda), based on leaf, flower, and fruit properties. The evolution of modern citrus cultivars and their diversity has been addressed (Swingle and Reece 1967). On the basis of morphological characteristics, studies have been carried out on the relationships between genera and species. This has led to the formulation of numerous classification systems. The most commonly used citrus classifications are by Swingle (Swingle and Reece 1967) and Tanaka (Tanaka and Kashio, 1977). In the genus *Citrus*, Swingle recognized only 16 species, whereas Tanaka recognized 162 species. The difference in these two systems reflected opposing theories on what degree of morphological difference justified species status and whether presumed hybrids among naturally occurring forms should be given species status (Soost and Roose, 1996).

In the classification of Swingle and Reece (1967) lemons and limes are classified as two separate species, *Citrus limon* (L.) Burm. F. and *C. aurantifolia* (Christm.) Swingle. In the classification of Tanaka (1954), limes and lemons are classified as 37 species. These conflicting classifications result from the total sexual compatibility between Citrus species and the frequent occurrence of apomixis due to nucellar polyembryony (Scora, 1975; Barrett and Rhodes, 1976), which led many taxonomists to consider interspecific hybrids fixed by apomixes or vegetative propagation (cuttings or grafting) as new species.

Taxonomic Classification of Acid Lime

Kingdom: Plantae

Phylum: Magnoliophyta

Class: Magnoliopsida

Order: Sapindales

Family: Rutaceae

Sub-Family: Aurantioideae

Tribe: Citrae

Sub-Tribe: Citrinae

Genus: Citrus

Species: Aurantifolia

Scientific name: *Citrus aurantifolia* Swingle

Modern Taxonomic Classification

The morphological and anatomical variations and the geographical place of origin were the key basis of previous citrus taxonomic classification. Later, Swingle (1943) suggested using chemical characters as the possible taxonomic marker like, using glycosides in combination with other usual morphological characters. Further, different scientist used several biochemical compositions of plant parts, such as leaves, flowers and fruits, flavonoids, leaves and rind oils, polyphenol oxidase catalysed browning of young shoots, leaf isozymes *etc.* (Nagy and Nordby, 1972; Tatum *et al.*, 1974; Malik *et al.*, 1974; Torres *et al.*, 1982; Hirai *et al.*, 1986). Taxonomic studies received a boost from the work of Barrett and Rhodes (1976). They did more elaborate phylogenetic work and evaluated a huge number of morphological and biochemical traits of tree, leaf, flower and fruit characteristics. Their study suggested that there were only three true species *i.e.*, *C. medica*, *C. reticulata*, and *C. grandis*. The other biotypes were derived from hybridization between these true species. Earlier also supported by Scora (1975). Lime (*C. aurantifolia*), *C. micrantha*, and *C. halmii* are also included in the list of 'true' citrus species by many researchers. Papeda is a group of Citrus species (*C. ichangensis*, *C. micrantha*, *C. latipes*, *C. celebica*, *C. hystrix*, and *C. macroptera*) having inedible fruit with acrid oil droplets in the juice vesicles. Understanding taxonomy, phylogenetic relationships, and genetic variability in citrus is critical for determining genetic relationships, characterizing germplasm, controlling genetic erosion, designing sampling strategies or core collections, establishing breeding programs, and registering new cultivars (Herrero *et al.*, 1996). However, these studies have not been able to clearly differentiate all the species. Hence, there is a need for additional taxonomic studies to further clarify the taxonomic distinctions (Gmitter *et al.*, 2009).

Genetically, the limes and lemons group of citrus fruits are more polymorphic including three different types of ploidy level namely, diploid, triploid and tetraploid varieties. With the discovery of DNA markers such as restriction fragment length polymorphisms (RFLPs), random amplified polymorphic DNA (RAPD), and sequence-characterized amplified regions (SCARs), have been widely used to look into the phylogenetic relationship of *Citrus* and its relatives clones to deduce into the genetic origins of the cultivated types. Recently, molecular marker studies have supported this hypothesis (Federici *et al.*, 1998; Nicolosi *et al.*, 2000; Moore 2001; Barkley *et al.*, 2006;Bayer *et al.*, 2009;Garcia-Lor *et al.*, 2012, 2013b;Ollitrault *et al.*, 2012;Curk *et al.*, 2014, 2015b). The phylogeny and genetic origin of important species of citrus have been investigated using molecular markers (Nicolosi *et al.*, 2000; Moore 2001; Berkeley *et al.*, 2006). Li *et al.* (2010)identified the paternal and maternal origins of cultivated *Citrus* such as sweet orange (*Citrus sinensis*), lemon (*C. limon*), and grapefruit (*C. paradisi*) using amplified fragment length polymorphism (AFLP) fingerprints, nuclear internal transcribed spacer (ITS), and three plastid DNA regions of 30 accessions of the cultivated citrus and their putative wild ancestors.

Their results indicated that bergamot (*C. aurantifolia*) and lemon were derived from citron (*C. medica*) and sour orange (*C. aurantium*), and grapefruit was a hybrid that originated from a cross between pummelo (*C. grandis*) and sweet orange. Rough lemon (*C. limon*) was likely as a parent of rangpur lime (*C. limonia*) and guangxi local lemon (*C. limonia*). Our data also demonstrated that sweet orange and sour orange were hybrids of mandarin (*C. reticulata*) and pummelo, while rough lemon was a cross between citron and mandarin. For Mexican lime (*C. aurantifolia*), our molecular data confirmed a species of Papeda to be the female parent and *C. medica* as the male. According to the recent publication, the maternal phylogeny of all cultivated *Citrus* species involved only four types of cytoplasm, whereas in the genus Citrus about six types of cytoplasm are encountered, and the nuclear genome involvement of the four basic Citrus taxa [*C. medica, C. maxima, C. micrantha* and *C. reticulata*] (Curk *et al.*, 2016) (Figure 4.1).

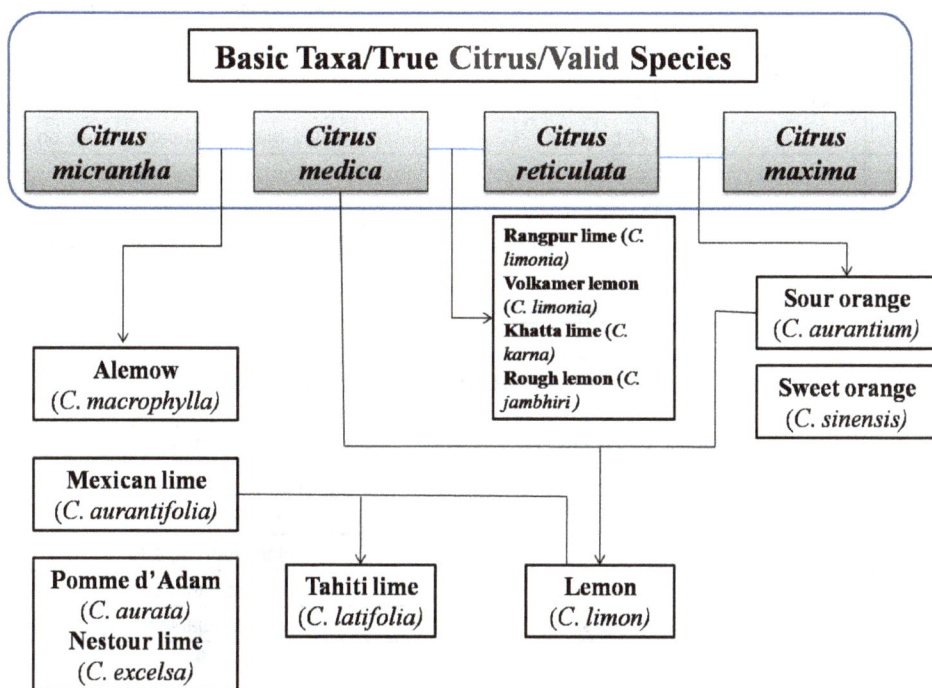

Figure 4.1: Origin of the Main Lime and Lemon Varietal Sub-groups (adapted from Curk *et al.*, 2016).

Genus *Citrus* and Species

Citrus is one of the world's leading fruit tree crop. In general, *Citrus* is believed to have originated in the tropical and subtropical regions of Southeast Asia, particularly eastern India, northern Myanmar and south-western China and then spread to other continents (Webber 1967;Chapot, 1975; Calabrese, 1992). At least for few citrus species, the NE region of India is known as one of the natural homes of citrus (Bose *et al.*, 2001). According to Bhattacharya and Dutta (1956), some citrus species such as, *Citrus ichangensis, C. indica, C. latipes* and *C. macroptera* are found in the NE states of India. The genus *Citrus* itself have a huge genetic variability comprising lots of valid species, their hybrids *etc*. It is clear from the taxonomic classification of genus Citrus, that how categorization of different species and their hybrids were done by W.T. Swingle (USA) and Tanaka (Japan).

Table 4.3a: Major Citrus Species of Horticultural Importance

Sl.No.	Major Citrus Groups	Members	Important Cultivars
I	Mandarin group	Citrus reticulata	Nagpur, Coorg, Khasi, Ponkan
		C. unshiu	Satsuma, Owari, Kara, Silver Hill
		C. deliciosa	Willow leaf, Kinnow, Wilking, Blinda
		C. nobilis	Kunembo, King
II	Orange group	C. sinensis	Mosambi, Blood Red Malta, Sathgudi, Valencia, Pineapple, Washington Navel
		C. aurantium	-
III	Pummelo-grapefruit group	C. grandis (earlier C. maxima)	Kao Pan, Buntan
		C. paradisi	Foster, Ruby, Marsh, Duncan, Thompson
IV	Acid group	C. limon	Eureka, Lisbon, Monachello, Bernia
		C. jambhiri	-
		C. aurantifolia	Kagzi lime, Rasraj, Jai Devi, Vikram, Pramalini
		C. limettoides	Mitha Nimboo

Table 4.3b: Wild, Semi-wild Species

Sl.No.	Major Citrus Groups	Common Name
I	Citrus indica	Indian Wild Orange
II	C. latipes	Khasi Papeda
III	C. macroptera	Melanesian Papeda
IV	C. ichangensis	Ichang Papeda
V	C. assamensis	Adajamir/Gajanimma

Table 4.3c: Related Genera

Sl.No.	Major Genera	Common Name	Species	Intergeneric hybrids
I	Poncirus	Trifoliate orange	Trifoliata	Citrange, Citrangequat, Citrangedin, Citrangor, Cicitrange, Citrumelo
II	Fortunella	Kumquat	Margarita, japonica, crassiflora, hindsii	Procimequat, Limequat, Orangequat

Table 4.4: The Varieties of Limes and their Recommended Areas in the Country

Sl.No.	Name of Variety	Varietal Description	Distribution
1	Kagzi lime	Very important cultivar of acid lime in India, but highly susceptible to citrus canker and tristeza virus	All over India
2	Pramalini	A selection from 'Kagzi' with more juice percentage and tolerance to Phytophthora.	Andhra Pradesh, Maharashtra, Gujarat, Bihar, Jharkhand, Karnataka, Tamil Nadu, Madhya Pradesh, Chattishgarh
3	Vikram	High yielding (46.40 kg/plant) than Kagzi (30-35per cent) and producing crops in September and May–June.	Andhra Pradesh, Maharashtra, Gujarat, Bihar, Jharkhand, Karnataka, Tamil Nadu, Uttar Pradesh, Madhya Pradesh, Chattishgarh, Uttarakhand
4	Jai Devi	Selection from PKM-1, which is a high yielding, with high juice content	Andhra Pradesh, Maharashtra, Gujarat, Bihar, Jharkhand, Karnataka, Tamil Nadu, Uttar Pradesh
5	Sai Sharbati	A high yielding (1000-1200 fruits/tree) with thin peeled, high juice content and large fruit size and some resistance to Phytophthora and Tristeza.	Andhra Pradesh, Maharashtra, Gujarat, Bihar, Jharkhand, Karnataka, Tamil Nadu, Uttar Pradesh, Uttarakhand
6	PKM-1	Evolved through intensive screening of acid lime alones at Periyakulam, TN. Fruits are bigger in size and juicy. Trees bear heavily.	
7	Phule Sharbati	High yielder (108.40 kg/plant)	Andhra Pradesh, Maharashtra, Gujarat, Bihar, Jharkhand, Karnataka, Tamil Nadu
8	Rasraj (IIHR Hybrid)		Karnataka
9	Balaji (Tenali Selection)	A clonal tree of Tenali, which is a free from citrus canker and bark eruption with high yielder and quality as well as rich in sugar and polyphenols.	Andhra Pradesh, Maharashtra, Madhya Pradesh, Chattishgarh
10	Seedless	A new selection made in Himanchal Pradesh. Matures late in the season. Fruits are oblong, thin peel, tight, yellow primrose colour at maturity.	
11	PDKV Chakradhar	A seedless and thornless variety, with spherical fruits rich in juice (60-65per cent) and vitamin C (118-140 mg/100 ml) content.	Andhra Pradesh, Maharashtra, Gujarat, Bihar, Jharkhand, Karnataka, Tamil Nadu

Sl.No.	Name of Variety	Varietal Description	Distribution
12	PDKV Lime	A Selection from existing local Kagzi lime, average fruit weight 32.6 to 52 g with 50-55 per cent juice content, TSS (7.4-7.8 per cent), acidity (8.0-8.4 per cent). The fruit is round to oblong, smooth and thin skin or papery rind, oval shape	Suitable for Vidarbha region of Maharashtra
13	PDKV Bahar	It is a large, round, juicy fruit with thin skin and an attractive and appealing colour. It grows on a tree with a large spreading canopy. The advantage with this variety is its yield capacity, which is much better than the existing varieties — about 36 tonnes yield per hectare. a. fruit size: Big, b. Yield : 30-35 t/ha, c. peel thickness:1.62mm, d. Juice: 53 per cent e. Acidity: 6.18per cent, f. Ascorbic acid: 31.47per cent	Suitable for Vidarbha region of Maharashtra
14	NRCC acid lime- 7	a. fruit size: 48 g, b. Yield : 54 t/ha, c. Juice: 50.50 per cent d. No. of fruits/plant: 4069, e. Growth habit: vigorous and spreading type	Developed from NRCC (now ICAR-Central Citrus Research Institute), Nagpur,Maharashtra
15	NRCC acid lime- 8	a. fruit size: 50 g, b. Yield : 59 t/ha, c. Juice: 51.53 per cent d. No. of fruits/plant: 4237, e. Growth habit: spreading type	Developed from NRCC (now ICAR- Central Citrus Research Institute), Nagpur,Maharashtra
16	Pusa Abhinav	A promising clonal selection having medium vigorous trees, dense foliage, and attractive bright yellow round shaped fruit, but moderately susceptible to canker.	Recently released from ICAR-IARI, New Delhi
17	Pusa Udit	A clonal selection which has a dense canopy, heavy yielder, fruits round in shape, medium in size (37.9 g), fruit length (43.9 mm), diameter (39.3 mm), peel thickness (1.1 mm), no. of seeds (8.3/fruit), fruits are bright yellow at ripening, juice (43.30per cent), TSS (8.5per cent) and acidity (6.9per cent).	Recently released from ICAR-IARI, New Delhi

Sources: https://www.pdkv.ac.in, http://www.iari.res.in.

Varietal Diversity of Acid Lime

Several varieties of acid lime are grown. Kagzi is the main variety in India due to its thin peel and is similar to 'Key' lime of Florida or 'Mexican' lime or 'West Indian' lime of Mexico. It is seeded, small fruited with profuse bearing.

Table 4.5: Specific Features of Cultivated Acid Lime Varieties grown in India

Specific Features		Variety
Tree growth habit	Erect	Chakradhar
	Spreading	Sai Sharbati, Vikram, Pramalini, Balaji, Phule Sharbati.
Spine density on the adult tree	Low (<10)	Chakradhar
	Medium (10- 15)	Sai Sharbati, Vikram, Pramalini, Balaji, Phule Sharbati
Petiole wings	Absent	Chakradhar
	Present	Sai Sharbati, Vikram, Pramalini, Balaji, Phule Sharbati.
Fruit weight (g)	Light (<40)	Chakradhar,Vikram, Pramalini, Pusa Abhinav, Pusa Udit
	Heavy (>40)	Sai Sharbati, Balaji, Phule Sharbati
Fruit diameter (mm)	Small (<41)	Chakradhar
	Medium (41 - 45)	Vikram, Pramalini, Balaji
	Large (>45)	Sai Sharbati, Balaji, Phule Sharbati
Fruit diameter (mm)	Short (<40)	Chakradhar
	Medium (40-45)	Vikram, Pramalini, Balaji
	Long (>45)	Sai Sharbati, Balaji, Phule Sharbati
Albedo colour	Greenish	-
	White	Sai Sharbati, Vikram, Pramalini, Balaji, Phule Sharbati, Chakradhar
Number of segments per fruit	8-10	Sai Sharbati, Vikram, Pramalini, Balaji, Phule Sharbati, Chakradhar
Fruit rind (epicarp) thickness (mm)	Thin (<2)	Sai Sharbati, Vikram, Pramalini, Balaji, Phule Sharbati, Chakradhar
Fruit juiciness (per cent)	Low (<40)	-
	Medium (40-50)	Vikram, Pramalini, Chakradhar, Pusa Udit
	High (>50)	Sai Sharbati, Balaji, Phule Sharbati, Pusa Abhinav
TSS (ºBrix)	High (>7)	Sai Sharbati, Vikram, Pramalini, Balaji, Phule Sharbati, Chakradhar, Pusa Udit
Titratable acidity (per cent citric acid)	Low (<5)	-
	Medium (5-6)	Chakradhar
	High (>6)	Sai Sharbati, Vikram, Pramalini, Balaji, Phule Sharbati, Pusa Abhinav, Pusa Udit
Seediness (Number of seeds/fruit)	<4	Chakradhar
	4-10	Sai Sharbati, Vikram, Pramalini, Balaji, Phule Sharbati, Pusa Udit
	>10	-

Source: http://www.plantauthority.gov.in/pdf/DAcidlime.pdf.

Ecological Adaptation

Acid lime is a tropical citrus crop; grow well under warm and slightly humid regions of the country. The region should be free from strong wind and frost as well as heavy rainfall. It thrives well at an altitude of 1000 msl. It is adapted to a wide range of edaphic conditions ranges from heavy clay to light sandy soils. Soil must be aerated with adequate drainage facility. Water table should not be deeper and soil pH upto 6.0 to 8.0 (Radha and Mathew, 2007). Trees of acid lime are susceptible to frost damage and hence not grown in frost prone areas. Warm and dry climate is suitable while humid climate with high rainfall increases problem of bacterial canker (Chadha, 2009).

Major Production Problems in Acid Lime and their Solutions

Citrus is considered as one of the most important tropical fruit crop in India. It plays a vital role in the fruit economy of the country next to mango. Citrus cultivation in India is facing plethora of production constraints *viz.*, non-availability of disease free planting material, bud wood transmissible diseases, general neglect, scarcity of water and poor management practices, cultivation in unsuitable soils. Beside this, *Citrus* spp. are prone to attack by more than 150 diseases and disorders caused by fungal, viral and few bacterial pathogens right from nursery level to bearing stage resulting in considerable yield losses (Lakshmi *et al.*, 2014). The major problem in acid lime cultivation is citrus bacterial canker caused by *Xanthomonas axanopodis* pv. citri (Chadha, 2009;Lakshmi *et al.*, 2014) followed by viral disease Tristeza (quick decline). A complex of twig die-back (caused by *Colletotrichum gloeosporioides*), canker, Phytophthora and Tristeza can be observed in declining orchards of acid lime (Chadha, 2009).

Cross protection of kagzi lime against Tristeza with mid strains successfully employed in Australia. In mid-eighties, emphasis was laid on production and distribution of cross protected plants by ICAR-IIHR, Bangalore and Acharya N.G. Ranga Agricultural University, Tirupati campus. As a result, partial success was reported to this technology. Use of resistant rootstocks or chemical control of Phytophthora, control of vectors of Tristeza, chemical control of canker and twig die-back are the standard practices. Adoption of canker resistant varieties and biological control for insect-pests and diseases has to be the approach to reduce costs and minimize pollution by chemicals (Chadha, 2009).

Morphology of Acid Lime

Citrus aurantifolia is a small shrubby tree, about 5 m tall. It is an evergreen and ever bearing tree that is densely and irregularly branched and possesses short and stiff spines (thorns). The leaves are alternate; elliptical to oblong-oviate (4-8cm×2-5cm) shaped and has a crenulate margin. The flowers are 3 cm in diameter and are yellowish white with a light purple tinge on the margin. The fruits are globose to ovoid berry of about 3 - 6 cm in diameter and sometimes have apical papilla. It is yellow when ripe but usually picked green commercially. The fruits and flowers appear throughout the year but are most abundant from May to September in the Northern hemisphere. The fruit peels are very thin with densely glandular segments with yellow-green pulp vesicles. The fruit juice is acidic and fragrant, sour as lemon juice but more aromatic. It is usually valued for its unique flavor compared to other limes. The seeds are small, plump, ovoid, pale, and smooth with white embryo (Sethpakdee, 1992; Okwu, 2008).

Tree

Key lime is a small, bushy tree, rarely taller than 12 feet (4.1 m). It has slender branches armed with short to medium length thorns. Spineless selections are more compact and upright in growth, have darker green foliage, and are characterized by low fruit production.

Root

Citrus root systems are important for tree anchorage and support. They also provide a mean of collection and transport of water and nutrients essential for tree growth and production. The ideal environment for citrus roots is a porous, medium-textured, well-drained soil, where water is easily available but not in excess. When such soil is irrigated, water distribution itself throughout the soil profile in the root zone or drains away, leaving no excess. The growth and production of citrus may be impaired either by an excess of water in the root zone or by a lack of easily available water. A deficiency of water in the root zone produces detrimental effects on root growth. As the soil dry, root growth becomes slower and eventually ceases (Alves *et al.*, 2012).

Leaves

The dense foliage consists of small, pale green, blunt pointed leaves with narrowly winged petioles (leaf stalks).

Flower

The flowers are held in axillary clusters (cymes) along the stems. The flowers are fragrant and small with white petals, a superior ovary, and 9 stamens. Flowering occurs throughout the year but mainly in the spring. Acid lime exhibits cyclical growth flushes throughout the year. In case of acid lime, the three blooming seasons are *Mrig bahar* (June-July bloom), *Hasta bahar* (September–October bloom) fruits fetch high price in the market, and Ambe bahar (January to February bloom).

Fruit

Fruit are small (3.8–5.1cm) in diameter, round to oval with 10–12 segments around a small, solid axis. The peel is thin, smooth and leathery, and greenish yellow to yellow at maturity. The pulp is greenish yellow, juicy, and highly acid with a distinctive aroma. There are 10 to 15 seeds, which are highly polyembryonic (two or more plants identical to the mother plant may be produced from one seed) [Davies and Albrigo, 1994].

Acid Lime Germplasm Collection in Genebanks

In India, ICAR-National Bureau of Plant Genetic Resources (ICAR-NBPGR) is a nodal agency, playing critical role in plant exploration and collection activities in various regions of the country and assembled large diversity in cultivated and wild *Citrus* species including acid lime. Total germplasm conservation in Fruit Gene Bank (FGB) is 1,285 accessions, including fruit crops, plantations crops *etc*. Out of which, *Citrus* spp. contributes 65 accessions (Table 4.6).

Table 4.6: Citrus Germplasm Collected at ICAR-NBPGR, New Delhi

Sl.No.	Citrus species	No. of Accessions
1	*Citrus jambhiri*	4
2	*Citrus sinensis*	19
3	*Citrus aurantifolia*	23
4	*Citrus medica*	5
5	*Citrus grandis*	3
6	*Citrus aurantium*	1
7	*Poncirustrifoliata*	1
8	*Citrus species*	9

Recently, ICAR-NBPGR, New Delhi undertook expedition and exploration in the year 2015.Thirty accessions of citrus comprising *Citrus aurantium* (1), *C. aurantifolia* (12), *C. grandis* (1), *C. jambhiri* (1), *C. limonia* (3), *C. maderaspatana* (1), *C. paradisi* (1), *C.sinensis* (7), *C. unshiu* (1) and *Citrus* sp. (2) were collected from parts of Chittoor, Nellore and Cuddapah districts of Andhra Pradesh and observed good variability in *C. aurantifolia* (Lime), *C. sinensis* (Satgudi) and *C. maderaspatana* (Kitchli). Lime variety Balaji was the most promising and sweet orange-Satgudi having high yield and sweetness is highly preferred by farmers. Kitchli a local uncommon citrus type is also grown in the area.About 60 accessions of *Citrus* grown at ICAR-CCRI, Nagpur were released after joint inspection at ICAR-NBPGR. Approximately 176 accessions of germplasm of non-orthodox seeds of different species of *Citrus* were collected, including 17 accessions of Kagazi nimboo (*Citrus aurantifolia*).

ICAR-CCRI, Nagpur

National Citrus Repository of ICAR-CCRI, Nagpur has a germplasm collection of 614 *Citrus* accessions including 62 exotic genotypes and 522 indigenous types. Only 4 accessions are of acid limes. About 14 accessions of acid lime were evaluated during 2014-15.

Genetic Diversity of Lime

Dubey *et al.* (2016) collected and characterized the pomological diversity of 53 lime and 12 lemon accessions, which were analyzed based on 11 quantitative traits from different states of the country like, Madhya Pradesh, Uttarakhand, Delhi, Rajasthan, Punjab and Haryana. The collected and characterized accessions were listed in Table 4.7.

Breeding Nature

India is a rich biodiversity of limes and lemons and well adopted to adverse agroclimatical zones of the country, and also grown in Homestead and kitchen gardens. Due to susceptibility of the most divesting diseases such as citrus canker and dieback, major emphasis is given to identify resistant/tolerant species/cultivars. Traditionally, genotype characterization and identification are based on phenotypic characters (Zamani, 2013). However, morphological/quantitative traits are more useful for assessing genetic diversity among morphologically distinguishable accessions.

Table 4.7: List of Lime and Lemon Accessions

Location	Accessions
Upper Gangetic plain zone	ALC-6, ALC-7, ALC-51, ALC-52, ALC-53, ALC-54, ALC-55, ALC-56, ALC-57, ALC-58, ALC-59, ALC-61, ALC-86, ALC-87, ALC-88, ALC-89, ALC-92, ALC-93, ALC-94, ALC-95, ALC-104, ALC-105, ALC-106, ALC-107, ALC-108
Trans Gangetic plain zone	ALC-3, ALC-4, ALC-5, ALC-11, ALC-12, ALC-13, ALC-16, ALC-17, ALC-21, ALC-22, ALC- 24, ALC-111, ALC-112, ALC-113 ALC-114, ALC-115, KagziKalan, Konkan Seedless, LS-1, LS-3, LS-4, LS-5, LS-8, LS-9,LS-10, LS-11, LS-12
Western dry zone	ALC-40, ALC-45
Central plateau and hill zone	ALC-96, ALC-97, ALC-98, ALC-99, ALC-100 ALC-101 ALC-102 ALC-103, LS-7
Western Himalayan zone	ALC-1, ALC-2

Acid lime (*Citrus aurantifolia*) is a cross pollinated fruit crop and has a high variation in fruit characters within the species. In citrus, most of the genotypes are selected on the basis of fruit character. Availability of a wide gene pool, species diversity and variability is utmost importance for promising genotypes selection for crop improvement programme.

Selection

In India, citrus germplasm collection and conservation started in middle of 19th century. North-eastern India is a rich diversity of citrus species, and first systematic work on citrus germplasm exploration, collection, conservation, citrus variability and species identification was made by Bhattacharya and Dutta (1956). As many as 17 Citrus species, their 52 cultivars and 7 probable natural hybrids are reported to have originated in the north-eastern region of India. The superior clones are selected and evaluated for different agroclimatic condition, mainly for adopting commercial cultivation. Clonal selection is an easy way to evolve promising cultivars in citrus industry. Evaluation of genetic resources for important characters is a most essential and useful steps before programming any breeding work to start with for introducing genetic improvement in any crop. The systematic evaluation of the biodiversity needs a thorough and well established approach.

In Eastern Nepal, a superior acid lime clones *viz.*, LD-23, LT-17, LD-49 and LM-44 identified from different agro-ecological *viz.*, terai, mid hill and high hill areas (Shrestha *et al.*, 2012). In India, acid lime breeding work has been initiated at Tirupati (AP), Periyakulam (TN), MAU., Parbhani (M.S.), Dr. PDKV., Akola (M.S.), MPKV.,

Rahuri (M.S.), ICAR-CCRI., Nagpur (M.S.), ICAR-IIHR., Bangalore and ICAR-IARI., New Delhi. In Maharashtra, Vikram and Pramalini (from MAU Parbhani), PDKV lime and PDKV bahar (from Dr. PDKV., Akola), Sai-Sharbati (Rahuri Sharbati) and Phule Sharbati (from MPKV., Rahuri), seedless and thornless strain of Chakradhar, cultivars were commercially cultivated in Maharashtra (Badge and Patil, 1989, Chakrawar and Rane, 1977; Paithankar *et al.,* 2015), and Jai Devi (PKM-1) and balaji were released from Horticulture Research and Training Institute, Periyakulam (Tamilnadu) and Citrus Research Station, Tirupati (AP), respectively. Some superior clones, also been identified at ICAR-AICRP on fruits (Citrus), Dr. PDKV., Akola for further evaluation *viz.,* genotype 20-5 and 15-1 (Kamatyanatti, *et al.,* 2015).

In Maharashtra, ICAR- CCRI, Nagpur initiated crop improvement programme for acid lime and indentified 12 elite clones based on qualitative and quantitative characters (*viz.,* KL - 2, KL - 4, KL - 7, KL - 9, KL - 11, KL - 12, KL - 15, KL - 16, KL - 21, KL - 22, KL - 23 and KL – 24) and they were field planted for further evaluation (http://nrccitrus.nic.in/index.php?c=pages and m=index and id=67). In, Arunachal Pradesh, a superior clone of acid lime ARL-1 was selected from the naturally existing variability (Dubey, 1998). Recently, collection, conservation and exploration of lime and lemon based on quantitative character was carried out in various parts of northern India *viz.,* Gangetic plain zone, Trans gangetic agroclimatic zone, Western dry Region, Central plateau and hill region and Western Himalayan region, some accessions of acid limes namely, ALC-1, ALC-4, ALC-6, ALC-12, ALC-53 ALC- 54, ALC-55, ALC-57, and ALC-58, and lemon accessions (LS-7, LS-8, LS-9, LS-10 and LS-12) found promising and are conserved *ex situ* in the citrus field gene bank of Indian Agricultural Research Institute, New Delhi (Dubey *et al.,* 2016). Acid lime flowers thought the year under tropical condition. The cultivar,Kadalipali-630 was reported for year round fruiting with better TSS and acidity, Bagabar-633 for better fruit quality, Basantpur-964 for thick skinned, excellent TSS and acidity with better fruit yield (19.5 kg/plant) and Odiapali-50 for higher yield (50.4 kg/plant) with twice flowering character (Sahoo *et al.,* 2010).

Acid lime is most suitable for dry climate and low rainfall areas of tropical and subtropical region, In West Bengal, Pati seedless and Kagzi Large are evaluated and commercially recommended for cultivation (Ghosh *et al.,* 2012). Due to more pleasant aroma of Gandharaj cultivar, fetches higher price in Kolkata and its adjoining markets.

Hybridization

Citrus aurantifolia fruits are small, spherical, and seedy fruit with true acid content. It's also known as West Indian or Key lime or Mexican lime. On the basis of molecular analysis suggest that, *Citrus aurantifolia* is probably originated by citron and *C. micrantha* Wester (Nicolosi *et al.,*2000).Acid limes are tropical species and well adapted to warm and humid areas of subtropics. There is a huge demand for development of high fruit quality, high acid content with low seed content for modern citrus industry. But also development of resistant varieties mainly against Asiatic citrus canker (*X. axonopodis* pv. citri), citrus tristiza virus (CTV), and witches broom of lime. Lime and lemons are also very sensitive to cold so, now a day's target

Table 4.8: Acid Lime Promising Clone/Selections and their Commercial Distribution in India

Sl.No.	Common name	Selections/Cultivars	Distribution
		North and North-Eastern India	
1	Acid lime (*C. aurantifolia* Swingle)	Abhayapuri lime, Karimgunj lime	Assam plains (Lower Assam and Barak valley, Tinsukia region)
2		Arunachal lime, Kaghzi-Nimbu	Arunachal Pradesh (East Siang District), Meghalaya (Khashi and Garo tribal region)
3		Kagzi lime	Kumaon and Garhwal hills of Uttarakhand, Himachal Pradesh (Kangra and Hamirpur Region) and parts of Uttar Pradesh and Madhya Pradesh (Tikamgarh area), Parts of Rajastan (Husangsar, Bikaner, Chattargarh, Khichiya and Nokiya area)
4	Acid lime	Pati seedless (erect and seedless), Pati hybrid (Bushy), Pati baramasi (dwarf), Kagzi large (spreading type), PKM-1 (spreading and drooping) and Gandharaj (dwarf and drooping)	Parts of West Bengal (*viz.,* Paschim Midnapore area, Bankura, Purulia, Birbhum)
5	Acid lime	Coorge lime, Sylhet lime, Kagzi lime	Part of Punjab, especially Abohar (Firozpur) region
		Southern India	
6	Acid lime (*C. aurantifolia* Swingle)	PKM-1, Sai-sarbati, Pramalini, Vikram, Tenali, Kasipentla, Jaidevi (PKM-1), Mungalipattu	Tamilnadu (Periyakulam, rainfed and irrigated conditions in the district of Trichy, Tirunelveli, Dindigul, Virudhunagar, Ramanathapuram, Madurai, Theni *etc.*)
7	Acid lime (*C. aurantifolia* Swingle)	Vikram, Pramalini, Sai Sarbati, Kagzi Kala, Tenali, PKM-1 and Tirupati	Andra Pradesh (Tirupati and Chittoor area) and parts of Telangana
8	Acid lime (*C. aurantifolia* Swingle)	Kagzi lime, Vikram, Pramalini, Sai Sarbati, Tenali and Coorg lime	Karnataka (Vijayapur, Bagalkote, Raichur and Dharwad area)
		Central and Western India	
9	Acid lime (*C. aurantifolia* Swingle)	PDKV Lime, PDKV Bahar, Sai Sarbati, Vikram, Pramalini, Chakradar, Tenali	Maharastra (Akola, Solapur,Ahmadnagar, Parbhani, Dhule, Jalgaon, Pune, Nasik, Nagpur, Amravati, Buldana, Wardha and Daud area)
10	Acid lime (*C. aurantifolia* Swingle)	Kagzi lime	Gujarat (Kheda, Mehsana, Baroda, Surendra Nagar, Gandhinagar area)

Sources: Malik *et al.,* 2012, Hazarika,2012, Upadhaya *et al.,* 2016, Tamata *et al.,* 2008, Murkute and Singh 2015,Kumar *et al.,* 2011 and Ghosh *et al.,* 2012.

for development of cold hardiness cultivars. Inbreeding depression is a characteristic of citrus hybrids from narrow crosses, while hybrid vigor is present is progenies from wider crosses (Soost and Roose, 1996).Natural hybridization is very common feature in citrus. Important intergeneric/intrageneric acid lime hybrids are:

1. Limequat

C. *aurantifolia* X F. *japonica* hybrid developed by Walter and Swingle at a U. S. Department of Agriculture Research station in Florida around 1909. Trees are evergreen and very thorny with spreading, small fruits with round to oval in shape and yellow green to yellow orange in color depending on the cultivar. Limequats are more cold tolerant to Mexican lime. The trees are now grown in Japan, Israel, Spain, Malaysia, South Africa, Armenia, the United Kingdom and the United States in California, Florida, and Texas. The fruit can be found, in small quantities, during the fall and winter months in the United States, India and Japan.

2. Perrine

Mexican lime X 'Genoa' lemon hybrid developed by Dr. Walter Swingle and colleagues in 1909. It is extensively cultivated in southern Florida on rough lemon rootstock. Bears, lemon-shaped fruit, with small nipple at apex, peel pale lemon – yellow, pulp pale greenish-yellow, in 10 to 12 segments having thin walls; tender, very juicy, high acidity, seeds usually 4 to 12, long-pointed. Trees are more tolerant to cold-sensitive than the lime; resistant to wither tip and scab but prone to gummosis and other bark diseases.

3. Rasraj

It's a hybrid between, Nepali round lemon (canker resistant) X acid lime (C. *aurantifolia*), resistant to bacterial canker, rind is thicker than lime, yellow fruit color with 55 g fruit weight, 70 per cent juice content and 12 seeds/fruit (http://www.iihr.res.in/varieties?page=9).

4. Lemonimes

C. *limon* X C. *aurantifolia* hybrids are found more resistant/tolerant to bacterial canker.

In Citrus, somatic hybridization plays an enormous potential for citrus improvement, due to occurrence of several biological barriers *viz.*, nucellar polyembryony, long juvenility and pollen/ovule sterility. Somatic hybridisation is performed (both physical and chemical method) by electric fusion or poly ethylene glycol (PEG) treatment. PEG-induced fusion has advantages because of low cost, high frequency of heterokaryons and no need of any special equipment (Grosser and Gmitter, 2011). For citrus somatic hybridization, the model of "embryogenic callus protoplast + mesophyll protoplast" was usually applied; since mesophyll protoplasts never divide and regenerate into plants, it is actually a half selection system, which is very conducive to identify mesophyll-parent type cybrids simply by their leaf morphology (Guo, 2013). Some reports of mesophyll parent type plants (Cybrids) regenerated *via* symmetric protoplast fusion in Kagzi lime/Acid lime (C. *aurantifolia* Swingle) are given in Table 4.9.

Table 4.9: Some Reported Somatic Hybrids of Acid Lime

Sl.No.	Somatic Hybrids	References
1	*C. deliciosa* 'willow leaf' + *C. aurantifolia* Swing.	Ollitrault *et al.*, 1996
2	*C. sinensis* 'shamouti'+ *C. aurantifolia* Swing.	Ollitrault *et al.*, 2000
3	*C. aurantifolia*+ *C. sinensis* G1	Ollitrault *et al.*, 2000
4	*C. Paradisi* ' star ruby'+ *C. aurantifolia* Swing.	Ollitrault *et al.*, 1996
5	*C. aurantifolia* Swing. + *C.aurantium*	Ollitrault *et al.*, 2000
6	*C. aurantifolia* Swing. + (*C. sinensis* X *P. trifoliata*) 'Carrizo'	Ollitrault *et al.*, 2000
7	*C. sudachi* ' shirai' + *C. aurantifolia* Swing.	Saito *et al.*, 1993
8	Key' lime (*C. aurantifolia* (Cristm.) Swing) + 'Valencia' orange,	Viloria *et al.*, 2005
9	*C.aurantifolia* cv. Macrophylla + Carrizo citrange (*C.sinensis* × *P. trifoliata*)	Dambier *et al.* (2011)

Cryopreservation

Citrus conservation *in vitro* has involved sequential culture of nodal stem segments (Marin and Duran-Vila, 1991). Cryopreservation of embryogenic callus can be used for long-term storage of totipotent lines and citrus genetic resources. Cultures are cryoprotected with 10 per cent (v/v) DMSO, frozen by slow cooling, stored in liquid nitrogen and warmed by fast thawing, after which they proliferate normally. Cryopreserved embryogenic cultures have been shown to be suitable for further proliferation (Perez *et al.*, 1998) and as a source for protoplast isolation, somatic hybridization and plant regeneration (Olivares-Fuster *et al.*, 2000).

Improvement of Fruit Quality in Kagzi Lime

In acid lime, consumers are mainly prepared seedless with high acidity, high juice content, TSS/Acidity ratio, and aroma with good colouration. All the citrus sp. and their cultivars have unique and diverse quality characteristics. In acid lime fruit size, fruit shape, aroma, acidity and ascorbic acid contents belong to fruit quality, so these characters to be considered for evaluation. Insects and animals are mainly attracted by fruit colors, which is attributed to water soluble (Flavonoids and betalains) and fat soluble (Carotenoids and chlorophylls) pigments. Flavanoids includes anthocyanins and colourless preanthocyanidin, mostly yellow flavonols which are derived from phenylalanine (Springob *et al.*, 2003). Anthocyanin are the group of pigments (glucosides and acylated derivatives) *viz.*, cyanidin, delphinidin, peonidin, malvidin etc., provide wide range of colors like yellow, red, blue, pink, black and magenta (Alcolea *et al.*, 2002). Most of the fruits at immature stage chlorophyll is a dominant and colorant, it act as a copigmentation for anthocyanin.

Previously, produced somatic hybrids that combine Key lime and 'femminello' lemon with Valencia or Hamlin sweet oranges are flowering and producing pleasant acid-type fruit (Grosser *et al.*, 1989; Tusa *et al.*, 1992). Somatic hybrids are colder hardy than cultivated lime and lemons, triploid hybrids are seedless, reduced fruit size and thinner rinds. Seedlessness in citrus spp. is mainly due to male sterility, female sterility, self incompatibility and some cultivars produce seedless fruits of parthenocarpy or rudimentary ovary (Washington naval) (Yamamato *et al.,* 1995). Sometimes natural mutation, bud mutation, irradiation (X rays and gamma rays) creating abnormal chromosomal division and cause seedlessness. In citrus, most of the seedless cultivars are developed by bud mutation and seedling selection (Deng *et al.,* 1996). The commercial kagzi lime cultivar, Chakradhar (seedless and thornless) is developed by selection (Badge and Patil, 1989) and some triploid limes *viz.,* Tahiti lime (2n=3x=27) (Morton, 1987) produce seedless fruits. Pickle is an important processed product of lime. For pickling, thick peeled and large fruits are more desirable. Paithankar *et al.* (2016) undertaken evaluation work at AICRP (Fruits) Dr. PDKV., Akola, and found Akola lime 3 (clone), most suitable for pickle making with desirable characters *viz.,* more peel thickness (0.28cm), highest peel percentage (21.52 per cent), less juicy (49.75 per cent), maximum rag per cent (28.73 per cent) and highest pickle recovery (67.50 per cent).

Several factors influencing the fruit quality are rootstock influence, agro-climatic conditions, cultivars and bearing season. In Central and Vidarbha region of Maharashtra, hasta bahar fruits are preferred for its quality and consumer demand. The influence of rootstock on fruit ripening and postharvest quality, reported that acid lime fruit grown on Rangpur lime rootstock reported longer shelf life with lower weight loss as compared with fruits grown on Troyer and Trifoliate stocks (Jawaharlal *et al.,* 1991).

Improvement for Biotic and Abiotic Stress Tolerance/Resistance in Kagzi Lime

Tristeza is a most devastated disease for citrus, it is caused by a closterovirus [citrus tristeza virus (CTV)] belongs to Closteroviridae family and caused through infected buds and by various aphid (*Aphis sp.*) species (Bar-joseph and lee, 1989). In limes, CTV transmitted by both Graft and aphid, and vein clearing, stem pitting, leaf chlorosis are most common symptoms (Ayllon *et al.,*1999, Liu *et al.,* 2012). Lemons are tolerance to CTV, and also slightly colder winter condition. The survey work has been conducted in Vidarbha, Marathwada and Western Maharashtra, during exploration most of the acid lime plants are affected by canker, except prominent acid lime cultivating areas (Beed area near Parli) (Singh and Singh 2003). Due to unaware of rootstock and non commercial citrus orchards, Phytophthora incidence has been reported up to 70 per cent of citrus growing parts of Maharashtra.

Citrus bacterial canker disease (CBCD) (*Xanthomonus axonopodis* pv. Citri) is also one of the major constraints in acid lime cultivation (Pal and Chand, 1983). Acid lime is very susceptible to bacterial canker, Although several acid lime clone/ selection or hybrids have been claimed either as resistant or tolerant from different regions *e.g.* RHR-L-49 (Sai-sarbati) (Desai *et al.,* 1999), Tenali (Madhavi *et al.,* 2000) and ALH-77 (lime x lemon hybrid) *i.e.* acid lime (as a female parent) and lemon (as a male parent *viz.,* Nepali oblong and Nepali round) hybrids were reported moderate to high resistant (Prasad *et al.,* 1997) and heritability resistant also reported by Koizumi (1982).

Acid lime anthracnose (*Colletotrichum acutatum*) is a fungal pathogen, this fungus earlier reported as *Gloeosporium limetticolum* Clausen (Timmer, 1978), on the basis of molecular characterization and pathogenicity tests it is confirmed to be *Colletotrichum acutatum* (Ruiz *et al.*, 2014).This disease is endemic to Mexican limes and Dominican thornless lime and does not affect other limes *viz.*, Tahiti lime. This fungus attacks almost all the parts (young leaves, twigs, flowers and flower bud, immature fruits) of the plant.

Witches' broom disease (*Candidatus Phytoplasma aurantifolia*) is a big threat to maxican lime growing countries mainly, Arab Emirates, Oman, Iran and other countries. The hesperidin, naringenin, and quercetin present in the leaves be used as a biomarker to identify disease infection at early stage in lime (Mollayi, *et al.*, 2016). If the plant is infected by the pathogen then increases the expression level of phenylalanine ammonia-lyase (PAL) and Chalcone synthase (CHS), followed by catechin, epicatechin biosynthesis in leaves so, this increased level of enzyme act as a defensive mechanism against the pathogen attack (Mollayi, *et al.*, 2016).

Citrus Scab (*Elsinoe fawcettii* Bitancourt) is caused by a fungus, and seriously affected in acid lime growing area, it attack almost all parts of the plant *viz.*, leaf, stem and fruits and deteriorate plant health. The scab and canker cause more than 50 per cent of yield loss (Ghosh and Ray, 2013). In citrus, more than 250 insect species are reported in India (Sreedevi and Rajulu., 2008). Acid lime is also infested by several pests mainly, Leaf minor, citrus butter fly, Thrips and aphids. These pest are considered to be a minor importance but sever incidence has been observed under specific agro-climatic condition (Butani, 1979). Field evaluation studies of 13 selected acid lime clones for thrips (*Scirtothrips* sp.) and mites (*Phyllocoptruta olievora* Ashmead) has been evaluated at Citrus Research Station (Petlur), Andra Pradesh, found that Selection-21,1 and Vikram for resistant to thrips, PKM1, Selection-1,7,25 for mites, respectively (Sreedevi and Rajulu., 2008).

Frost is the major problem for cultivation of Kagzi lime in northern India mainly Kumaon and Garhwal hills, Nanital of Uttarakhand and parts of Himachal Pradesh and Utter Pradesh so, there is a need to identify and exploit superior germplasm for breeding programme. In northern India, (Tamata *et al.*, 2008) reported three kagzi lime genotypes (*viz.*, IC319045, IC319068 and IC259667) among that IC319045 reported as a frost resistant and can be used for parent hybridization.

Nutritional Composition of *C. aurantifolia* Fruit Juice

The nutritional content of raw *C. aurantifolia* fruit juice per 242 g (1 cup) of the juice includes various nutrients which are presented in Table 4.10 [USDA Nutrient Database 2013].

Domestic Uses of *Citrus aurantifolia*

Acid Lime Juice

Citrus fruits are consumed globally in the form of fresh fruit or are processed into citrus products and citrus-by-products. Approximately, $1/3^{rd}$ of total citrus production is utilized for processing (Okwu, 2008). *C. aurantifolia* juice is used in beverages as flavorings agent (Yano *et al.*, 1999); and it is also used by juice producing companies to produce freshly squeezed orange juice or frozen concentrated orange juice. In addition, *C. aurantifolia* fruits can be processed to obtain other food products such as marmalades, jams, sorbet, pickles, jellies, candies; sugar boiled and dehydrated citrus products (Okwu, 2008).

Table 4.10: Nutritional Composition of Acid Lime Juice

Nutritional Composition	Concentration (Fruit juice per 242 g)
Water	220 g
Total calories	60.5 KJ
Protein and amino acids	1.0 g
Total carbohydrate	20 g
Dietary fibre	1.0 g
Sugar	4.1 g
Total fat	0.2 g
Vitamin A	121 IU
Vitamin C	172 mg
Vitamin E	0.5 mg
Vitamin K	1.5 mcg
Thiamine	0.1 mg
Niacin	0.3 mg
Vitamin B6	0.1 mg
Folate	24.2 mcg
Pantothenic acid	0.3 mg
Choline	12.3 mg
Betaine	0.5 mg
Calcium	33.9 mg
Iron	0.2 mg
Magnesium	19.4 mg
Phosphorus	33.9 mg
Sodium	4.8 mg
Zinc	0.2 mg
Copper	0.1 mg
Selenium	0.2 mg
Ash	0.8 mg

Essential Oils

Another and very important by-product of acid lime fruit is citrus essential oils. These are volatile oils obtained from the citrus fruit peels' sacks. It is used in the food industry to add flavor to drinks and foods. They are also a major component for the pharmaceutical industry for the preparation of drugs, soaps, perfumes, hair cream, body oil and other cosmetics as well as for home cleaning products (Ferguson, 1990).

Medicinal Value of Acid Lime

Citrus aurantifolia is an important medicinal and food plant cultivated worldwide. It is known for its nutritional composition and numerous health benefits. The plant is used in traditional medicine as an antiseptic, antiviral, antifungal, anthelmintic, astringent, diuretic, mosquito bite repellent, for the treatment of stomach ailments, constipation, headache, arthritis, colds, coughs, sore throats and used as appetite stimulant. These health benefits of *Citrus aurantifolia* are associated with its high amounts of photochemical and bioactive compounds such as flavonoids, limonoids, phenols, carotenoids, minerals and vitamins (Enejoh *et al.*, 2015).

The decoction of pounded leaves is drunk for stomach ache, used as eye wash and to bath feverish patient. Poultice of leaves are applied to ulcer wounds, used for skin disease and also applied to abdomen after child birth. Infusion of *C. aurantifolia* leaves have been given to treat fever with jaundice, sore throat and oral thrush (Kunow, 2003). A decoction of the flower is believed to help induce sleep for those with insomnia (Kunow, 2003;Khan *et al.*, 2010).

Citrus is rich in flavonoids and the most abundant flavonoids in *C. aurantifolia* extracts include: apigenin, rutin, quercetin, kaempferol, nobiletin, hesperidin, hesperitin, and neohesperidin (Jayaprakasha *et al.*, 2008;Loizzo *et al.*, 2012). The flavanoids have strong inherent ability to modify the body's reaction to allergens, viruses and carcinogens. They show anti-allergic, anti-inflammatory, antimicrobial and anti-cancer activity (Okwu, 2005). It has been demonstrated that quercetin, one of the most active flavonoids possess significant anti-inflammatory activity because of direct inhibition of several initial processes of inflammation. Quercetin also showed remarkable anti-tumor properties and may have positive effects in combating or helping to prevent cancer, prostatitis, heart diseases, cataracts, allergies/inflammations and respiratory diseases such as bronchitis and asthma. Carotenoids are also found in citrus and are believed to reduce the incidence of age-related macular degeneration, the leading cause of blindness in human after the age of sixty five (Seddon *et al.*, 1994). They play essential roles as sources of vitamin A. The most active role is protection against serious disorders such as cancer, heart diseases and degenerative eye diseases. It is an antioxidant and acts as regulators of the immune system. Carotenoids commonly found in citrus are β-carotene, lutein, zeaxanthin and cryptoxanthin (Mangels *et al.*, 1993). Citrus is one of the main sources of Vitamin C (Okwu, 2006). Ascorbic acid in the body aids in iron absorption from the intestines. It is required for connective tissue metabolism especially the scar tissue, bones and teeth. It is necessary as an anti-stress and protector against cold, chills and damp. It prevents muscle fatigue and scurvy. It is needed for normal wound healing (Okwu, 2007). The production of collagens is also dependent on Vitamin C. It helps in the promotion and restoration of skin (Roger, 2002).

Conclusion

The Acid lime is a cross pollinated thus; wide genetic variation exist in different agro climatic conditions. At present situation of citrus genetic diversity is alarming as enormous destruction in the natural habitat is taking place to fulfill various requirements of mankind. Special drives need to be launched in the remote areas to maintain a parallel repository of germplasm *ex situ*, so that if diversity is extincted the species will be available for future utilization. At the same time every attempt should be made to conserve the purity of the existing germplasm *in situ*. Since some wild and rare citrus species (like, *Citrus indica*) accessions failed to conserve in filed gene bank. This will protect one alternative to other, if any one out of the two approaches failed. Such effort must be more confined to collection sites where seemingly the primitive types may still be found in an uncontaminated and pristine form. The inaccessible remote area and the forest possess great potential for primitive and wild citrus types, need our immediate attention. The unique area where a few citrus types tolerate adverse conditions (biotic and abiotic), may be thoroughly surveyed and collected as a rare collection for future utilization.

The future of acid lime depends on availability of disease and insects resistant varieties with good storability. Acid lime cultivars/hybrids with less seeds content, higher juice content, acidity (per cent), and higher yield with more shelf life will boost acid lime productivity and availability of quality material. Seedlessness/ triploid is another important character in citrus processing industry so, seedless cultivars (*viz.*, Chakradhar) need to be improved and commercialized. Some of the resistant cultivars of *citrus sp.* (*e.g.* Lime and Lemon cultivars) can be utilized as sources for resistance to diseases and pests, the genetic resources of acid lime need to be explored and evaluated systematically to identify desirable parental lines worth use in acid lime breeding. Besides the cultivated species acid lime the semi-cultivated and wild species need to be searched to derive genes of interest and to broaden the genetic base of the breeding material. The diseases namely Bacterial canker, anthracnose, citrus scab and virus diseases like, Citrus tristeza virus which are of high significance in major acid lime growing areas (tropical and subtropical areas). The insects like, aphids, which is both a pest and vector for acid lime viruses also needs priority in breeding. Abiotic stresses particularly both drought, frost (Subtropical and Himalayan region) and excess moisture are major limiting factors in acid lime cultivation. With changing climate these problems are likely to aggravate. Hence, existing acid lime accessions need to be evaluated for various agro climatic regions for yield, quality, abiotic and biotic stresses, and which can be used for further breeding programme.

References

Anonymous. 2015. Indian Horticulture Database. National Horticulture Board, Gurgaon.

Alcolea, J., Cano, A. and Acosta, M. 2002. Hydrophilic and liphophilic antioxidant activities of grape. *Nahrung* **46**: 353-356.

Alves Júnior, José, WijeBandaranayake, Larry R. Parsons and Evangelista, Adao W.P. 2012. Citrus root distribution under water stress grown in Sandy soil of central Florida. *Engenharia Agrícola* **32** (6): 1109-1115.

Al-Abadi, S.Y., Al-Sadi, A.M., Dickinson, M., Al-Hammadi, M.S., Al-Shariqi, R., Al-Yahyai, R.A., Kazerooni, E. A and Bertaccini, A. 2016. Population genetic analysis reveals a low level of genetic diversity of '*Candidatus Phytoplasma aurantifolia* 'causing witches' broom disease in lime. *Springer Plus* **5** (1): 1701.

Ayllon, M. A., Rubio, L., Moya, A., Guerri, J. and Moreno, P. 1999. The haplotype distribution of two genes of citrus tristeza virus is altered after host change or aphid transmission. *Virology* **255**: 32–39

Badge, T. R. and Patil,V.S. 1989. Chakradhar lime-A new thornless and seedless selection in kaghzi lime (*Citrus aurantifolia* Swingle). *Annals of Plant Physiology* **3** (1): 95-97.

Bar-Joseph, M. and Lee, R. F. 1989. Citrus tristeza virus. AAB *Descriptions of Plant Viruses No.* 353.

Barrett, H. C. and Rhodes,A.M. 1967. A numerical taxonomic study of affinity relationships in cultivated Citrus and its close relatives. *Systematic Botany* 105-136.

Barkley, N. A., M. L. Roose, R. R. Krueger and Federici,C.T. 2006. Assessing genetic diversity and population structure in a citrus germplasm collection utilizing simple sequence repeat markers (SSRs). *Theoretical and Applied Genetics* **112** (8): 1519-1531.

Bayer, R. J., D. J., Mabberley, C., Morton, C. H., Miller, I. K., Sharma, B. E., Pfeil, S., Rich, R., Hitchcock and Sykes,S. 2009. A molecular phylogeny of the orange subfamily (Rutaceae: Aurantioideae) using nine cpDNA sequences. *American Journal of Botany* **96** (3): 668-685.

Bhattacharya, S.C. and Dutta,S. 1956. Classification of citrus fruits of Assam. Government of India Press, Delhi Science Monograph. 20: 110.

Bose, T.K., Mitra, S.K. and Sanyal, D., 2001. *Fruits: tropical and subtropical. Volume 1* (*No.Ed.* 3).Naya Udyog.

Butani, D. K. 1979. Insects and fruits. Periodical Expert Book Agency, New Delhi-2

Calabrese, F. 1992.The history of citrus in the Mediterranean countries and Europe. *In Proc Int Soc Citricult* **1**: 35-38.

Chadha, K.L. 2009. R&D initiatives in acid lime for growth and prosperity. *Indian Journal of Horticulture* **66** (3): 291-294.

Chakrawar, V.R. and Rane, D.A. 1977. Bud mutations in Nagpur mandarin oranges. *Res. Bull. MAU* **1**: 22-23.

Chapot, H., 1975. The citrus plant. *Citrus. Basle: Ciba-Geigy Agrochemicals, Ciba-Geigy Ltd*, pp.6-13.

Curk, F., G., Ancillo, A., Garcia-Lor, F., Luro, X., Perrier, J. P., Jacquemoud-Collet, L., Navarro and Ollitrault,P. 2014. Next generation haplotyping to decipher nuclear genomic interspecific admixture in Citrus species: analysis of chromosome 2. *BMC Genetics* **15** (1), p.152.

Curk, F., G., Ancillo, F., Ollitrault, X., Perrier, J.P., Jacquemoud-Collet, A., Garcia-Lor, L., Navarro and Ollitrault,P. 2015b. Nuclear species-diagnostic SNP markers mined from 454 amplicon sequencing reveal admixture genomic structure of modern citrus varieties. *PloS one* **10** (5), p.e0125628.

Curk, F., F., Ollitrault, A., Garcia-Lor, F., Luro, L. Navarro and Ollitrault,P. 2016. Phylogenetic origin of limes and lemons revealed by cytoplasmic and nuclear markers. *Annals of Botany* **117** (4): 565-583.

Dambier, D., Hamid, B., Giovanni, P., Yildiz, A., Yann, F., Zina, B., Beniken, L., Najat, H., Bruno, P., Raphael, M., Turgut, Y., Luis, N. and Patrick, O. 2011. Somatic hybridization for citrus rootstock breeding: an effective tool to solve some important issues of the Mediterranean citrus industry. *Plant Cell Rep.* **30**: 883–900

Davies, F.S. and Albrigo, L.G., 1994. Environmental constraints on growth, development and physiology of citrus. *DAVIES, FS; ALBRIGO, LG Citrus. Wallingford: CAB International*, pp.52-82.

Deng, X. X., Guo, W. W. and Sun X. H. 1996. Progress of seedless citrus breeding in China. *Acta horticulture sinica* **23** (3): 235-240.

Desai, V.T., S.A. Ranpise, C.V. Pujari and Raijadhav,S.B. 1999. "Sai Sarbati" promising acid lime cultivar for western Maharashtra. *Proc. Natl. Symp. Citric.,* Nov. 17-19, 1997, Nagpur, Maharashtra. pp. 38-41.

Dubey, A.K. 1998. ARL-1: an acid lime variety for Arunachal Pradesh. ICAR News, New Delhi. pp 13–14

Dubey, A.K., Sharma, R.M., Awasthi, O.P., Srivastav, M. and Sharma, N. 2016. Genetic diversity in lime (*Citrus aurantifolia* Swing.) and lemon (*Citrus limon* (L.) Burm.) based on quantitative traits in India. *Agroforestry Systems* **90** (3), pp.447-456.

Enejoh, O. S., I. O., Ogunyemi, M. S., Bala, I. S., Oruene, M. M., Suleiman and Ambali,S.F. 2015. Ethnomedical Importance of *Citrus aurantifolia* (Christm) Swingle. *Pharma Innovation Journal* **4**: 1-6.

FAOSTAT. 2014. FAO statistics division 2014. http: //faostat.fao.org/ (accessed on 12 May 2017).

Federici, C. T., D. Q. Fang, R. W. Scora and Roose,M.L. 1998. Phylogenetic relationships within the genus Citrus (Rutaceae) and related genera as revealed by RFLP and RAPD analysis. *Theoretical and Applied Genetics* **96** (6): 812-822.

Ferguson, U. 1990. Citrus fruits processing. Horticultural Science, Florida, 117-118.

Garcia-Lor A, F., Luro, L., Navarro and Ollitrault,P. 2012. Comparative use of InDel and SSR markers in deciphering the interspecific structure of cultivated citrus genetic diversity: a perspective for genetic association studies. *Molecular Genetics and Genomics* **287**: 77–94.

Garcia-Lor, A., F., Curk, H., Snoussi-Trifa, R., Morillon, G., Ancillo, F., Luro, L., Navarro and Ollitrault,P. 2013b. A nuclear phylogenetic analysis: SNPs, indels and SSRs deliver new insights into the relationships in the 'true citrus fruit trees' group (Citrinae, Rutaceae) and the origin of cultivated species. *Annals of Botany* **111** (1): 1-19.

Ghosh S. N., B. Bera and Ray,S. 2012. Evaluation of acid lime cultivars in laterite zone of West Bengal. *Journal of Crop and Weed* **8** (1): 31-33.

Ghosh S. N. and Ray,S. 2013. Management of canker and scab in acid lime for better yield and plant health. *Environment and ecolosy* **31** (2A): 772-774.

Gmitter, F. G., J. R., Soneji and Rao,M.N. 2009. Citrus Breeding. (In) Jain, S.M. and Priyadarshan, P.M. (*Eds*). Breeding plantation tree crops: tropical species. *Springer* 3-50p.

Grosser, J. W., G. A. Moore and Gmitter,F.G. 1989. Interspecific somatic hybrid plants from the fusion of 'Key lime (*Citrus aurantifolia*)' with Valentia sweet orange (*Citrus sinensis*) protoplasts. *Scientia Hort* **39**: 23-29.

Grosser, J. W. and Gmitter, F.G. 2011. Protoplast fusion for production of tetraploids and triploids: applications for scion and rootstock breeding in citrus. *Plant Cell Tiss. Org. Cult.* **104**: 343–357.

Guo, W. W., Shi-Xin Xiao and Xiu-Xin, Deng. 2013. Somatic cybrid production *via* protoplast fusion for citrus improvement. *Scientia Horticulturae* **163**: 20–26.

Hazarika, T. K. 2012. Citrus genetic diversity of north-east India, their distribution, ecogeography and ecobiology. *Genet Resour Crop* **59**: 1267–1280.

Herrero, R., M. J. Asins, E. A. Carbonell and Navarro,L.1996. Genetic diversity in the orange subfamily Aurantioideae. I. Intraspecies and intragenus genetic variability. *Theoretical and Applied Genetics* **92** (5): 599-609.

Hirai, M., Kozaki, I. and Kajiura, I. 1986. Isozyme analysis and phylogenic relationship of citrus. *Japanese Journal of Breeding*, **36** (4), pp.377-389

Hodgson, R.W. 1961. Taxonomy and nomenclature in citrus. (In) *Proc. 2nd Conf. Intern. Organ. Citrus Virol* (pp. 1-7).

http: //www.iari.res.in

https: //www.pdkv.ac.in

http: //www.plantauthority.gov.in/pdf/DAcidlime.pdf

Jayaprakasha, G. K., Mandadi, K. K., Poulose, S. M., Jadegoud, Y. and Patil, B. S. 2008. Novel triterpenoid from *Citrus aurantium* L. possesses chemopreventive properties against human colon cancer cells. *Bioorganic Medical Chemistry* **16** (11): 5939-5951.

Jawaharlal, M., Thangaraj, T. and Irulappan, I. 1991. Influence of rootstock on the post-harvest qualities of acid lime fruit. *South Indian Hort* **39**: 151–152.

Kamatyanatti, M., Nagre, P. K., Vikas, R. and Murlimanohar, B. 2016. Evaluation of Acid Lime (*Citrus aurantifolia* Swingle) genotypes during *hasth bahar* for growth, yield and quality attributes. *7th Indian Horticulture Congress-2016*, IARI, New Delhi.

Khan, I.A. 2007 (Ed). *Citrus genetics, breeding and biotechnology*.CABI.London

Khan, I. A. and Abourashed,E.A. 2010. Leung's Encyclopedia of Common Natural Ingredients. John Wiley and Sons Publication, New Jersey. 422-423.

Koizumi, M. 1982. Resistance of citrus plants to bacterial canker disease. *Proc Int. Soc. Citriculture* 402-405.

Kumar, M., S. Parthiban S; Saraladevi,D. and Aruna,P. 2011. Evaluation of acid lime (*citrus aurantifolia* Swingle) cultivars for yield attributes. *The Asian Journal of Horticulture* **6** (2): 442-444.

Kunow, M. A. 2003. Maya Medicine: Traditional Healing in Yucatan. UNM Press, New Mexico, p117.

Lakshmi, T. N., Gopi, V., Sankar, T G., Sarada, G., Lakshmi, L.M., Ramana, K.T.V. and Gopal, K. 2014. Status of diseases in sweet orange and acid lime orchards in Andhra Pradesh, India. *International Journal of Current Microbiology and Applied Science* **3** (5): 513-518.

Li, X., R., Xie, Z., Lu and Zhou,Z. 2010. The origin of cultivated citrus as inferred from internal transcribed spacer and chloroplast DNA sequence and amplified fragment length polymorphism fingerprints. *Journal of the American Society for Horticultural Science* **135** (4): 341-350.

Liu, Y., Guoping, W., Zeqiong, W., Fan, Y., Guanwei, W. and Ni, H. 2012. Identification of differentially expressed genes in response to infection of a mild Citrus tristeza virus isolate in *Citrus aurantifolia* by suppression subtractive hybridization. *Scientia Horticulturae* **134**: 144–149.

Loizzo, M. R., Tundis, R., Bonesi, M., Menichini, F., De Luca, D., Colica, C. and Menichini, F. 2012. Evaluation of Citrus aurantifolia peel and leaves extracts for their chemical composition, antioxidant and anti-cholinesterase activities. *Journal of Science, Food and Agriculture* **24** (12): 1893-18937.

Madhavi, M., Seshadri, K.V., Reddy, G. Subbi, Reddy, M.R.S., Gopal, K. and Rao, R. 2000. Tenali acid lime – a high yielding canker resistant acid lime clone. *Hi-tech Citrus Management – Proc. Intn. Symp.Citriculture,* Nov. 23-27, 1999, Nagpur, Maharashtra. (S.P.Ghosh and Shyam Singh, Eds.), pp. 977-981.

Malik, M. N., Scora, R. W. and Soost, R. K. 1974. Studies on the origin of the lemon. *Hilgardia* **42**: 361–382

Malik, S.K., Chaudhury, R., Kumar, S., Dhariwal, O.P. and Bhandari, D.C. 2012. NBPGR publication, Citrus genetic resources in India- present status and management. pp-78-94.

Mangels, A. R., Holden, J. M. and Beecher, G. R. 1993. Carotenoid content of fruits and vegetables, an evaluation of analytic data. *Journal of American Dietetic Association* **93**: 284-296.

Marin, M. L. and Duran-Vila, N. 1991. Conservation of citrus germplasm *in vitro. Journal of the American Society for Horticultural Science* **116** (4): 740-746.

Mollayi, S., Mohsen, F., Faezeh, G., Hassan, Y. and Aboul-Enein, A. G. 2016. Study of catechin, epicatechin and their enantiomers during the progression of witches' broom disease in Mexican lime (*Citrus aurantifolia*). *Physiological and Molecular Plant Pathology* **93**: 93-98

Moore, G. A. 2001. Oranges and lemons: clues to the taxonomy of Citrus from molecular markers. *Trends in Genetics* **17** (9): 536-540.

Morton, J. 1987. Tahiti lime. (In) Fruit of warm climates. (Ed.) J.F. Morton, pp. 172-175, Southern Book Service, Miami.

Murkute, A. A. and Singh,I.P. 2015. Citrus improvement through selection and mutagenesis constraints and opportunities. *International Journal of Tropical Agriculture* **33** (3): 2361-2366.

Nagy, S. and Norby, H.E. 1972. Long-chain hydrocarbon profiles of grapefruit juice sacs. *Phytochemistry*, **11** (9): 2789-2794.

Nicolosi, E., Z. N. Deng, A. Gentile, S. La Malfa, G. Continella and Tribulato,E. 2000. Citrus phylogeny and genetic origin of important species as investigated by molecular markers. *Theoretical and Applied Genetics* **100** (8): 1155-1166.

Okwu, D. E. 2005. Phytochemicals, Vitamins and Mineral Contents of Two Nigerian Medicinal Plants. *International Journal of Molecular Medicine and Advance Sciences* **1**: 375-381.

Okwu, D. E. and Emenike, I. N. 2006. Evaluation of the Phytonutrients and Vitamins Contents of Citrus Fruits. *International Journal Molecular Medicine and Advance Sciences* **2** (1): 1-6.

Okwu, D. E. and Emenike, I. W. 2007. Nutritive Value and Mineral Content of Different Varieties of Citrus Fruits. *Journal of Food Technology* **5** (2): 92-1054.

Okwu, D. E. 2008. Citrus fruits: a rich source of phytochemicals and their roles in human health: a review. *International Journal of Chemical Science* **6** (2): 451-471.

Ollitrault, P., Dambier, D., Sudahono and Luro, F. 1996. Somatic hybridization in Citrus : some new hybrid and alloplasmic plants. *Proc. Int. Soc. Citricult* **2**: 907-912

Ollitrault, P., Dambier,D., Froelicher, Y., Carreel, F., D'Hont, A., Luro, F., Bruyere,S., Cabasson, C., Lotfy, S., Joumaa, A., Vanel, F., Maddi, Treanton, K. and Grisoni, M. 2000. Somatic hybridisation potential for citrus germplasm utilisation. *Cahiers Agric* **9**: 223-236

Olivares-Fuster, O., Asiáns, M. J., Duran-Vila, N. and Navarro, L. 2000. Cryopreserved callus, a source of protoplasts for citrus improvement. *The Journal of Horticultural Science and Biotechnology*, **75** (6): 635-640.

Ollitrault, P., J.,Terol, A., Garcia-Lor, A., Bérard, A., Chauveau, Y., Froelicher, C., Belzile, R., Morillon, L., Navarro, D., Brunel and Talon,M. 2012. SNP mining in C. clementina BAC end sequences; transferability in the Citrus genus (Rutaceae), phylogenetic inferences and perspectives for genetic mapping. *BMC Genomics* **13** (1): 13.

Radha, T. and Mathew, L. 2007.Peter,K V (Ed) Fruit crops (Vol. 3). New India Publishing Agency,New Delhi.

Paithankar, D. H., P. Nagre and Kale,V.S. 2016. Evaluation of Acid Lime Genotypes for Pickle.*7th Indian Horticulture Congress-2016*, IARI, New Delhi.

Paithankar, D. H., P. K. Nagre, Ekta Bagde and Sadawarte,A.K. 2015. New Promising Variety of Acid Lime: PDKV Bahar, a Prolific Bearer. *PKV Res. J.* **39**: (1 and 2)

Pal, V. and Chand, J. N. 1983. Preliminary evaluation of Dettol for *Xanthomonas citri* control in citrus. *Ann. Appl. Biol* **102**: 60-61, Suppl.

Perez, R.M., Galiana, A.M., Navarro, L. and Duran-Vila, N. 1998. Embryogenesis in vitro of several Citrus species and cultivars. *The Journal of Horticultural Science and Biotechnology* **73** (6): 419-429.

Prasad, M.B.N.V., Roopali, S., Rekha,A. and Ramesh, C. 1997. Evaluation of lemon cultivars and acid lime X lemon hybrids for resistance to *Xanthomonas axonopodis* pv. Citri. *Scientia Horticulturae* **71**: 267-272.

Ray, P. K. 2002. Breeding tropical and subtropical fruits. Springer Science and Business Media.

Roger, G. D. P. 2002. Encyclopedia of medicinal plants. Education and Health Library Editorial, 153-154.

Ruiz, A., Cynthia, C., Parra John, V. Da Graça, Bacilio Salas, Nasir; Malik,S.A. and Madhurababu, K. 2014. Molecular characterization and pathogenicity assays of *Colletotrichum acutatum*, causal agent for lime anthracnose in Texas. *Revista Mexicana De Fitopatología* **32** : 1

Saito, W., Ohgawara, T., Shizimu, J., Ishii, S. and Kobayashi, S. 1993. Citrus cybrid regeneration following cell fusion between nucellar cells and mesophyll cells. *Plant Sci* **93**: 195-201.

Sahoo, H. S., Das, B. K. and Srichandan, S. 2010. Clonal selection of kagzi lime. *Environment and Ecology* **28** (1): 117-119

Scora, R.W. 1975. On the history and origin of citrus. *Bulletin of the Torrey Botanical Club* **102** (6): 369–375.

Seddon, J. M., Ajani, A. U. and Sperduto, R. D. 1994. For the Eye Disease Case-Control Study Group, Dietary Carotenoids, Vitamin A, C, and E and Advanced Age-Related Macular Degeneration. *Journal of America Medical Association* **271**: 1413-1430.

Sethpakdee, R. 1992. *Citrus aurantifolia* (Christm. and Panzer) Swingle. (In) R.E. Coronel., and E.W., Verheij. (Eds.): Plant Resources of South-East Asia. Edible fruits and nuts. *Prosea Foundation, Bogor, Indonesia* **2**: 126- 128.

Shrestha, R. L., Durga D. D., Durga M. G., Krishna P. P. and Sangita S. 2012. Study of fruit diversity and selection of elite acid lime (*Citrus aurantifolia* Swingle) genotypes in Nepal. *American Journal of Plant Sciences* **3**: 1098-1104

Singh. I. P and Shyam Singh. 2003. Citrus genetic diversity in central india. *Indian Journal of Plant Genetic Resources* **16**: 3.

Soost, R. K. and Roose,M.L. 1996. Citrus. (In) Janick, J. and Moore, J.N. (Eds.), Fruit breeding, Vol. I: Tree and Tropical Fruits. John Wiley and Sons Inc., New York, Chichester, Brisbane, Toronto, Singapore. 257–323p.

Sreedevi, K. and Rajulu,B.G. 2008. Screening of acid lime (*Citrus aurantifolia* Swingle) clonal selections for resistance to thrips, *Scirtothrips sp.* and rust mite, *Phyllocoptruta oleivora* (Ashmead).*Pest Management in Horticultural Ecosystems* **14** (1): 16-19.

Springob, K., Nakajima, J. and Yamazaki, M. 2003. Recent advances in the biosynthesis and accumulation of anthocyanins. *Natural products report* **20** (3): 288-303.

Swingle, W.T. 1943. The botany of Citrus and its wild relatives of the orange subfamily (Family Rutacae, subfamily Aurnatiodideae). *Webber, HJ and Batchelor, LD, The Citrus Industry* **1**.

Swingle, W.T. 1948. New taxonomic technique in studying wild relatives of major crop plants illustrated by citrus. In *American Journal Of Botany* (**35** (10): 798-798). Ohio State Univ-Dept Botany 1735 Neil Ave, Columbus, Oh 43210: Botanical Soc Amer Inc.

Swingle, W.T. and Reece, P.C. 1967. The botany of Citrus and its wild relatives. (In) The Citrus Industry, (Eds.) Reuther, W., H.J. Webber and L.D. Batchelor), Univ. California Press, Berkeley **1**: 190-430.

Tamata, S., Regar, K.C., Verma, S.K., Arya, R.R. and Rager, R. 2008. Germplasm evaluation of Kagazi lime (*Citrus aurantifolia*) at Bhowali (Nainital). *New agriculturist* **19** (1, 2): 147-149.

Tanaka, T. 1954. Species problems in citrus. Japanese Society for the Promotion of Science, Ueno, Tokyo, 152 p.

Tanaka, T. 1969. Misunderstanding with regards citrus classification and nomeclature. *Bulletin of the University of Osaka Prefecture. Ser. B, Agriculture and biology*, **21**: 139-145.

Tanaka, M. and Kashio, T., 1977. Biological studies on *Amblyseius largoensis* Muma (Acarina: Phytoseiidae) as a predator of the citrus red mite, *Panonychus citri* (McGregor) (Acarina: Tetranychidae). *Bulletin of the Fruit Tree Research Station. Series D. Kuchinotsu (Japan)*.

Tatum, J. H., Berry, R. E. and Hearn, C. J. 1974. Characterization of citrus cultivars and separation of nucellar and zygotic seedlings by thin layer chromatography. In *Proc. Florida State Hort. Soc* **87**: 75-81.

Timmer, L.W. 1978. Identification of lime anthracnose in Texas. *J. Rio Grande Hort. Soc* **32**: 35-38

Torres, A.M., Soost, R. K. and Mau-Lastovicka, T. 1982. Citrus isozymes Genetics and distinguishing nucellar from zygotic seedlings. *Journal of Heredity* **73** (5): 335-338.

Tusa, N., Grosser, J.W., Gmitter, F.G. and Louzada, E.S. 1992. Production of tetraploid somatic hybrid breeding parents for use in lemon cultivar improvement. *HortScience* **27**: 445-447.

Upadhaya, A., Chaturvedi, S.S. and Tiwari, B.K. 2016. Utilization of wild Citrus By Khasi and Garo tribes of Meghalaya. *Indian Journal of Traditional Knowledge* **15** (1): 121-127.

Viloria, Z., Grosser, J.W. and Bracho, B. 2005. Immature embryo rescue, culture and seedling development of acid citrus fruit derived from interploid hybridization. *Plant Cell, Tissue and Organ Culture* **82**: 159–167

Webber, H. J. 1967. History and development of the citrus industry. (In) The Citrus Industry (Eds.: Reuther, W., H.J. Webber and L.D. Batchelor). University California Press, Berkeley **1**: 1-39.

Yamamato, M., Matsumoto, R and Yamodo, Y. 1905. Relationship between sterility and seedless in citrus. *Journal of Japanese society of horticultural science* **64** (1): 23-29.

Yano, M., Kawaii, S., Tomono, Y., Katase, E. and Ogawa, K. 1999. Quantification of flavonoid constituent in citrus fruits. *Journal of Agricultural Food Chemistry* **47**: 3565- 3571.

Zamani, Z., Adabi, M. and Khadivi-Khub, A. 2013. Comparative analysis of genetic structure and variability in wild and cultivated pomegranates as revealed by morphological variables and molecular markers. *Plant Syst Evol* **299** (10): 1967–1980

Ziegler, L.W. and Wolfe, H.S. 1961. The kinds of fruits. *Citrus growing in, 331,* pp.12-19.

Chapter 5

Banana in Tripura

Sukhen Chandra Das and K.S. Thingreingam Irenaeus

Department of Horticulture,
College of Agriculture, Tripura – 799 210,
E-mail: sukhenchandra@rediffmail.com

Banana (*Musa* sp.) is the second most important fruit crop next to jackfruit in Tripura and is cultivated in about 13,274.0 ha area with an annual production of 1,30,085.20 tones with a productivity of 8.9 tones ha^{-1} which is far below the national productivity (37 tones ha^{-1}). The national area under banana is 802.6 thousand ha with an annual production of 29724.6 thousand tones. Banana has been growing in Tripura since several years comprising of different local genotypes and a good potential exists in the state for its exploitation and commercialization. The genetic variation among the local genotypes offers a huge scope for selection of superior genotypes and other improvement programmes. At present most of the banana genotypes are found growing naturally in hill slopes and are being maintained in home gardens by farmers to a small extent. Tripura is considered to be the natural home of many species and cultivars of banana and it is reservoir of many local genotypes. Due to favourable agro-climatic conditions, fertile and acidic soil with good depth and abundant rainfall favour the growth of various types of bananas and they are growing wild and semi-wild in the state. Diverse forms of banana have been growing in hill slopes and plains. This chapter includes report from study on collection, conservation and evaluation of different local banana genotype at College of Agriculture, Tripura, India during 2009-2015. The available germplasm were evaluated for plant characters like plant height, pseudo-stem girth, number of leaf at flowering and harvest, leaf length and breadth, life cycle, and yield. Fruit bunch and finger characters like bunch weight, fruit length, number of hands/bunch, number of fingers/bunch, number of hands/bunch, number of finger/hand, fruit weight, fruit length, fruit diameter, fruit pulp weight, peel weight, TSS and other related characters were recorded. Some of the local genotypes are Shabri Kela (AAB), Samai Kela or Gopi Kela or Bangla Kela (ABB), Champa Kela (AAB),

Attia kela (BB), Katch kela (ABB), Kanai Bashi (AA), Red Banana (AAA), Mizo-Cavindish (AAA), Ram Kela *etc.* Among all local genotype, Shabri (AAB), Samai (ABB) and Champa (AAB) showed better performance with good quality of fruit for fresh consumption and medium to tall stature plants.

Banana originated from South-East Asia with *M. acuminata* and *M. balbisiana* as its ancestral species. Though a great diversity has been observed for acuminata in Malaysia, Philippines, Indonesia and north- Eastern region of India, India harbours a greater diversity for balbisiana and acuminata-balbisiana bispecific clones (AB, AAB, ABB and BB/BBB). In India, More than 90 distinct clones have been identified at different field gene banks located in various locations (Anon, 1995).

Morphotaxonomic Characterization

Assigning Genomic Status

Assigned genomic group using the Simmond and Shepherd' (1962) 15 character score card. Here the plant is scored from 1 to 5 as per description available in Table 5.2. All the trails mentioned under Musa acuminate were given the score of 1 and those coinciding with Musa balbisiana were scored as 5. The variations between these two extreme characters were given the intermediary score based on the experience. The total score ranged from 15-75 depending upon the genomic groups. The total scores thus obtained were compared with the score card (Table 5.2) and corresponding genomic status was assigned to a particular genotype.

Assignment of Ploidy Level

Meiotic chromosome number is an important character for determining ploidy status. The ploidy status was determined using morphological studies like stomatal density (no. of stomata per unit leaf area) and stomatal measurements (length and breadth) and specific leaf weights.

Banana Classification

Based on the utility, banana and plantains are classified as dessert bananas, cooking bananas and beer bananas. Dessert bananas are those, which are consumed raw when they are fully ripe and are characterized by soft texture, high sugar and a pleasant aroma. These are easily digestible and do not need cooking prior to consumption. While cooking bananas are starchy and invariably cooked to make them palatable. A small category of cooking bananas can be allowed to ripe before consuming as raw while many are not palatable. They are referred, as true plantains with dual utility, by Swennen and Vuylsteke (1987). The last category of beer bananas can neither be cooked nor eaten at ripeness owing to their bitter and astringent pulp and also utilized for the preparation of juice and other alcoholic beverages.

Linnaeus was the first person to give scientific nomenclature for banana, based on morpho-taxonomic traits, as *Musa paradisiaca* Linn in 1953. He referred this to the plantains with long slender fruits which are starchy and needed-cooking prior to cooking. Simmonds and Shepherd (1955) presented the theory of origin of edible bananas. The new classification system using 15 differentiating traits and the score card gave a fine-tuning to the Musa classification scheme.

Different forms or subspecies of *M. acuminata* were found in S.E. Asia where their distribution overlapped with pure balbisiana clones especially in Malaysia, Philippines, and NE India *etc*. This led to natural introgression resulting in acuminata-balbisiana hybrids. Depending on the relative contribution of these two species various genomic combination AA, AAA, AB, AAB, ABB, BB, BBB, ABBB have developed. The ploidy level ranged from diploid, triploid and tetraploid denoted by the number of letters of acuminata and balbisiana.

Taxonomy

The family Musaceae comprise of two genera, *viz.*, Ensete and *Musa* with about 50 species in this family. Ensete in an old declining genus. This originated in Asia and spread to Africa. It has about 6-7 species, of which *E. ventricosa* is reported to be grown in Ethiopia as a food crop. The name Musa is derived from the Arabic word Mouz. It has about 40 species of perennial, stooling or rhizomatous herbs and is found in South-East Asia and the pacific. The genus Musa is divided into the following sections namely *Callimusa* and *Australimusa* which have basic chromosome number of 10, and *Eumusa* and *Rhodochlamys* having the number x as 11.

Eumusa

The section Eumusa has 13-15 species. Eumusa is the only source of present day edible bananas. They are characterized by robust pseudostem, horizontal, angular or pendulous bunches. All the edible bananas are believed to have originated from two species, *Musa acuminata* designated by A genome and *Musa balbisiana* designated by B genome. The edible cultivated parthenocarpic bananas, except for the Fe'i bananas belong to this section which are derived from *M. accuminata* and inter specific crosses with this species and *M. bulbisiana*. *M. basjoo* in Japan yields fibre which may be made into textiles. The present contribution of each genome has resulted in various combinations *viz.*, AA, AAA, AB, AAB, ABB, BB, BBB and ABBB *etc*. Here two letters represent a diploid, 3 letters a triploid and four letters a tetraploid. The recent molecular studies have indicated the involvement of *Australimusa* member's genomes namely *M. textilis* and *M. schizocarpa* represented by T and S genomes. The occurance of 3 genomes namely A, B and T for accuminata, *bulbisiana* and *textilis* has also added a new dimension to the evaluation. For a better identification and classification of a cultivar or a land race, a basic knowledge about their evaluation, taxonomic status and description of traits is essential. For this purpose, the banana descriptor published by IPGRI/INIBAP is utilized for distinguishing the important traits.

Rhodochlamys

This section has 5-7 species and is found growing from India to Indonesia. Parthenocarpy is absent in this genus. *M. ornata* and *M. veluntia* and most numbers of Rhodochlamys are grown as ornamental plants. They are distinguished by their slender status; erect inflorescence and uniseries fruits covered in bright coloured bracts. They do not exhibit self incompatibility and are female fertile.

Table 5.1: Details of Species and Distribution of Genus Musa and Ensete.

Genus	Basic Chromosome No.	Section	Distribution	Species	Uses
Ensete	9		W. Africa	E. surerbum	Fibre
			P.N. Guinea	E. glaucum	Vegetable and Medicinal
			India	E. ventricosum	
				E. gillettii	
				E. homblei	
				E. perrieri	
Musa	10	Australimusa	Queenaland	M. textilis	Fibre
			New Caledomia	M. maclayi	Fruit and Vegetables
			Philippines	M. lolodensis	
				M. peekelii	
				M. fehi	
		Callimusa	Indo-China	M. coccinea	Ornamental
			Indonesia	M. violascens	
				M. gracilis	
	11	Eumusa	Asia	M. accuminata	Fruit, Vegetable Fibre and Medicinal
			Africa	M. bulbisiana	
			S. America	M. schizocarpa	
			Others	M. itinerans	
		Rhodochamys	India	M. velutina	Ornamental
			Indo- China	M. ornata	
				M. laterita	

Calllimusa

This section consists of many non-domesticated members with lot of ornamental value. They multiply fast with their rhizomatous stems, put forth erect inflorescence with bright coloured bracts.

Australimusa

This section has 5-7 species and is distributed in the regions from Queensland to the Philippines. The most important species are fiber yielding *M. textiles*, *M. fehi*, and *M. maclayi*. Fruits are parthenocarpic and female sterile. The distribution of *M. textiles* is mostly in S.E. Asia. The recent finding of involvement of *M. textiles* in the evaluation of present–day bananas has evoked much interest.

Incertae Sedis

Members in this section includes *M. ingens* and *M. baccarri* (Purseglove, 1975).

Several Latin names have been used till recently in the botanical nomenclature of the banana, three of the earliest employed were *M. paradisica*, *M. cavendishii* and

M. sapientum. These are no longer favoured and have been superseded by a genome nomenclature for cultivars in recognition of their derivation from two wild species *M. accuminata* and *M. bulbisiana.*

Chromosome Number

Edible bananas have 2n= 22, 33 or 44 chromosomes; the basic haploid number is 11 so that these cultivars are respectively diploid, triploids and tetraploids.

1. Musa (x = 10, 11 rarely 7/9)
2. Eumusa (x = 11, 2n = 22 in wild spp. 22, 33, 44 in cultivars)
3. Rhodoclamys (x = 11, 2n = 22)
4. Australimusa (x=10, 2n = 20)
5. Incertae sedis *M. ingens* (x =7; 2n = 14) *M. baccarii* (x = 9, 2n = 18)

Morphology

Banana is a large herb belonging to the family *Musaceae* under the order Zingiberales. The true stem of banana is referred as rhizome with roots and vegetative buds. Bananas are basically clumped in growth habit owing to its sympodial rhizome. The central stem is a pseudostem forrmed out of closely packed leaf sheaths embedding the growing tip. In general leaf is uniformly green with various shapes except in few cvs. such as Nendran and Cavendish clones where young leaves has purplish pigmentation. Rate of production of leaf, referred as phyllochron, varies from 7-10 days under tropical conditions. Appearance and orientation of mature leaves is an important discriminating trait of ploidy state. Diploids have narrow, slender and erect leaves clustered around the crown. Triploids have normal spreading leaves while tetraploids have wider, thick, leathery and drooping leaves. Foliage is large, oblong or elliptic, becoming as much as 9 feet long and affect wide. Approximately 44 leaves will appear before the inflorescence. The inflorescence is a spike originating from the tip of the corm. Female flowers, with inferior ovaries, occupy the lower 5 to 15 rows on the stalk, with neuter or hermaphrodite flowers in the center, and males at the top. Male flowers and bracts are shed one day after opening.

The orientation of the bunch depends on the extent of positive geotropism expressed by the plant. Based on that, they are referred as i) **Pendulous** - the inflorencence/bunch develops downwards parallel to the pseudostem *e.g.* Cavendish clones (AAA) of Eumusa, *Ensete glaucum* of the genus *Ensete* ii) **Sub horizontal** - the bunch is put forth at an angle to the pseudostem. *e.g.* Silk, Mysore clones. iii) **Horizontal** - the bunch is held at almost 90° to the pseudostem *e.g.* Poovan, Pome. iv) **Erect** - the bunch looks like a continuation of the pseudostem axis and the inflorescence expresses negative geotropism axis. Here the male bud faces the sky. Members of Rhodochlamys, Callimusa, Australimusa sections express this trait *e.g., M. ornate, M. velutina* etc. of the selection Rhodochlamys *M. coccinea* of Callimusa, *M. Fe'i* of Australimusa.

The flowers are of three types, female flowers, which emerge in the beginning and develop into fruits, hermaphrodite flowers, which appear immediately after the female phase and finally the male flowers. The flowers consist of three sepals, three petals, two outer whorl stamens, two inner stammen, and a tri-locular, inferior ovary. The adaxial petal is not opposed by an inner whorl stamen, resulting in only two stamens in this whorl. Fruit is epigynous berry borne in hands of up to 20 numbers with 5-13 hands per spike. Fruits appear as angled, slender, green fingers during growth reaching harvest maturity in 90-120 days after flower opening. Banana blooms year round, but most flowering is concentrated from February to August. The longevity of female flowers is greater than that of male flowers and opening of both types of flowers occurs at any time during the day, but is concentrated in the morning. The natural volume and rate of female flowers are significantly higher than that of male flowers. Bananas of the Cavendish group are triploids and therefore completely sterile, fruit is set parthenocarpically. Floral morphology suggests that wild bananas are bat pollinated in their native range. Insects such as bumble bees (*Bombus eximius* and *B. montivolans*), honey bees (*Apis cerana* and *A. florea*), and wasps (*Vespa mandarinia*) are the primary floral visitors and show a preference for female flowers. The bracts covering the flowers are important trait used in identification. The colour, shape, apex, shoulder ratio, type of lifting, nature of colour fading and finally the bract scars decide the genomic grouping of an accession through morpho-taxonomic characterization. In some cases, bract lifts up and turns back and remains, as in case of Plantains. The former situation is referred as reflex and revolute while the later as reflex. The female flowers develop into fruits and biserially arrange on each nodal cushion. Each cluster is referred as a hand and individual fruit in a hand is called a finger. Male flowers fall after 1-4 days of bract lifting and the dried nodal cushion remains as a scar. The appearance of the scare is another factor contributing to the genomic status of a particular accession. In banana, two distinct flower colours are reported *viz.*, cream and pink and all members take different shades of these two. The fruit development in banana is characterized by vegetative parthenocarpy, the development of pulp without pollination. In case of wild seeded bananas, pollination precedes normal fruit development without which fruits remain unfilled and shriveled. Seeds in case of banana exhibit a lot of variation with respect to size, shape and colour. The size varies from 4mm to 20 mm (4mm-6mm in normal wild cultivars, 15-20mm in Ensete species), shape from spherical to triangular to ovoid. Colour from brown to pitch black. Seed coat may be warty or smooth. Simmonds and Shepherd (1955) devised a method of indicating the relative contributions of these two wild species to the constitution of any cultivar through a scoring technique for distinguishing *M. accuminata types* from those of *M. bulbisiana*. They identified 15 diagnostic characters score card (Table 5.2). The modified score card was developed by Silayoi and Chom Chalow (1987). A modified scorecard has been proposed (Singh and Uma, 1996).

For each characters in which the cultivar agreed with wild *M. accuminata* the score of 1 was given and for *M. bulbisiana* the score of 5 was given and intermediate expressions of the characters were assigned score of 2, 3 or 4 according to their intensity. According to this scoring technique, the scores range from 15 (15x1) for *M. accuminata* to 75 (15x5) for *M. bulbisiana*.

Table 5.2: Taxonomic Scoring for Banana Cultivars by Simmonds and Shepherd (1995)

Characters	Musa acuminata	Musa balbisiana
Pseudostem colour	More or less heavily marked with black or brown blotches	Blotches slight or absent
Petiolar canal	Margin erect or spreading with scarious wings below, not clasping	Margins enclosed, not winged below, clasping pseudostem
Peduncle	Usually downy or hairy	Glabrous
Pedicel	Short	Long
Ovules	Two regular rows in each loculus	Four irregular rows in each loculus
Bract shoulder	Usually high (ratio <0.28)	Usually low (ratio>0.30)
Bract curling	Bracts roll after opening	Bracts lift but do not roll
Bract shape	Lanceolate or narrowly ovate, tapering sharply from the shoulder	Broadly ovate, not tapering sharply
Bract apex	Acute	Obtuse
Bract colour	Red, dull purple or yellow outside, pink dull purple or yellow inside	Distinctive brownish purple outside, bright crimson inside
Bract scars	Prominent	Scarcely prominent
Colour fading	Inside bract colour fades to yellow towards base	Inside bract colour continues to base
Free tepal of male flower	Variably corrugated below	Rarely corrugated
Male flower colour	Creamy white	Variably flushed with pink
Stigma colour	Orange or rich yellow	Cream, pale yellow or pale pink

Table 5.3: Genomic Group by Stover and Simmonds (1987)

Genomic Group	Score
AA Diploid	15-23
AAA Triploid	15-23
AAB Triploid	24-46
AB Diploid	49
ABB Triploid	59-63
ABBB Tetraploid	67

Table 5.4: Genomic Group by Silayoi and Chomchalow (1987)

Genomic Group	Score
AA/AAA	15-25
AAA	26-46
ABB Triploid	59-63
ABBB Tetraploid	61-69
BB/BBB	70-75

Table 5.5: Modified Score Card for Assigning Tentative Genomic Groups

Genomes	Score Card of		
	Simmonds and Shepherd (1982)	Silayoi and Chom Chalow (1987)	Singh and Uma (1996)
AA/AAA	15-23	15-25	15-25
AAB	15-23	26-46	26-45
AB	49	-	46-49
ABB	59-63	59-63	59-65
ABBB	67	-	66-69
BB/BBB	-	70-75	70-75

Banana in Tripura

Tripura is a tiny state in North East Region of India with a total geographical area 10,491 sq. km. Seventy per cent (70 per cent) of total geographical area of this state is under forest and only 27 per cent of total geographical area is under cultivation. Tripura is one among the other northeastern states with rich and diverse banana resources (Table 5.6) while a lot of wild types still exist unidentified. In Tripura, banana (*Musa sp.*) is a natural crop and most important fruit crop of Tripura next to jackfruit with rich source of vitamins and minerals and is very important for nutritional and health security of rural and urban area of the state. Unopened male bud, pseudosheath and pseudostem (Figure 5.1) of both balbisiana and acuminata bananas are used as vegetable by different tribal communities and rural people of Tripura, *viz.*, Delong, Tripuri, Reang, Debbarma, Mok, Chakma, Uchoi, Halam and others. The different tribal community use for preparing local dishes like 'Gudok' with or without other local vegetables and is a favourite among the tribals in Tripura. The unopened male bud cost around 10-12 rupees and five (5) feet long pseudosheath from the base cost around 15-20 rupees. The diverse agro-climatic condition, fertile soil, slightly acidic, sub-topical climate and abundant rainfall area enriched with undulating tilla land type with varying degree of slope play a vital role to check soil erosion and can stand even with minimum orchard care. It grows abundantly and luxuriantly as wild in forests areas in different parts of the state (Figure 5.2) and expresses their original characters.

Important banana types are Shabri Kela (AAB), Samai Kela (ABB) and Champa Kela (AAB), grown extensively in hill slop and plain parts of the state. Area under Banana cultivation in Tripura is 13,274.0 ha with an annual production is 130085.20 MT with a productivity of 9.8 t/ha. However, maximum (70 per cent) area of banana cultivation is under Shabri Kela (AAB) compared to Samai/Gopi/Bangla Kela (ABB) and Champa Kela (AAB). Shabri Kela (AAB) and Mizo-Cavindish (AAA) are very popular and excellent (in respect of flavour, texture, aroma and TSS) cultivars of Tripura for fresh consumption with medium plant stature. Many other local and traditional banana are available in the state namely Katch Kela (ABB), Attia Kela (BB), Ram Kela and Kanai Basi (AA) and many more seeded banana are available, whose pseudosheath is edible and used by tribal communities, rural and urban

Table 5.6: Diversity of Wild Species and Types of Banana in North-eastern India

Genus	Species	Subspecies/Types	Section	States
Musa	acuminata	-	Eumusa	Tripura
	balbisiana	type Athiakol		
		type Bhimkol		
	ornata	-	Rhodochlamys	
Ensete	glaucum	type medium	Eumusa	
Musa	acuminata	ssp. *burmannica*	Eumusa	Assam
		Type Kaziranga		
	balbisiana	type Athiakol		
		type Bhimkol		
		type Rissue		
Musa	acuminata	type Kaziranga	Eumusa	Mizoram
		type Variegated		
	balbisiana	type Athiakol		
		type Bhimkol		
	rosacea	-	Rhodochlamys	
	rubra	-		
Ensete	glaucum	type medium	Eumusa	
Musa	acuminata	ssp. *burmannica*	Eumusa	Meghalaya
		type Kaziranga		
	balbisiana	type Athiakol		
		type Bhimkol		
		type Rissue		
	nagensium	-		
	sikkiemensis	-		
Musa	acuminata	type Rigitchi	Eumusa	Nagaland
		ssp. *burmannica*		
	balbisiana	type Paglapahad		
		type Themenglong		
		type Phirima		
	nagensium	-		
Musa	acuminata	type Kaziranga	Eumusa	Manipur
	balbisiana	type Themenglong		
		type Athiakol		
		type Bhimkol		
Musa	acuminata	type Kaziranga	Eumusa	Arunachal
	itinerans	-		Pradesh
	aurantiaca	-	Rhodochlamys	
	rosacea	-		
	ornata	-		
	velutina			

Figure 5.1: Unopened Male Flower Buds and Pseudosheaths Used as Vegetables.

Figure 5.2: Wild Bananas growing in Forests of Tripura.

A: Hezamara block, West Tripura district; B: Re-emergence of new shoots after the plants was chopped for jhuming in Chaumanu, block, Dhalai district; C: Mandai block, West Tripura; D: Gandacherra, Dhalai district.

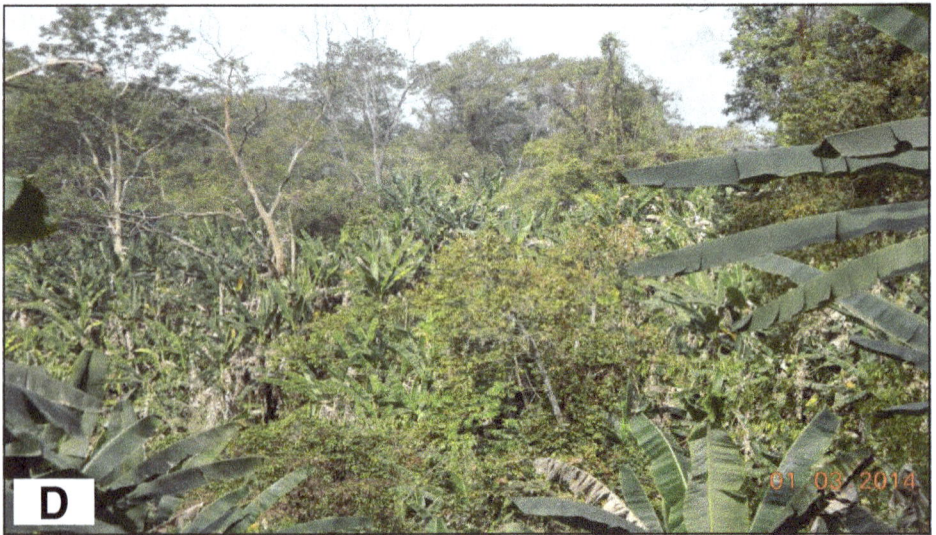

people of the state. Seven to eight types of important local and traditional genotypes of banana are available like Shabri Kela (AAB), Samai/Gopi/Bangla Kela (ABB), Champa Kela (AAB), Mizo-Cavindish (AAA), Katch Kela (ABB), Kanai Basi (AA), Attia Kela (BB) and Ram Kela, are grown in different parts of the state.

Description of Common Bananas in Tripura

Some of the important characters of some local and wild banana genotypes have been recorded in (Tables 5.7 and 5.8).

Table 5.7: Phenological Parameters of some Important Local Bananas of Tripura.

Name of Genotypes	P. Level	Genome	Crop Cycle Duration (Days)	Pseudostem		No. Leaf		No. of Suckers Pt⁻¹	B. Wt (Kg)	No. of Hands	No. of Fingers		TSS (°Brix)
				Height (cm)	Girth (cm)	S	Ha				H	B	
Shabri Kela	3x	AAB	416	343	70.0	13	8	4	15.5	8	13	95	24.20
Champa Kela	3x	AAB	410	315	72.0	12	9	4	25.0	20	20	400	22.00
Samai Kala	3x	ABB	453	4.55	80.0	13	9	6	32.5	14	20	273	27.00
Katch Kela	3x	ABB	426	355	76.0	11	9	4	21.5	7	15	105	00.00
Red Banana	3x	AAA	553	384	78.0	14	9	2	12.0	6	14	80	19.50
Mizo Cavendish	3x	AAA	387	290	62.0	14	10	2	09.0	8	14	110	20.00
Kanai Bashi	2x	AA	250	130	34.0	8	6	2	07.0	5	12	56	24.00
Attia Kela	2x	BB	393	345	78.0	15	13	5	26.5	8	14	109	20.50
Ram Kela	2x	BB	455	410	60.5	13	10	6	9.5	12	14	162	20.00

P: Ploidy; S: Shooting; Ha: Harvesting; Pt: Plant; H: hand; B: Bunch; Wt: Weight.

Table 5.8: Finger Characters of some Important Local Bananas of Tripura

Name	Finger Length (cm)	Finger Girth (cm)	Pedicel Length (cm)	Pedicel Girth (cm)	Pulp Wt (g)	Peel Wt. (g)	Fruit Wt. (g)	Fruit Pulp Colour	Bunch Orientation
Shabri Kela	19.50	14.50	1.50	4.80	91.00	37.00	126.0	Creamy white	Pendulous
Champa Kela	13.50	14.00	1.50	4.00	73.00	12.00	85.0	Yellow white	Pendulous
Samai Kala	13.00	13.50	2.50	4.00	93.00	29.00	122.0	Creamy white	Oblique
Kanai Bashi	18.00	9.80	1.60	3.60	60.00	25.00	90.0	Creamy white	Horizontal
Katch Kela	26.50	16.70	5.50	5.00	00.00	00.00	262.0	Creamy white	Oblique
Red Banana	11.50	14.00	1.10	3.50	87.00	38.00	125.9	Yellow orange	Pendulous
Mizo Cavendish	21.00	13.00	2.00	4.50	132.00	48.00	180.0	Creamy white	Pendulous
Attia kala	19.50	16.50	3.50	3.50	155.00	52.00	207.0	Brownish white	Oblique
Ram Kela	10.50	14.50	2.5	3.80	78.00	14.00	92.0	Brownish white	Horizontal

Shabri Kela (AAB)

This banana is similar to Rashthali banana. It is an excellent table type of banana in respect of appearance, flavour, texture, aroma, pulp colour and TSS. There is a huge market demand with four (4) fingers costing Rs. 50.0 to 60.0. One good banana bunch can be sold in the market at 700.0-1000.0 rupees. It is commercially grown in Tripura and West Bengal. It is similar to Marthaman in West Bengal. The plant can easily be identified by green pseudostem with slightly brownish and blackish blotches, reddish margin of the petiole and leaf sheath and at young plant pseudostem have reddish blotches. The plant is medium in stature with 2.5-3.5 m height and 60-70 cm circumference at the base. The duration of the crop is 13 to 14 months (416–435 days). The flowering started from 9 months onwards and flowering to maturation of fruits takes three and half to four months (102-113 days). The average bunch weight is 15 to 16 kg having 8 to 9 hands, each hand with 12-14 fingers or fruits with 95-120 fruits/bunch (Figure 5.3) with pendulous bunch orientation. Fruits are pale green throughout their development and turn attractive yellow on ripening. The number of leaf at shooting varies from 12-13 and at harvesting it varies from 8-9. The length of the fruit is 17 to 20 cm with 14-15 cm girth at the middle. The fruit skin is very thin and pulp is good in taste with fine flavour, texture, aroma and overall acceptance. The TSS range from 24-25° Brix with fine texture and TSS is increase up to 26-27° Brix during dry month of December to February. The pulp is firm, mealy and creamy white coloured, susceptible to the physiological disorder 'hard lumps' (unripe tissues) and fruits easily detach at maturity. Sometimes, fruits splitting during maturation are another common physiological disorder due to long spell of dry weather. It is susceptible to Panama wilt (*Fusarium oxysporum* f.sp. cubense race-1). It is also slightly susceptible to yellow sigatoka leaf spot disease. The plant produces 4-5 suckers for completion of its life cycle. The production of one leaf to another leaf interval takes seven (7) days. In Tripura, banana can be grown throughout the year in the hill slopes and plains. During December to February, temperature is very low *i.e.*, 7-15°C and plant is under water stress condition but plants can survive under such stressed condition. The recommended spacing in hilly slope is 2 x 2 m² accomodating 2,500 plant per hectare area with a yield of 40,000 kg per hectare. Keeping quality of fruits is good. Fruits can be kept for 5 to 7 days under normal condition. It is suitable for banana wine preparation as well.

Champa Kela (AAB)

This is a Poovan type of banana. It is grown throughout the state from hill slope to plain area round the year with very good local demand. The cost of 20 bananas ranges from 40-50.0 rupees. One good Banana bunch can fetch 400.0-600.0 rupees in the market. This is the second most important Banana in the state. It is a very hardy type of banana grown in the state and farmer are getting good yield with minimum care. It can be grown on a wide range of soil and climatic conditions with wide adaptability, free from pests and diseases. The fruit is small, peel on ripening turns yellow in colour (Figure 5.4A) and is thin. There is slight reddish pink colouration of the other side of mid rib when leaves are young. It is highly amenable for ratooning. However, it is slightly susceptible to banana bunchy top

Figure 5.3
A and B: Bunches of Harvested Shabri (AAB) banana; C: A Hand of Shabri banana.

virus when grown as a ratoon in the same location repeatedly. In many parts of Tripura, it is grown semi-perennially with one plant followed by one ratoon. The plant is tall and robust with a height of 2.5 to 3.5 m and 60-75 cm circumference at the base. The plant produces 4-5 suckers for completion its life cycle. The number of leaf at shooting varies from 12-13 and at harvesting it varies from 9-10. The crop duration is 13–14 months (410 days). The average bunch weight is 20-25 kg and it has 20-25 hands, each hand have 18-25 fingers or fruits with 350-450 fingers or fruits/bunch with pendulous bunch orientation (Figures 5.4B, C and D). TSS range from 21-22°Brix and TSS is increased up to 23-24°Brix during winter months. The bunch is compact, length of the fruit is 13.5-14.5 cm and girth of fruits in the middle is 13-14 cm. The fruits are medium sized with bottle necked tip. The fruits are pleasantly acidic sweet in taste with firm flesh. The bunch is medium sized with closely packed fruits and has good keeping quality. The fruit skin is medium to thick. It is tolerant to nematodes and *Fusarium* wilt. The recommended spacing in hilly slope is 2 x 2 m² with 2500 plant per hectare area and yield of 50,000 kg per hectare. Keeping quality of fruits is good; 6 to 8 days under normal condition. It is suitable for banana wine preparation. The cultivar is very hardy compared to Shabri Kela with more tolerance to severe water and temperature stress.

Gopi/Samai/Bangla Kela (ABB)

This banana is similar to Karpuravalli type and is commercially grown in Tripura. It is an important component in most of the homestead gardens and is indispensably used for puja purpose. This is the third most important banana in the state with a cost of 60-70.00 rupees per 20 bananas. One good Banana bunch can be sold in the market at 600.0-700.0 rupees. Stem is light green with purplish tinge, 3.5 to 4.5 m tall, leaves are large and 70-80 cm circumference at the base. The

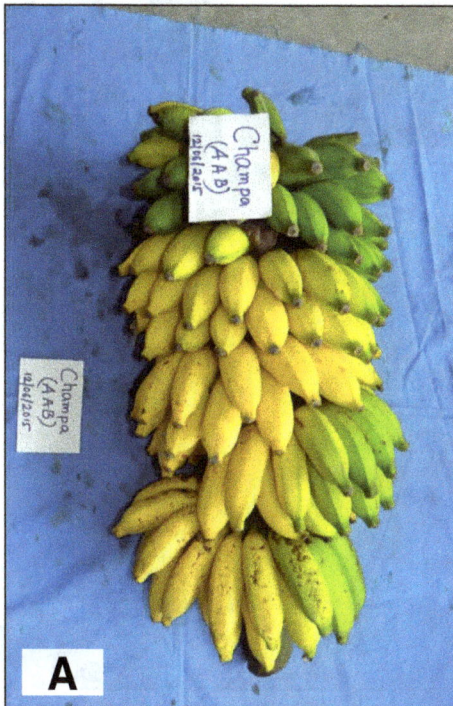

Figure 5.4: A: Ripening bunch of Champa Kela (AAB).

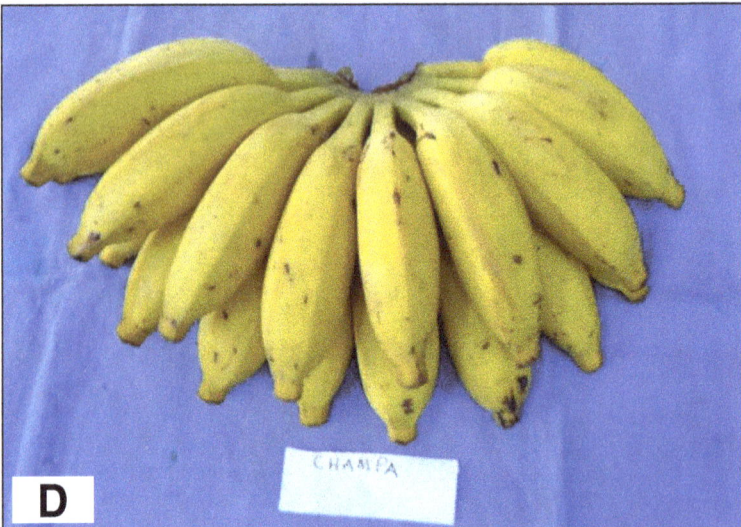

Figure 5.4: B and C: Fully ripened bunches of Champa Kela;
D: A hand of Champa Kela.

duration of the crop is 15-16 months (453 days). Flowering starts from 10-11 month onwards and takes four to five months (125-135 days) after flowering for maturation. The average bunch weight is 25 to 35 kg, having 11 to 14 hands with 15-20 fingers or fruits in each hand and 250-275 fruits/bunch (Figure 5.5), and oblique bunch orientation. Skin yellow with ashy coating, pulp cream coloured, crisp, sweet with a pleasant taste and flavour. The number of leaf at shooting varies from 11-13 and at harvesting it varies from 8-9. The length of the fruit is 13 cm with 13.5 cm girth at the middle. The fruit skin is very thin and fruits are very attractive yellow colour at the time of ripening, good taste with fine flavour, texture and aroma. Fruits have one or two rare seeds. The TSS ranges from 26-27°Brix. The plant has a more suckering habit wit 7-8 suckers for one life cycle. The production of one leaf to another leaf interval (phyllochron) is eight days. This banana can grow throughout the year in the hills and plains. The recommended spacing in hilly slope is 2.5 x 2.5 m^2 with 2000 plant per hectare area with a yield of 50,000 kg per hectare. Keeping quality of fruits is good (8 to 10 days under normal condition). It is also suitable for banana wine preparation. Plants can survive stress condition during winter months (December-February). This cultivar is hardy, can survive without water for

Figure 5.5: Bunches of Samai/Gopi/Bangla (ABB) Banana.

3-4 months and is free from almost all the major pests and diseases like pseudostem weevil, nematodes, *Fusarium* wilt, bunchy top, sigatoka *etc.*

Katch Kela/Sabzi Kela (ABB)

Katch Kela is one the most important genotype suitable for use as vegetable and used for preparing various dishes by local people. Cost of four (4) bananas is 20.0 rupees and a good banana bunch is sold in the market at 400.0-500.0 rupees. It is a very popular banana grown in homestead garden to meet the vegetable requirement in rural and semi urban areas. A popular local dish known as Gudok is a favourite among the different tribes of Tripura prepared from this banana flower. Variation exists among this type of banana and distinctly three types of Kach Kela has been differentiated. Generally, the plant is tall, robust, with 3.0-3.5 m height and 70-76 cm circumference at the base and leaves are light green. The plant is hardy and tolerant to various stresses (biotic and abiotic). The duration of the crop is 13 to 14 months (426 days). The flowering commence from 9-10 months onwards and takes three and a half month (90-110 days) from flowering to fruit maturation. The bunch is not compact, average bunch weight is 18 to 25 kg having 6 to 7 hands, and each hand has 13-15 fingers or fruits averaging 90-105 fruits/bunch (Figure 5.6). Bunch orientation is oblique. The number of leaf at shooting varies from 11-12 and at harvesting it varies from 8-9. The length of the fruit is 26 to 26.5 cm with 16.70 cm girth at the middle of the fruits. Stigma persistent at maturity, ripe fruits turn yellow, pulp is rich yellow-orange coloured and tough. The plant produce 4-5 suckers for completion its life cycle. The interval for the production of one leaf to another leaf is eight days. This type of banana can grow throughout the year in the hill slope as well as plain areas. The recommended spacing in hilly slope is 1.8 x 1.8 m² with plant accommodation of 3086 numbers per hectare with a average yield 50,000 kg per hectare. The fruits have prominent ridges at maturity. The keeping quality is very good and can be kept for 10 -15 days under normal condition.

Red Banana/Amritsagar (AAA)

The colour of the pseudostem, petiole, midrib and fruit rind is purplish red. The fruit size is good, slightly curved with a blunt apex (Figure 5.7). The plant is medium to tall stature with 3.5-4.0 m (3.84 m) height and 78 cm circumference at the base. The duration of the crop is 18 to 19 months (553 days). The flowering starts from 14 month onwards and takes four months (120 days) after flowering to reach maturity. Average bunch weight is 9-12 kg with 5-6 hands; each hand has 12-14 fingers or fruits with 80 fruits/bunch and has pendulous bunch orientation. The number of leaf at shooting varies from 12-13 and at harvesting it varies from 9-10. The length of the fruit is 11.50 cm with 14.00 cm girth at the middle. Ripe fruits are reddish; pulp is a rich yellow-orange in colour and tough, stigma persistent at fruit maturity. The plant produces 1-2 suckers for completion of one life cycle. This type of banana can grow throughout the year in the hills and slope. The fruit is costly and a fruit/finger is priced around 10-15 rupees which amounts to around 500.0-600.0 rupees per bunch. The recommended spacing in hilly slope is 2.5 x 2.5 m² with an average of 2000 plant per hectare, and average yield is 15,000-17,000 kg per hectare.

Figure 5.6: Different Types of Kach Kela (ABB) Bunch and their Respective Hands.

Figure 5.7: Red Banana (AAA); Bunch (above left) and Hands (below),
and Plant with Bunch and Flower (above right).

Kanai Bashi (AA)

The plant is diploid and dwarf in stature with 1.2 m to 1.5 m height and 30-35 cm circumference at the base. The duration of the crop is 7 to 8 months (236–250 days). It is a short duration crop with the flowering starting from 5 months after planting and flowering to maturation lasts for three months (89-90 days). The plant has horizontal bunch orientation and average bunch weight is 5 to 7 kg with 4 to 5 hands, each hand has 10-12 fingers or fruits with 53-56 fruits/bunch (Figure 5.8). The number of leaf at shooting varies from 8-9 and at harvesting it varies from 6-7. The length of the fruit is 15.0-18.0 cm with 9.5 cm girth at the middle. The fruits are very attractive yellow in colour at the time of ripening, possess good taste and aroma, TSS range from 23-24°Brix. The pulp texture is fine, firm with creamy yellow colour. The plant produces 1-2 suckers for completion of one life cycle. The recommended spacing in hilly slope is 1.25 x 1.25 m accommodating 4444 plants per hectare area with an average yield of 22,500 kg per hectare. The plant is very hardy in nature with very less infestation by pests and diseases. Plant is resistant to *Fusarium* wilt and nematodes. Keeping quality of fruits is moderate. Fruits can stay for 4 to 5 days under normal condition.

Mizo-Cavendish (AAA)

The plant is tall, fruits large, curved, skin thick and greenish, flesh soft and sweet with pleasant aroma at ripening. The entire pseudostem has blackish blotches. The height of the plants ranges from 2.90-3.10 m and circumference at the base is about 59.00-62.20 cm. It is found growing naturally in the border of Mizoram and Tripura and is locally known as Mizo-Cavendish. A short duration crop and life cycle is completed is in 12-13 months (387 days). The flowering starts from 8-9 months onwards and flowering to maturation of fruits it takes three to four months (100-120 days). The average bunch weight is 9.0-10.0 kg having 5-7 hands, each hand with 12-14 fingers or fruits and 90-100 fruits/bunch with pendulous bunch orientation (Figure 5.9). The number of leaf at shooting varies from 12-14 and at harvesting it varies from 10-11. The length of the fruit is 21.0 cm with 13.0 cm girth at the middle. The fruits turn yellow colour at the time of ripening. The TSS ranges from 20-21°Brix, the pulp is firm and creamy white in colour with very fine texture. The plant produces 6-7 suckers for completion of its life cycle. The recommended spacing in hilly slope is 1.8 x 1.8m^2 with 3086 plant per hectare area; yield is 22,500 kg per hectare. The plant is very hardy in nature with very less infestation by pest and disease.

Attia Kela (BB)

The plant is diploid, tall and is seeded. Pseudostem green with slight black blotches and leaves are green. The height of the plant is 3.0 m to 3.45 m and 78.0 cm circumference at the base. The duration of the crop is 13-14 months (393 days). The flowering starts from 9 months onwards and flowering to harvesting of fruits takes four months (100-115 days). The average bunch weight is 26.5 kg having 7 to 8 hands; each hand has 13-14 fingers or fruits with 109 fruits/bunch (Figure 5.10). Bunch orientation is oblique. The number of leaf at shooting varies from 13-15 and at harvesting it varies from 12-13. The length of the fruit is 19.5 cm with

Figure 5.8: Kanai Bashi (AA).
A: Plant with a bunch and male flower; B: Ripened bunch; C: Plant with distinct horizontal bunch orientation.

16.5 cm girth at the middle. The fruits are yellow in colour at the time of ripening and fingers are heavily seeded, TSS is 20°Brix. The plant produces 5-6 suckers to complete its one life cycle. Local people use for fresh consumption and is also used

Figure 5.9: Mizo Cavendish (AAA) Banana.
A: Plant with a bunch and male bud; B: Ripened hands and bunch.

for different local puja festivals. Traditionally it is consumed as a medicine by the pregnant women. This type of banana is available in forest as well as plain areas but is not commercially cultivated.

Ram Kela (BB)

It is a tall type banana with heavily seeded character. Unopened male bud and pseudostem are used as vegetable by different tribal communities and rural people of Tripura, *viz.*, Halam, Tripuri, Reang, Debbarma, Mok, Chakma, Uchoi and others. The unopened male bud cost around 10-12 rupees and is used for preparing local dishes like Gudok which is a favourite among the tribals in Tripura. The height of the plants varies from 4.0 m to 4.10 m. and 60.5 cm circumference at the base. The duration of the crop is 14-15 months (455 days). The flowering starts from 11 months onwards and complete its maturation in three and half months after flowering. The average bunch weight is 9-10.5 kg having 11-12 hands; each

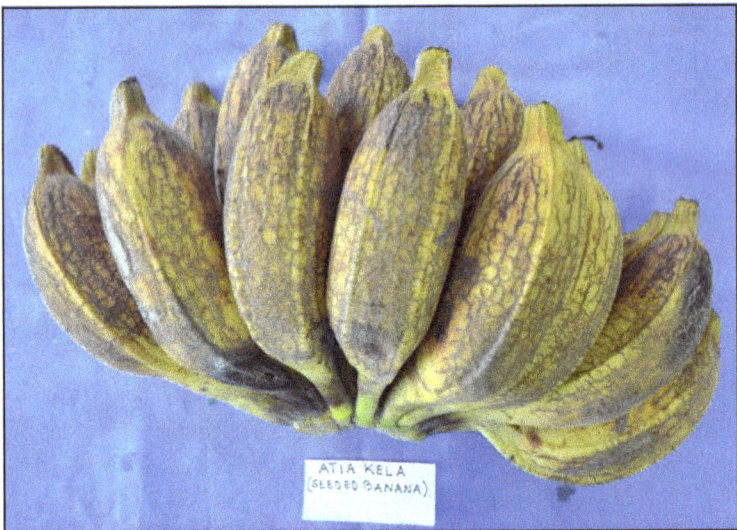

Figure 5.10: Bunches, Hand and LS of Attia Kela (BB).

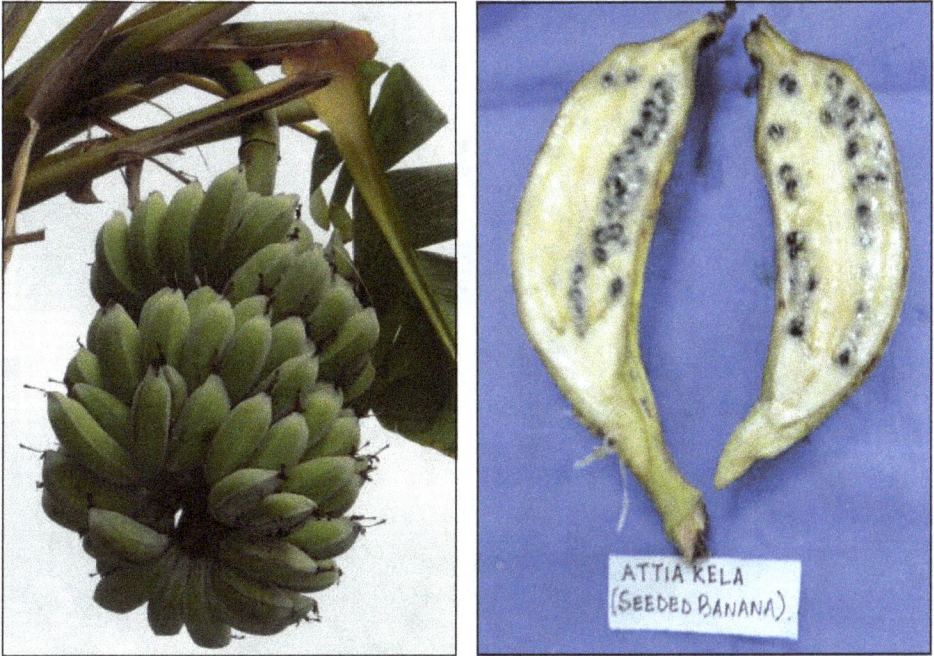

Figure 5.10: Bunches, Hand and LS of Attia Kela (BB).

hand has 12-14 fingers or fruits with 162 fingers or fruits/bunch (Figure 5.11) with horizontal bunch orientation. Fruits turn yellow on ripening. The number of leaf at shooting varies from 13-14 and at harvesting it varies from 9-10. The length of the fruit is 10.5 cm with 14.5 cm girth at the middle. TSS ranges from 20-21°Brix. Plant produces 6-7 suckers in its one life cycle. Pseudostem has a whitish green ash like layers, leaves are dark green and orients erect towards the sky. This banana is very hardy, water stress tolerant and not infected with any pests and diseases. It is resistant to nematodes and *Fusarium* wilt and it is widely grown in the hilly areas.

Musa bulbisiana Banana (BB)

Morpho-taxonomic classification of wild genotypes identified seeded BB clones. Bhimkol is an important seeded banana of Tripura, Assam, Nagaland and other north-Eastern states. Seed and pulp are used as food for medicinal purpose. Highly tolerant to different pests and diseases like bunchy top, panama wilt, leaf spot and different nematodes. Another wild banana is Attiakol with yellow green pseudostem. Seeds are very fertile and germination is very easy. The plants are very tall with 4-5 m height and 62.5 cm circumference at the base. The duration of the crop is 13-14 months (430 days). The flowering starts from 11 months onwards and complete its maturation in three and half months after flowering. The average bunch weight is 8-9.5 kg having 7-8 hands; each hand has 11-12 fingers or fruits with 99 fingers or fruits/bunch (Figure 5.12) with horizontal bunch orientation. Fruits turn yellow on ripening. The number of leaf at shooting varies from 12-13 and at

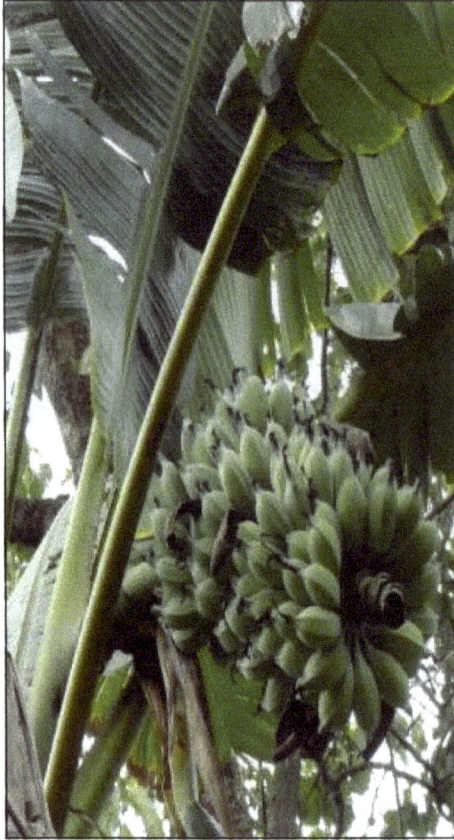

Figure 5.11: Ram Kela (BB).

harvesting it varies from 8-9. The length of the fruit is 9.5 cm with 13.0 cm girth at the middle. TSS ranges from 21-22°Brix. Plant produces 6-7 suckers in its one life cycle. Pseudostem has a whitish green ash like layers, leaves are dark green and orients erect towards the sky. This banana is very hardy, water stress tolerant and not infected with any pests and diseases. It is resistant to nematodes and *Fusarium* wilt and it is widely grown in the hilly areas.

Musa acuminata (AA)

Wild *M. acuminata* bananas are available in the forest of Tripura and are seeded. Approximately 10-15 per cent area of forest is under this type of seeded banana. The height of the plants varies from 2.0 m to 2.50 m and 35.0 cm circumference at the base. The duration of the crop is 12-13 months (375 days). The flowering starts from 9 months onwards and complete its maturation in three and half months after flowering. The average bunch weight is 3-4.5 kg having 4-5 hands; each hand has 11-12 fingers or fruits with 62 fingers or fruits/bunch (Figure 5.13) with horizontal bunch orientation. Fruits are very small and turn yellow on ripening. The number of leaf at shooting varies from 12-13 and at harvesting it varies from 7-8. The length

Figure 5.12: Different Wild *Musa balbisiana* (BB) Plants, Flowers and Fruits.

of the fruit is 6.5 cm with 9.0 cm girth at the middle. TSS ranges from 20-21°Brix. Plant produces 5-6 suckers in its one life cycle. Pseudostem has a whitish green ash like substance with black patches, leaves are dark green and orients erect towards the sky. This banana is very hardy, water stress tolerant and not infected with any pests and diseases. It is resistant to nematodes and *Fusarium* wilt and it is widely grown in the hilly areas.

Ensete Banana

It is found in forest of Tripura, Mizoram, Hills of Assam, Arunachal Pradesh, Nagaland and other parts of North-East. The name of the species *Ensete glaucum*. In Tripura, this banana is also called as Elephant Banana by the local tribes due to the resemblance of the flowering bunch to that of the elephant trunk (Figure 5.14A). Plant is single stemmed, monocarpic herb, non-stoloniferous. The plants do not produce any sucker and propagation is only by means of seed which are quite bold, hardy and black in colour, irregularly angulated and compressed. Seeds are used as medicine by the tribals of Tripura by preparing a garland and worn by jaundice

Figure 5.13: Wild *Musa acuminata* (AA) Plants with Flower and Fruit.

patients. The individual pseudosheath is consumed by different indigenous tribes of Tripura *viz.*, Halam, Tripuri, Reang, Debbarma, Mok, Chakma, Uchoi and other tribes as their favourite vegetable dish. The height of pseudostem is usually 2.0-2.5 m, conspicuously swollen at the base with 80-100cm in circumference and psuedostem has black patches (Figure 5.14). Leaves are clustered and crowned just above the ground level. The leaves are dark green, dull and spreading, 1.0-1.5 m long, and have symmetrically pointed laminar bases. Leaves exhibit gradual transition from foliage to bracts. Petiole widely opened, petiolar base has spreading green margins. Plant has high water content in the pseudostem due to which it is a favourite for Elephants. It is said that elephants search for this banana species when thirsty and consume this pseudostem to quench their thirst. Pseudostem is dark and dull green in colour with profuse ash like coating. The bunch has 5-8 hands of closely packed fruits; male flowers are persistent and are hidden under the persistent bracts (Figure 5.14). Male bud is ovoid with distinct yellow-green coloured strips on the outer face of the bract. The fruits are obovoid, dull green ash coated with prominent tip.

Ornamental Banana

Ornamental banana (*Musa ornate*) under section Rhodochlamys (Table 5.6) was observed growing naturally in some parts of Tripura and other northeastern states like Nagaland and Sikkim. The fruits, flower and pseudostem are red in colour. Leaf and bunch orients erect towards the sky (Figure 5.15). The height of the plant is 1-1.5 m and pseudostem has 30-35 cm circumference at the base.

Figure 5.14: Ensete Banana in Tripura.
A: Plant with mature bunch (male hidden under persistent bracts); B: Flower; C: Plant with emerging bunch.

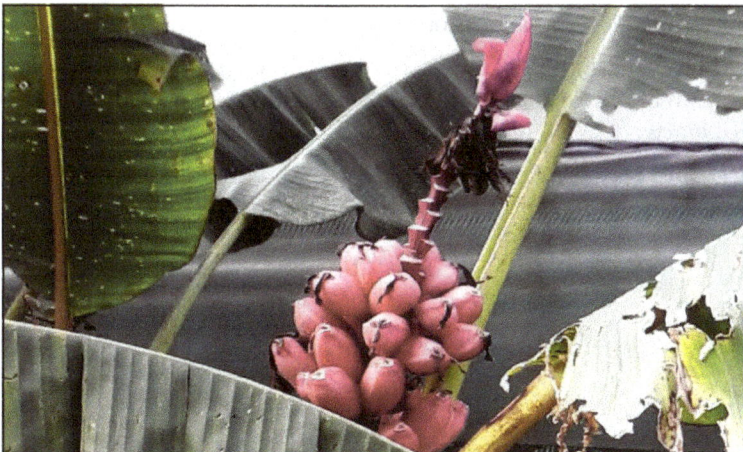

Figure 5.15: Ornamental Banana, *Musa ornate*.

References

Anonymous, 2014. Annual report 2013-2014. Department of Horticulture, Government of Tripura, India.

Bhakthavatsalu, C.M. and Sathiamoorthy, S. 1979. Fruits, 34: 99-105.

IBPGR. 1984. Revised Banana Descriptors. IBPGGR Secretariat, Rome.

IPGRI-INIBAP/CIRAD, 1996. Descriptor for banana (Musa spp.) INIBAP, Parc Scientifique Agropolis, Montpellier, France. p-55.

Purseglove, J.W. 1975. Tropical Crops: Monocotyledons. The English Language Book Society and Longman, Pp. 345-376.

Silayoi, B. and Chomchalow, N.1987. In: Proc. Int. Workshop (Eds. G.J. Persely and E.A. De Longhe), Cairns. Australia, 1986. ACIAR Proc. No. 21: 157-160.

Simmonds N.W. 1962. The evolution of the bananas. Longmans, Green and Co, London.

Simmonds, N.W. 1948. The effects of ploidy upon the leaf of *Musa*. Ann. Bot., 12: 441-453.

Simmonds, N.W. 1953. Seggregations in some diploid bananas. J. Genet., 51: 458-469.

Simmonds, N.W. 1954. Isolation in *Musa*, sections *Eumusa* and *Rhodochlamys,* Evolution, 8: 65-74.

Simmonds, N.W. 1966. Bananas. Longman, London.

Simmonds, N.W. and Shepherd, K. 1955. The taxonomy and origin of the cultivated bananas. J. Linn. Soc. Bot., 55: 302-12.

Simmonds, N.W. and Shepherd, K. 1987. A tentative key for identification and classification of Indian Bananas. National Research Centre for Banana (ICAR), Thiruchirapalli, India, 2001.

Simmonds, N.W. 1952. Experiments on the pollination of seeded diploid bananas. J. Genet., 51: 32-40.

Simmonds, N.W. 1966. Bananas. Longmans, London, p.512.

Singh, H.P. and Uma, S. 1997. Collection, conservation and characterization of Musa germpasm at NRCB. Proc. All India Co-opdinated Research Project Meeting, UAS, Dharwad, 27-30 November, 1997.

Stover, R.H. and Simmonds, N.W. 1987. Bananas, Longman Edition, London.

Chapter 6

Pear

Amit Kumar¹, Tawseef Rehman Baba¹,
Nirmal Sharma² and Manmohan Lal¹

¹*Division of Fruit Sciences,*
SKUAST-Kashmir, Shalimar, Srinagar (J&K)
²*Division of Fruit Sciences*
SKUSAT-Jammu, Chatha, Jammu (J&K)
khokerak@rediffmail.com

Pear (*Pyrus communis* L.) is a truly wondrous hardy fruit, widely grown in the temperate regions of the world, with varied size, shape, texture, and flavors. The long-lived trees attain great size and are relatively easy-to-grow. It is well-known temperate fruit popular throughout recorded history in the West and the East. Yet, in art and literature, as well as pomology, the pear plays second fiddle to the apple, suggesting that the problems of the pear continue to bedevil its champions. Therefore pear improvement should be undertaken so that it can take its rightful place in the pantheon of pome fruit.

Pear, a typical fruit of temperate climates, with delicate pleasant taste and smooth, has a wide acceptance throughout the world. By its shape, it inspires designers and architects. The fruit pleases generations; already in 1661, Jean-Baptiste de La Quintinie, lawyer and botanist, responsible for the gardens of the Versailles palace, passionate about the cultivation of pears, wrote in reports: "It must be confessed that, among all fruits in this place, nature does not show anything so beautiful nor so noble as this pear. It is pear that makes the greatest honor on the tables." As a fruit of typical temperate regions, its origin and domestication is considered at two different points, China and Asia Minor until the Middle East. It is the fifth most widely produced fruit in the world, being produced mainly in China, Europe, and the United States. Pear belongs to rosaceous family, being a close "cousin" of the apple, but with some particularities that make this fruit special with a delicate flavor.

Pear tree is also an important ornamental and is beloved in Asia where pear is considered a sign of good luck. Yet, we cannot ignore the fact that world production of pear is only about one-quarter that of apple, indicating that the appreciation of pear has not attained the universality or the depth of appeal of its better known relative. The perfect fruit at its optimum state of maturity and stage of ripeness, to produce the perfect proportion of texture, flavor, acidity, and sweetness. The pear can be a stately ornamental and the 'Bradford' pear and other selections *of Pyrus calleryana*, are admired as a street tree for its elegant, pyramidal form, red fall color, and white flowers. We, who love this fruit - fresh, cooked, spiced, fermented, dried, liquid or even grown in a bottle and smothered by brandy - are convinced that all attempts to overcome any defects, natural or imposed, are worthy of the struggle and we are proud to be a part of this sublime activity. Pear cider is usually made from cultivars of *P. nivalis* and is called perry.

Europeans prefer soft flesh, "pyriform" pears that must be ripened to come to optimum quality while the pears of the Asia are round and crisp and do not require softening. Both qualities, although very different, are delectable in their own way. It is not surprising, therefore, that the pear should permeate the cultures that consume it. In European winter pears this eating quality is achieved by ripening after harvest. *Pyrus koehnii*, an evergreen species native to Taiwan, is planted in California and Florida. But, alas, the foul scented flowers, a distinguishing characteristic of *Pyrus*, may be objectionable in mass plantings.

Table 6.1: Area and Production of Pear

Country	2013	
	Area (ha)	Production (tones)
China	12,70,000	1,73,00,751
USA	19,840	7,95,557
Italy	34,241	7,43,029
Argentina	28,420	7,22,324
Turkey	34,430	4,61,826
Spain	24,200	4,25,700
South Africa	11,148	3,43,203
India	38,500	3,40,000
Netherlands	8,509	3,27,000
Japan	14,600	2,94,400
Republic of Korea	13,740	2,82,212

Source: FAO, 2013.

The world production of pear is second only to apples among the deciduous pear tree fruits. About 2,52,03,751 tons of pears were produced in 2013 (FAOSTAT, 2015), up over 2,35,80,845 tons from 2012. Asia produced the most pears (1,94,52,003 t), followed by Europe (30,06,895 t), America (18,55,844 tons), Africa (7,51,383 tons) and Oceania (1,37,626 tons). In Asia, China was the largest producer with the largest producer with 68.64 per cent of the world volume, followed by Turkey

(1.83 per cent), India (1.34 per cent) and Japan (1.17 per cent). The major producers in Europe are Italy (2.94 per cent), Spain (1.69 per cent) and France (1.30 per cent). In America, the highest producers were USA (3.16 per cent) followed by Argentina (2.86 per cent) and Chile (0.89 per cent). Production in Oceania was the highest in Australia (0.43 per cent).

Pyrus is genetically quite diverse, with considerable variability in morphology and physiology adaptations. (Knight, 1963; Westwood, 1982; Lombard and Westwood, 1987; Bell *et al.*, 1996). European and Asian pears breeders have utilized this variability to develop high quality cultivars with larger size and attractive appearance that are well adopted local conditions.

The European pears are distinguished by the juiciness, delicate flavour and aroma, while the Oriental (Asian or nashi) pears are known for their crispness and sweet flavour. North American breeders have had to focus more on disease resistance and cold hardiness than the Europeans, although the spread of the bacterial disease fire blight throughout Europe is forcing the breeders there to concentrate more on disease resistance (Bell *et al.*, 1996).

Brief History of Genus *Pyrus*

Ancient Greek poet Homer praised pears as one of the 'gifts of God'. This prehistoric fruit has been under cultivation both in Europe and Asia for a long ago. In Europe, pears have been cultivated as early as 1000 B.C. (Hedrick *et al.*, 1921). Sand pear (Japanese and Chinese species) has been domesticated as edible fruit and cultivated in Asia for more than 3000 years (Lombard and Westwood, 1987). Early developments took place in Greece and Rome around 200-300 BC. The description of varieties and methods of vegetative propagation were well documented. Advances in pear culture and varietal improvement were brought only in the 18th century.

Antiquity

The first mention of the pear is found in Homer's (9th century BC) epic poem, *The Odyssey*, confirming that the pear was cultivated in Greece as early as three thousand years ago. The pear is included as one of the "Gift of the God" which grew in the garden of Alcinöus, the King of the Phaeacians, a legendary country. Marcus Procius Cato (234–149 BC), known as Cato the Censor, is the author of the famous agricultural manual *De Agri Cultura* (*De Re Rustica*), the oldest extant specimen of a treatise in Latin prose. He writes extensively on pomological subjects and describes six types of pear. The treatise *De Re Rustica* of Marcus Terentius Varro (116–27), written in the form of a dialogue, emphasizes the dependence of the commonwealth on a sound agriculture. He gives directions for grafting pear, including inarching, and discusses storage of the fruit.

Medieval

The introduction of the pear in France is unknown. It may have been independently domesticated as two cultivated species grow wild or it may have been introduced by the Greeks who founded Marseille in 600 BC, but is more likely that the pear was introduced by the Romans. Charlemagne (742–814) the ruler of the Franks in the 9th century is credited with establishing the first collection of pear in France. The *Capitularies* or "lists of laws" includes comments on pear cultivation in the king's orchards; orchardists are commanded to plant pears of distinct kinds for distinct purposes: *Plant pear trees whose products, because of pleasant flavor, could be eaten raw, those which will furnish fruits for cooking, and finally, those which mature late to serve for use in winter.*

Renaissance

The great Leonardo Da Vinci (1452–1519) reveled in collecting cryptic puns, aphorisms, fables, prophecies, jests, mottoes, and fantastic tales. *Codex Arundel 67* contains several botanical fables including one involving the laurel, the myrtle, and the pear:

The laurel and the myrtle on seeing the pear tree being cut down, cried out in a loud voice: 'O pear tree where are you going? Where is the pride that you had when you were laden with ripe fruit? Now you will no longer make shade for us with your thick foliage.' The pear tree replied: 'I am going with the husbandman who is cutting me down and who will take me to the workshop of a good sculptor, who by his art will cause me to assume the form of the god Jove, and I shall be dedicated in a temple and worshipped by men in place of Jove. While you are obliged to remain always maimed and stripped of your branches [while] men shall set around me in order to do me honor" (Embode, 1987).

Modern Time

The modern history of the pear in Belgium, France, England and Central and Western Europe, as well as America is well covered in Hedrick (1921) and readers are referred to this great work for a detailed review of more recent pomological history. The pear has a great tradition in France and there appears to be an explosion of diversity from the 16th to 19th centuries. Thus, in 1540, Charles Estienne lists sixteen pears with brief descriptions in an agricultural work entitled *Seminarium*. De Serres, known as the French father of agriculture, describes various types in *Le Theâtre d'Agriculture* (1608).

Le Lecier, an attorney of the King at Orleans collected fruits and describes 254 pears in his catalogue of 1628. Subsequently, 197 sorts are described by Merlet in 1667, 67 by La Quintinye in 1690, 119 by Duhamel in 1768, 102 by the Chartreuse fathers in 1775, 120 by Tollard in 1851, and 238 by Noisette in 1833. Leroy in 1867 claimed 900 cultivars with 3000 names. Synonymy has continued to be a problem with pear. 'Bartlett', the most famous pear in North America, was discovered in 1770 in England and named 'Williams Bon Crétien' and is still known as 'Williams' in Europe.

Genus *Pyrus* was classified into more than twenty primary diploid species distributed over Europe and Asia (Layne and Quamme, 1975; Zohary and Hopf, 1988) and at least six naturally occurring inter specific hybrids (Bell *et al.*, 1996) during 18th and 19th century. Out of these recognized species, pear is classified into three main groups on the basis of origin and commercial fruit production *viz.*, European pear (*Pyrus communis* L.), Japanese pear (*P. pyrifolia* Burm.) and Chinese pear (*P. bretschneideri* Rehd. and *P. ussuriensis* Maxim). Furthermore, the genus is also divided into two native groups, *i.e.* occidental and oriental pears (Lee, 1948;

Layne and Quamme, 1975). The occidental pears include over twenty species mostly found in Europe, Northern Africa, Asia Minor, Iran and Central Asia. The oriental pears include 12 to 15 species distributed from the Tian-Shan and Hindu Kush mountains eastward to Japan (Rubtsov, 1944). Oriental pears are divided into five groups *i.e.* Ussurian pear, Chinese white pear, Chinese sand pear, Xinjiang pear (*P. sinkiangensis*) and Japanese pear. Four of these are grown in China and one in Japan (Teng and Tanabe, 2004). Among occidental pears, *P. communis* (the common pear) is the major cultivated species widely distributed throughout the Europe, North and South America and Africa (Bell, 1990). The advancements in pear varieties have been seen in Belgium, France and England, and great diversity of germplasm was accumulated there (Watkins, 1976; Layne and Quamme, 1975; Bell *et al.*, 1996). Pear was first introduced in Europe and in their colonial states by seed. However, these were propagated clonally as varieties which were later introduced. 'Bartlett', 'Anjou', and 'Bosc' are examples of European varieties that are still being grown commercially. The cultivation of pears spread quickly throughout these colonies. There apparently was a rather broad range of genetic diversity in these fruits as each state accumulated a wide variety of local pear varieties that had been selected from chance seedlings.

Origin and Distribution

The genus *Pyrus* has probably originated in Central Asia, the mountainous regions of Western and Southern China, from Asia Minor to India and further diversified and moved both in Eastern and Western directions from primary centre of origin (Watkins, 1976). Speciation has occurred mainly in Eastern and Central Asia in the Himalayas, Caucasus, Asia Minor and Eastern Europe. Distribution of wild species of *Pyrus* extends from the Balkans in Europe, through Caucasus, Turkmenistan, Altai Mountains, Siberia to China and Japan (Vavilov, 1951; Zagaja, 1970 and 1977). Some species of this genus although naturalized in America but America is not a native home of this genus (Rehder, 1986). Vavilov (1951) proposed three centres of origin (centre of biodiversity) for cultivated pears *i.e.* Chinese centre, Central Asiatic centre and Near Eastern centre. In addition to these, Zeven and Zhukovsky (1975) proposed the fourth centre of diversity as European Siberian centre.

Chinese Centre

The primary gene centre comprises the regions of North and Central China, Japan and Korea. The important species originated in this tract are *P. pyrifolia*, *P. ussuriensis*, *P. betulaefolia* and *P. calleryana*. It is further evidently proved that most of the Japanese pear cultivars/varieties belonged to *P. pyrifolia* which have been domesticated from wild *P. pyrifolia* found in Japan (Kikuchi, 1948). However, the progenitors of these pear varieties probably came from China and Korean peninsula since long time (Shimura, 1988). Asian species Ussuri pear (small round fruit), most advanced varieties of Japanese and Korean pears, are derived from sand pears (Lombard and Westwood, 1987; Teng *et al.*, 2002). However, the Japanese pear (*P. pyrifolia* Nakai) and Chinese pear (*P. bretschneideri* Rehd.) are the main cultivated species in Asia (Bell *et al.*, 1996).

Central Asiatic Centre

Western Tian-shan, Pamir-Alai, Tajikistan, Uzbekistan, North-western India and Afghanistan are included in this centre. The indigenous species are *P. communis*, *P. salicifolia*, *P. regeli* and *P. pashia* (Vavilov, 1951).

Near Eastern Centre

This centre included Asia Minor and Caucasus mountains tracts and the species included to this centre are *P. communis, P. syriaca* and *P. caucasica* (Vavilov, 1951).

European Siberian Centre

According to Zeven and Zhukovsky (1975) *P. communis, P. nivalis, P. salvifolia*, and *P. cordata* are widespread to this centre. *P communis* (syn. *P. domestica*) has been grouped into two species *i.e. P. pyraster* and *P. nivalis*. The former grows in Central Europe and West Asia and the latter in western Switzerland and France. Almost all the commercial varieties of European pear predominating throughout the world except China and Japan are descendents of *P. communis*. The ancestors of this species are considered to be *P. communis* var. *Pyraster*, *P. communis* var. *Caucasica* and probably *P. nivalis* (Lombard and Westwood, 1987).

Table 6.2: Places of Origin of Primary *Pyrus* Species

Species	Origin
Pyrus amygdaliformis Vill	Greece, Sardinia, Turkey, Yugoslavia
Pyrus betulifolia Bunge	Central North China, South Manchuria
Pyrus callen/ana Decne.	Central and Soudi China
Pyrus communis L.	West and South-east Europe, Turkey
Pyrus communis var.pyraster Burgsd	South-east Europe
Pyrus communis var. caucasia Fed.	South-east Europe
Pyrus cordata Desv.	Algeria
Pyrus cossonii Rehd.	Japan
Pyrus dimorphylla Mak.	South-east Europe,Turkey, West Russia
Pyrus fauriei Scheid.	Korea
Pyrus gharbiana Trab.	Morocco
Pyrus glabra Boiss.	South-east Europe,Turkey
Pyrus hondoensis Kik. and Nak.	Japan
Pyrus koehnei Schneid	South China, Taiwan
Pyrus marnorensis Trab.	Morocco
Pyrus nivalis Jacq.	West, Central and South Europe
Pyrus pashia D. Don	India, Nepal, Pakistan
Pyrus pseudopashia Yu	Morocco
Pyrus pyrifolia (Burm.) Nak.	China, Japan, Korea, Taiwan
Pyrus regelii Rehd.	Afghanistan
Pyrus salicifolia Pall.	North Iran, South-west Russia

Species	Origin
Pyrus syriaca Broiss.	Israel, Lebanon, Libya, Syria, Tunisia
Pyrus ussurienisis Maxim	North China, Korea, Manchuria, Siberia
Pyrus xerophilus Yu	North-west China

Source: Bell, 1990.

Interspecific hybridization was probably involved in the origin of domesticated forms of *Pyrus communis*, the species originally cultivated in Europe and the Asian species *P. ussuriensis* and *P. pyrifolia*. A naturally occurring interspecific hybrid *P. bretschneidcri* is thought to have arisen where the range of *P. pyrifolia* and *P. betulifolia* Bunge overlap. Many of the domesticated cultivars of "ussuri" pears grown in China are undoubtedly the result of interspecific hybridization of *P. ussuriensis* and *P. pyrifolia*. Wild populations of *P. communis* var.*pyraster* and/or *P. caucasia* Fed. are tire probable ancestors of the cultivated forms of *P. communis*, and there is some evidence that *P. nivalis* Jacq. was also involved (Challice and Westwood, 1973). A number of non-primary species designations also appear in die literature, which may be botanical varieties, subspecies, or "arboretum hybrids" not known to occur in natural populations. Almost all the commercial cultivars of European pear predominating throughout tire world except China and Japan arc descendents of *Pyrus communis*. The ancestors of this species are considered to be *Pyrus communis* var. *pyraster*, *Pyrus communis* and probably *Pyrus nivalis*. *Pyrus nivilis*, snow' pear, is mainly utilized for making pear cider in Europe (Lombard and Westwood, 1987). Where ranges of species overlap, inter specific hybridization can occur and pure species can be difficult to locate if population size is small. This is the case with the evergreen pear, *P. kawakanni* Hayata, which is found at only a few' sites on Taiwan and southeast China, where it occurs together with *P. pyrifolia*. The probable interspecific *Pyrus hybrids* and their distribution is given in Table 6.3 and different *Pyrus* species and hybrids originating in different parts of world are tabulated in Tables 6.4–6.6.

Table 6.3: Interspecific Hybrids of *Pyrus*

Species/Parentage		Distribution
Naturally occurring		
P. x *bretschneideri* Rehd.	*P. ussuriensis* x *P. betulifolia*	Northern China
P. x *phaeocarpa* Rehd.	*P. betulifolia* x *P. ussuriensis*	Northern China
P. x *serrulata* Rehd.	*P. pyrifolia* x *P. callen/ana*	Central China
P. x *complexa* Rubtzov	Unknown	Caucasus
P. x *salicifolia* DC	*P. communis* x *P. nivalis*	Europe, Crimea
P. x *canescens* Sparch	*P. nivalis* x *P. salicifolia*	Unknown
Probable arbortea or artificial hybrids		
P. x *lecontei* Rehd.	*P. communis* x *P. pyrifolia*	Unknown
P. x *michauxii* Bose ex Poiret	*P. amygdaliformis* (?) x *P.nivalis*	Unknown
P. x *uyematsuana* Makino	*P. dimorphophylla* x *P. liondoensis*	Korea

Table 6.4: *Pyrus* Species and Hybrids from Asia

Species	Site of Origin	Crop
Pyrus alnifolia (S. and Z.) Franch. and Sav.	Russian Far East, China, Japan, Korea, Taiwan	*
Pyrus armeniacifolia T. T. Yu	China	*
P. aucuparia var. *randaiensis* Hayata	Taiwan	*
Pyrus baccata L.	Russia, Mongolia, China, Korea	*
Pyrus baccata var. *aurantiaca* Regel	Russia, Mongolia, China, Korea	*
Pyrus baccata var. *himalaica* Maxim.	China, Bhutan, India, Nepal	*
Pyrus baccata var. *mandshurica* Maxim.	Russia, China, Japan, Korea	*
Pyrus betulifolia Bunge	China, Laos	*
Pyrus x *bretschneideri* Rehder	China	*
Pyrus calleryana Decne.	China, Korea, Taiwan, Vietnam	USA, Canada
Pyrus calleryana var. *dimorphophylla* (Makino) Koidz.	Japan	
Pyrus calleryana var. *fauriei* (C. K. Schneid.) Rehder	Korea	
Pyrus calleryana var. *koehnei* (C. K. Schneid.) T. T. Yu	China	
Pyrus cathayensis Hemsl.	China	
Pyrus delavayi Franch.	China	
Pyrus discolor Maxim.	China	*
Pyrus doumeri Boiss.	China, Taiwan, Laos, Vietnam	*
Pyrus folgner (C. K. Schneid.) Bean	China	*
Pyrus foliolosa Wall.	Burma, Bhutan, India, Nepal, China	*
Pyrus glabra Boiss.	Iran	*
Pyrus gracilis Siebold and Zucc.	Japan	*
Pyrus harrowiana Balf. f. and W. W. Sm.	China, India, Nepal, Burma	*
Pyrus heterophylla Regel and Schmalh.	Kyrgyzstan, Tajikistan, China	*
Pyrus hondoensis Nakai and Kikuchi	Japan	*
Pyrus x *hopeiensis* T. T. Yu	China	*
Pyrus hupehensis Pamp.	China, Taiwan	*
Pyrus indica Wall.	South Asia and Far East Asia	*
Pyrus japonica Thunb.	Japan	*
Pyrus keissleri (C. K. Schneid.) H. Lev.	China, Myanmar	*
Pyrus kansuensis Batalin	China	*
Pyrus lanata D. Don	Afghanistan, India, Nepal, Pakistan	*
Pyrus matsumurana Makino	Japan	*
Pyrus nussia Buch.-Ham. ex D. Don	Far East, South Asia	*
Pyrus x *phaeocarpa* Rehder	China	*

Species	Site of Origin	Crop
Pyrus pohuashanensis Hance	Russia, China, Korea	*
Pyrus prattii Hemsl.	China	*
Pyrus prunifolia Willd.	China	*
Pyrus pseudopashia T. T. Yu	China	*
Pyrus pyrifolia var. *pyrifolia*	China, Laos, Vietnam	*
Pyrus ringo Wenz.	China, Korea	*
Pyrus ringo var. *kaido* Wenz	China	*
Pyrus scabrifolia Franch.	China	*
Pyrus scalaris (Koehne) Bean	China	*
Pyrus x *serrulata* Rehder	China	*
Pyrus sieboldii Regel	China, Japan	*
Pyrus sikkimensis Hook. f.	China, Bhutan, India	*
Pyrus sinensis var. *maximowicziana* H. Lev.	Korea	*
Pyrus x *sinkiangensis* T. T. Yu	China	*
Pyrus spectabilis Aiton	China	*
Pyrus taiwanensis Iketani and H. Ohashi	Taiwan	*
Pyrus ussuriensis Maxim.	Russia, China, Japan, Korea, Brazil	Brazil
Pyrus x *uyematsuana* Makino	Japan, Korea	*
Pyrus vestita Wall. ex G. Don	China, Bhutan, India, Nepal, Myanmar	*
Pyrus vilmorinii (C. K. Schneid.) Asch. and Graebn.	China	*
Pyrus xerophila T. T. Yu	China	*
Pyrus yunnanensis Franch.	China, Myanmar	*
Pyrus zahlbruckneri (C. K. Schneid.) Cardot	China	*
Pyrus tschonoskii Maxim.	Japan	*
Pyrus tschonoskii Maxim.	Japan	*
Pyrus cydonia L.	Iran, Armenia, Azerbaijan, Russia, Turkmenistan	*
Pyrus germanica (L.) Hook. f.	Middle East and Northern Asia	*
Pyrus korshinskyi Litv.	Afghanistan, Tajikistan, Uzbekistan	*
Pyrus kumaoni Decne.	Middle East, Far East and South Asia	*
Pyrus salicifolia Pall.	Iran, Armenia, Turkey, Arzebaijao	*
Pyrus trilobata (Poir.) DC.	Israel, Lebanon, Turkey, Bulgaria, Greece	*
Pyrus turkestanica Franch.	Kyrgyzstan, Tajikistan, Turkmenistan, Afghanistan	*

Source: Silva *et al*., 2014.

Table 6.5: *Pyrus* Species and Hybrids Originating in Europe and Southern Africa

Species	Geographic Distribution-Site of Origin	Crop
Pyrus aria (L.) Ehrh.	Canary Islands, North Africa, All of Europe	*
Pyrus aria (L.) Ehrh. var. *cretica* Lindl.	North Africa, Middle East, Central Europe Oriental and Southern and Turkmenistan	*
Pyrus aucuparia var. *dulcis* (K) A. and G.	All Europe	North America
Pyrus boissieriana Buhse	Azerbaijan, Turkmenistan, Iran	*
Pyrus korshinskyi Litv. subsp. *bucharica* (Litv.) B. K.	Former Soviet Union	*
Pyrus bulgarica Kuth. and *Sachokia* (*Pyrus* x *nivalis* Jacq.)	Western Europe, Central Eastern and Southern	*
Pyrus caucasica Fed.	Eastern Europe and Central Greece	*
Pyrus chamaemespilus (L.) Ehrh.	Western Europe, Central Eastern and Southern	*
Pyrus communis L.	All Europe	Eastern Europe Central, South and West, and South America
P communis var. *cordata* (Desv.) H.f.	UK, Portugal, Spain, France	*
P communis subsp *gharbiana* (T.) Maire	Algeria, Morocco	*
P communis subsp. *marmorensis* (Trab.) Maire	Morocco	*
P communis subsp. *pyraster* (L.) Ehrh.	Western Europe, Central Eastern and Southern	*
Pyrus x *complexa* Rubtzov	Former Soviet Union	*
Pyrus cossonii Rehder	Algeria	*
Pyrus crataegifolia Savi	Turkey, Albania, Serbia, Greece, Italy, Macedonia	*
Pyrus cuneifolia Guss.	Central Eastern Europe, South and Central	*
Pyrus decipiens Bechst.	All Europe and North Africa	*
Pyrus domestica (L.) Sm.	Algeria, Cyprus, Eastern Europe Central West and Meridional	*
Pyrus elaeagrifolia Pall.	Turkey, Ukraine, Albania, Bulgaria, Greece, Romania	*
Pyrus elaeagrifolia subsp. *kotschyana*	Turkey	*
Pyrus germanica (L.) Hook. f.	Middle East, Eastern Europe, Central, Southern and Northern Asia	*
Pyrus gharbiana Trab.	Algeria, Morocco	*
Pyrus intermedia Ehrh.	All Europe	*
Pyrus malus subsp. *paradisiaca* (L.) Schubl. and G. Martens	Western, Eastern, and Central Europe and Greece	*
Pyrus minima Ley	UK	*

Species	Geographic Distribution-Site of Origin	Crop
Pyrus nebrodensis Guss.	Italy - Sicily	*
Pyrus pinnatifida Ehrh.	All Europe	*
Pyrus praemorsa Guss	South of Italy, France	*
Pyrus sachokiana Kuth.	Georgia	*
Pyrus spinosa Forssk.	Central Eastern Europe, South, and Central	*
Pyrus sudetica Tausch	Western Europe, Central Eastern and Southern	*
Pyrus syriaca Boiss.	Caucasus and Middle East Region	*
Pyrus torminalis (L.) Ehrh.	North Africa, Middle East, South Caucasus, whole Europe	*
Pyrus trilobata (Poir.) DC.	Turkey, Bulgaria, Greece, Israel, Lebanon	*

Source: Silva *et al.*, 2014.

Table 6.6: *Pyrus* Species and Hybrids Originating in the Americas

Species	Place of Origin	Crop
Pyrus americana DC	Greenland, USA, Canada	*
Pyrus angustifolia Aiton	USA, Canada	*
Pyrus arbutifolia (L.) L. f.	USA	*
Pyrus arbutifolia (L.) L. f. var. nigra Willd.	USA	Northern and Eastern Europe Center
Pyrus coronaria L.	Canada, USA	*
P. coronaria var. *ioensis* Alph. Wood	USA	*
Pyrus diversifolia Bong.	USA, Canada	*
Pyrus floribunda Lindl.	USA, Canada	Korea, Russia, Sweden, Czech Republic, Slovakia, Germany, Latvia, Bulgaria
Pyrus fusca (Raf.) C. K. Schneid.	USA, Canada	*
Pyrus sanguinea Pursh	Canada, USA	*

Source: Silva *et al.*, 2014.

Taxonomy

The name *pear*, from the Anglo Saxon *pere* or *peru hu*, is derived from Latin pera or pira; thus, *poire* in French, *peer* in Dutch, *paere* in Danish, *paron* in Swedish and *pera* in Spanish and Italian. The German and Danish name is *Birne* and in Greece as *acras* as wild type and *apios* as cultivated pear with diploid number of chromosomes (2n = 2x = 34) belongs to family *Rosaceae* subfamily Pomoideae and order Rosales.

The Rosaceae family, where the Pyrus gender belongs, has a basic chromosome number as x =17, which is fair if compared with other species of Rosaceae, where x = 7 or x = 9. Of the three hypotheses that emerged from the 1920s to explain the event, the most accepted theory [8] suggests an allotetraploid or allopolyploid from the cross between two primitive forms of Rosaceae family, Prunoideae with $x = 8$ and Spiraeoideae with $x = 9$. This theory was based on the observation of predominance of univalent (unpaired chromosomes) and not from multivalent chromosomes during meiosis. Subsequently, isozyme studies supported this theory [9]. Most cultivated pears are diploid ($2x = 34$), but there are a few polyploid cultivars of *P. communis* and *Pyrus* x *bretschneideri*. According to some authors [9], the speciation of *Pyrus* occurred without a change in chromosome number. It is believed that gender *Pyrus* originated during the Tertiary period (65 to 55 million years ago) in the mountainous regions of western China where a very large number of species of the gender Pomoideae and Prunoideae are concentrated. Chromosome number and ploidy level of different *Pyrus* species is given in Table 6.7.

Table 6.7: Chromosome Number and Ploidy Level in *Pyrus* Species

Species	Chromosome Number	Ploidy Level
Pyrus betulifolia	2n=2x=34	2x
Pyrus callen/ana	2n=2x=34	2x
Pyrus bretschneideri	2n=2x=34,2n=2x=51,2n=4x=68	2x,3x,4x
Pyrus sinkiangensis	2n=2x=34,2n=3x=51,2n=4x=68	2x,3x,4x
Pyrus xerophilus	2n=2x=34	2x
Pyrus serrulata	2n=2x=34	2x •
Pyrus phaeocarpa	2n=2x=34	^~2x
Pyrus pashia	2n=2x=34	2x
Pyrus pyrifolia	2n=2x=34,2n=3x=51	2x,3x
Pyrus ussuriensis	2n=2x=34,2n=3x=51	2x,3x
Pyrus amicniacaefolia	2n=3x=51	3x
Pyrus hopeiensis	2n=2x=34	2x

Source: Shengua and Chengquan, 1994.

Taking into consideration the areas of distribution of the various genres of Pomoideae, it is likely that the common ancestor of these was widely distributed in that territory during the Cretaceous or Paleocene and prior to the Tertiary. Evidence suggests that pear dispersion and speciation followed the mountain ranges to both the east and the west [10, 11]. In this period, only few traces of leaves in some localities from eastern Europe and the Caucasus were found, as the village of Parschlug, Austria, and the Kakhetia mountains, where *Pyrus theobroma* fossils were found. Whereas in eastern Georgia, Horizon Akchagyl, Azerbaijan, and Turkey, *Pyrus communis* L. Fossil leaves were also found. In postglacial records, traces of fruits were found in lacustrine deposits in Switzerland and Italy [12]. It is believed that the process of domestication followed what is currently seen in the Caucasus, where one can find many types of pear trees that grow abundantly [13].

There are two domestication centers and primary origin of the genus *Pyrus*: the first is located in China, the second located in Asia Minor to the Middle East, in the Caucasus mountains, and a third secondary center located in Central Asia [14, 15].

The number of cataloged species varies greatly according to the interpretation of each author, 20 to 75 species [16]. There are 23 wild species cataloged, all native to Europe, temperate Asia, and northern mountainous regions of Africa [7, 17, 18]. Pears are classified into three groups according to the number of carpels and fruit size: small fruits that have two carpels known as Asian pears, large fruits with five carpels, and fruits with three to four carpels that are hybrids of fruits mentioned above. Asian pears have a crisp texture, while the European pear has a buttery and juicy texture, with characteristic flavor and aroma.

The total number of species included in the genus is not definite; however, Zeven and Zhukovsky (1975) mentioned about 60 species. Challice and Westwood (1973) made detailed taxonomic studies of genus *Pyrus* on the basis of both chemical and botanical characters and categorized 22 primary species. Beside these, the rest are non-primary species may be botanical varieties, subspecies or interspecific hybrids (Westwood, 1982; Bell and Hough, 1986). These twenty two primary species exist in all temperate regions in more than 50 countries (Bell *et al.*, 1996). Some of the primary species of Bell and Hough (1986) differ from that of Challice and Westwood (1973). *P. pashia* and *P. kawakamii* were also considered as primary species by Challice and Westwood (1973). These 22 primary species were further classified on the basis of geographical distribution (Bell and Hough, 1986) as given in Table 7.8.

Table 7.8: Groups of *Pyrus* Species

Group	Species
European Group	*Pyrus communis, P. cordata, P. caucasia, P. nivalis.*
North Africa Group	*P. syriaca, P. gharbiana, P. memorensis.*
West Asian Group	*P. syriaca, P. elaeagrifolia, P. amygdaliformis, P. salicifolia, P. glabra, P. regelii.*
East Asian Group	*P. pyrifolia, P. kanseunsis, P. ussuriensis, P. hondoensis.*
Asian Group	*P. calleryana, P. betulifolia, P. faurici, P. dimorphophylla, P. koehnei.*

Botany

Morphological characters of *Pyrus* species is given in Table 6.9 and described below:

Tree Growth Habit

All species of the genus Pyrus are deciduous. European pear (*P. communis*) trees are pyramidal in shape with medium tall size (Janick, 1977 and 2006), while Japanese (*P. pyrifolia*) and Chinese pear (*P. bretschneideri* Rehd. and *P. ussuriensis* Maxim) are tall, vigorous and spreading in nature. On the other hand, *P. pashia* is comparatively medium in size with open-headed tree. However, tree height is generally influenced by soil fertility, cultural practices such as pruning and type of rootstock.

Table 6.9: Some Morphological Characters of *Pyrus* Species

Species	Tree Size (m)	Fully Grown Leaf Margins	Growing Leaves	Leaf Length (cm)	Peel Surface	Calyx Type	Fruit Diameter (cm)	Carpel No.
Pyrus caucasia	5-6	Crenate	Simple	4.81	Smooth	Persistent	4.70	5
Pyrus communis	5-6	Crenate	Simple	4.95	Smooth	Persistent	3.80	5
Pyrus cordata	2-3	Crenate	Simple	7.87	Russet		1.50	2-3
Pyrus nivalis	3-4	Crenate	Simple	4.87	Smooth	Persistent	5.60	5
Pyrus amygdaliformis	1-2	Entire or slight crenate	1-lobed	6.52	Smooth	Persistent	2.60	
Pyrus elaeagrifolia	3-4	Entire or slight crenate	Simple	570	Smooth	Persistent	2.40	5
Pyrus gharbiana	3-4	Crenate	Simple	5.15				
Pyrus longipes	3-4	Crenate	Simple	3.77	Russet	Deciduous (some persistent)	1.70	3.4
Pyrus mamorensis	-	Crenate	Simple	4.58	-		-	-
Pyrus syriaca	1-2	Crenate	Simple	4.78	-	Persistent	-	5
Pyrus pasltia	3-4	Crenate	2-lobed	7.99	Russet	Deciduous	2.39	4-5
Pyrus regelii	1-2	Crenate	Lanciniate	580	Smooth	Persistent	2.0-3.0	5
Pyrus salicifolia	1-2	Entire	Simple	5.10	Smooth	Persistent	1.90	5
Pyrus bctulifolia	5-6	Coarse serrate	1-lobed	5.74	Russet	Deciduous	0.88	2 (3)
Pyrus calleryana	3-5	Crenate	Simple	660	Russet	Deciduous	1.06	2
Pyrus dimorphophylla	3-1	Crenate serrate	2-lobed	7 16	Russet	Deciduous	1.24	2
Pyrus fauriei	1-2	Crenate	Simple	4.35	Russet	Deciduous	1.34	2
Pyrus liondoensis	3-4	Fine serrate, setose	Simple	7.57	Smooth	Persistent	2.80	5
Pyrus pseudopashia	3-4	Crenate	1-lobed	6.42	Smooth	Persistent	4.26	5
Pyrus kawakamii	1-3	Crenate	3-lobed	6.14	Smooth	Persistent	1.04	4 (3)
Pyrus pyrifolia	3-5	Fine serrate, setose	Simple	10.65	Russet	Deciduous	4.10	5
Pyrus ussuriensis	1-3	Coarse serrate setose	Simple	8.23	Smooth	Persistent	3.75	5

Leaf

The leaves are alternatively arranged, simple, 2-12 cm long, glossy green of some species densely silvery-hair in others. Leaf shape varies from broad oval to narrow lanceolate, orbicular ovate to elliptic or crenate-serrated (Paganova, 2003). The leaves of *P. communis* are orbicular ovate to elliptic, crenate-serrated and glabrous when old but new growth and inflorescence is pubescent. *P. pyrifolia* leaves are ovate-oblong, dark green, pubescent, elliptical and comparatively large in size. *P. pashia* have crenate, ovate to lanceolate leaves (Parmar and Kaushal, 1982).

Flower

P. communis has white flowers mostly born in corymbs. *P. pyrifolia* has white flowers that appear prior to the emergence of leaves. On the other hand, *P. pashia* flowers are white and fragrant with cross-compatible (self incompatible) (Parmar and Kaushal, 1982). Generally, pear species has perfect flowers on old spur, inflorescence with 6-8 flowers in umbel-like racemes, petals white or rarely pink, spread sepals, 20-30 pink, 6 red or purple anthers with 2 to 5 free style and 2 ovules per locule. The flowers of European pear are white rarely tinted yellow or pink, 2-4 cm diameter and have five petals and size and colour also resembled with apple flowers (Garriz *et al.*, 1998).

Fruit

The morphological characteristics of fruit in *Pyrus* genotypes are variable with respect to genetics and environments. Generally, all species have pome fruits along with variable fruit shape like oblong, bulbous or pyriform, round and pear shaped with persistent or deciduous calyx. Kajiura and Suzuki (1980) and Paganova (2003) described variability in fruit shapes as round or apple shaped and ovate to oblong in various pear genotypes. Fruit size is generally 0.5 20 cm long with or without grit cell. Ground colour changes from green to yellow or red during maturation. Browning russet is mainly due to increase in humidity. Ripening season is generally from June to November. Precisely, *P. communis* fruits vary in shape mostly oblate/ bulbous and pyriform, calyx persistent, fleshy pedicels and pulp with gritty. The pulp is melting and buttery in texture. *P. pyrifolia* fruits are round and *P. pashia* with small, dark brown fruits, which become soft and sweet to some extent when ripened but gritty and astringent with poor quality (Parmar and Kaushal, 1982). This species is commonly used as a rootstock for commercial varieties in India and Pakistan.

References

Bell, R.L. 1990. Pears (*Pyrus*). (*In*) Moore, J.N. and Ballington, J.R. Jr. (Eds.). Genetic Resources of Temperate Fruit and Nut Crops - I. International Society for Horticultural Science, Wageningen, The Netherlands, pp. 655-697.

Bell, R.L. 1996. Pears. (*In*) Fruit Breeding. J. Janick and J. N. Moore (Eds.), Tree and Tropical Fruits, pp. 441–514, John Wiley and Sons.

Bell, R.L. and Hough, L.F. 1986. Interspecific and intergeneric hybridization of Pyrus. *HortScience* 21: 62-64.

Bell, R.L., Quamme, H.A., Layne, R.E.C. and Skirvin, R.N. 1996. Pears. (In) Janick, J. and Moore, J.N. (Eds.). Fruit Breeding, Vol. 1: Tree and Tropical Fruits. John Wiley and Sons, New York, pp. 441-514.

Challice, J.S. and Westwood, M.N. 1973. Numerical and taxonomical studies of genus Pyrus using both chemicals and botanical characters. *Botanical Journal of Linnaeus Society* 67: 121-148.

Emboden, W.A. 1987. Leonardo da Vinci on plants and gardens. Dioscorides Press, Portland Oregon. p. 182.

Erhardt, W.; Gotz, E.; Bodeker, N. and Seybold, S. Zander. 2002. Handworterbuch der Pflanzennamen 1, Eugen Ulmer Verlag, Stuttgart, Germany.

FAO. 2015. www.fao.org.in

Fideghelli, C. 2007. Origine ed evoluzione," inIl Pero, R. Angelini, (Eds 1), Bayer/ CropScience, Milano, Italy.

Garriz, P.I., Colavita, G.M. and Alvarez, H.L. 1998. Fruit and spur leaf growth and quality as influenced by low irradiance levels in pear. *Scientia Horticulturae* 77: 195-205.

Hedrick, U.P. 1921. The pears of New York. J. B. Lyon, Albany, New York.

Hedrick, U.P., Howe, G.H., Taylor, O.M., Francis, E.H. and Tukey, H.B. 1921. The pears of New York. 29th Annual Report, Department of Agriculture. JB Lyon Co. Printers, Albany, New York.

Janick, J. 1977. 'Honeysweet' Pear. *HortScience* 12: 357.

Janick, J. 2006. 'H2-169' (Ambrosia™) Pear. *HortScience* 41: 467.

Kajiura, I. and Suzuki, S. 1980. Variations in fruit shapes of Japanese pear cultivars: Geographic differentiations and changes. *Japanese Journal of Breeding* 30: 309-328.

Kikuchi, A. 1948. Speciation and taxonomy of Chinese pears. Collected Records of Horticultural Research, Faculty ofAgriculture, Kyoto University, 3: 1-11.

Knight, R.L. 1963. Abstract bibliography of fruit breeding and genetics to 1960. *Malus* and *Pyrus*. Tech. Comm No. 29 Commonwealth Bureau of Horticultural and Plantation Crops. Commonwealth Agriculture Bureau, East Malling.

Layne, R.E.C. and Quamme, H.A. 1975. Pears. (*In*) Advances in Fruit Breeding. J. Janick and J.N. Moore. (Eds.). pp. 38–70. Purdue University Press, West Lafayette, Ind, USA

Lee, S.H. 1948. A taxonomical survey of the oriental pears. *Proceedings of the American Society for Horticultural Science* 51: 152-156.

Lombard, P.B. and Westwood, M.N. 1987. Pear rootstocks, pp: 145-183. (*In*) Rootstocks for Fruit Crops. Rom, R.C. and Carlson, R.F. (Eds.). John Wiley and Sons, New York, USA

Paganova, V. 2003. Taxonomical reliability of leaf and fruit morphological characteristic of the Pyrus L. taxa in Slovakia. *HortScience* 3: 98-107.

Parmar, C. and Kaushal, M.K. 1982. *Pyrus pashia* Buch. Kalyani Publishers, New Dehli, India, pp. 78-80

Rehder, A. 1986. Manual of Cultivated Trees and Shrubs, 2nd edn. Dioscorides Press, Portland, pp. 401-406.

Rubtsov, G.A. 1944. Geographical distribution of the genus Pyrus: Trends and factors in its evaluation. *American Nature* 78: 358-366.

Sax, K. 1931. The origin and relationships of the pomoideae. Journal of the Arnold Arboretum 12: .

Shimura, I. 1988. Nashi (Pear). (*In*) Heibonsha's World Encyclopedia, 36 [in Japanese]. Heibonsha, Tokyo, pp. 354–372.

Teng, Y. and Tanabe, K. 2004. Reconsideration on the origin of cultivated pears native to East Asia. *Acta Horticulturae* 634: 175–182.

Teng, Y., Tanabe, K., Tamura, F. and Itai, A. 2002. Genetic relationships of Pyrus species and cultivars native to East Asia revealed by randomly amplified polymorphic DNA markers. *Journal of the American Society for Horticultural Science* 127: 262–270.

Vavilov, N.I. 1951. The Origin, Variation, Immunity and Breeding of Cultivated Plants. Ronald Press, New York.

Vavilov, N.I. 1992. Origin and Geography of Cultivated Plants 15, Cambridge University Press, Cambridge, UK.

Watkins, R. 1976. Cherry, plum, peach, apricot and almond. (*In*) Simmonds, N.W. (Eds.). Evolution of Crop Plants. Longman, London, pp. 242–247.

Weeden, N. and Lamb, R.C. 1987. Genetics and linkage analysis of 19 isozyme loci in apple. *Journal of the American Society of Horticultural Science* 112: 865–872.

Westwood, M.N. 1982. Pear germplasm of the new national clonal repository: its evaluation and uses. *Acta Horticulturae* 124: 57–65.

Zagaja, S.W. 1970. Temperate zone fruits. (*In*) Frankel, O.H. and Bennet, E. (Eds.). Genetic Resources in Plants: Their Exploration and Conservation. Blackwell Scientific Publications, Oxford, pp. 327-333.

Zagaja, S.W. 1977. Fruits of North East China. *Fruit Science Report* 4: 1-8.

Zeven, A.C. and Zhukovsky, P.M. 1975. Dictionary of cultivated plants and their centres of diversity. Wageningen: Centre for Agricultural Publishing and Documentation.

Zielinski, Q. B. and M. M. Thompson, M.M. 1967. Speciation in *Pyrus*: chromosome number and meiotic behavior. Botanical Gazette 128 pp. 109–112.

Zukovskij, P.M. 1962. Cultivated Plants and Their Wild Relatives, Farnham Royal, London, UK.

Chapter 7

Cluster Bean

Vikas Kumar, R.B. Ram and Chhatarpal Singh

*Department of Applied Sciences (Horticulture) and
Department of Environmental Microbiology,
Babasaheb Bhimrao Ambedkar University, Lucknow – 226 025, U.P.
E-mail: vs1744@gmail.com*

Among all the beans grown in India, cluster bean (guar) [*Cyamopsis tetragonoloba* (L.) Taub.] is cultivated to a lesser extent. The research work done on different aspects of this crop is also meager. It is mainly grown for green vegetable, dry seeds and also as a forage crop. Cluster bean has several health benefits in both vegetable and powder form (guar gum). Guar gum has also been shown to be useful in weight loss and diabetes treatment. Guar pods are rich in soluble dietary fibre and lowers blood cholesterol levels. Guar gum is a common ingredient in fibre-rich drinks marketed as health drinks and weight-loss drinks. Low in calories, the cluster bean (guar) contains vitamin C, vitamin K, vitamin A, dietary fibre, folate, iron, manganese and potassium. The vitamin K is important for maintaining strong bones and proper development of foetus. The nutritional values per 100 gm of raw cluster bean is one of the most important and potential vegetable cum industrial crop grown for its tender pods for vegetable purpose and for endospermic gum (30-35 per cent) (Kumar and Singh, 2002). The tender pods are used as vegetable and in the southern parts of India they are dehydrated and stored for use. It is a nutritious fodder for livestock and the seeds are also fed to the cattle. Besides, the crop can be used for soil improvement and as a medicine. The mucilaginous seed flour is valued as guar gum (galactomanan) and this gum is used in textile, paper, cosmetic and oil industries throughout the world and is a useful absorbent for explosive (Smith, 1976).

Guar is basically a crop that is cultivated mostly in the arid and semi arid areas as it is drought resistant. That is why the Southern Asian continent suits well to the cultivation of this crop especially the Indian subcontinent. The powder made after refining the gum obtained from the plant makes an important raw material in many industries. This powder has some unique characteristics like grease resistance, thickening agent, capacity to bind water, high viscosity and the capability to function in low temperatures which makes it a highly popular in those sectors. Among other by products of guar, guar gum powder is the main marketable commodity. The world's total production of cluster bean is around 7.50 to 10.00 million tons at every year. The production list of guar is dominated by India as a leading producer of this crop. The consumption pattern of guar seeds is largely influenced by the demands from the petroleum industry of United States of America and the oil fields in the Middle East as the derivative products of these seeds are quite useful in the petroleum drilling industries. United States alone constitute to around 40 thousand tons of guar and its derivatives demand. Also, in rest of the world, the trend of consumption has increased with time that has lead to the introduction of this crop in many countries. The Major guar producing countries in the world are India, Pakistan, Sudan, USA, South Africa, Brazil, Malawi, Zaire and Australia. India leads the list of the major guar producing countries of the world contributing to around 75 to 80 per cent in the world's total production of around 7.5 to 10 million tons. Pakistan follows India in the list with 10-15 per cent share in the world's total produce. Thus, the total area of cluster bean in India is 4.25 million hectare and production is 2.41 million tones with productivity of 0.57 million tones/ha in 2015. The most important by-product of this crop *i.e.* guar gum is obtained through the processing of endosperm of the seeds 5 of guar. This product is vastly produced in the countries such as USA, Germany, China, Italy, South Africa, and United Kingdom though these countries are not really indulged in the production of guar as a crop. (Anonymous, 2012). The other major producers of Guar are Pakistan, USA, South Africa, Malawi, Zaire and Sudan (Yadav *et al.*, 2014). The area under cultivation and the level of production has an increasing trend for guar crop in India. It is expected that the area under cultivation of guar in India in the year 2015 will be around 36 lakh 6 hectares and the production will reach to around 17 to 18 lakh metric tones (Yadav *et al.*, 2014).

History

Cluster bean is considered to be originated by domestication of the African wild species, *C. senegalensis* which appears to be the ancestor of the *C. tetragonoloba*. The domestication process could have been taken place in the dry areas of the north western region of the Indo-Pakistan Subcontinent (Hymowitz, 1972). It was cultivated as a minor crop in India during ancient times as a vegetable and feed for cattle. Guar was introduced in USA in 1903 for experimentation in the southwest region where the climate is hot and has long growing seasons to suit its adaptation (Hymowitz and Matlock, 1963). The objective was to use cluster bean as a soil improving legume and forage for cattle. Before World War II the carob (locust bean) seed (*Ceratonia siliqua*) from Mediterranean was used to extract carob gum for extensive use in paper industry. During World War II the supply of imported carob seed from the Mediterranean region was cut off, as a result search for domestic source of galactomanan gum was initiated in USA by Institute of Paper Chemistry.

This study revealed guar as an alternative source for galactomanan (Anderson, 1949). Further, studies were done on milling the guar seeds for gum production and application in the manufacturing of paper revealed the beneficial effects of guar gum on paper processing. This information helped in the adoption of guar gum in the different manufacturing process.

Origin and Distribution

Guar is native to India where it is grown principally for its green fodder and for the pods that are used for food and feed. Guar was introduced into the United States from India in 1903. Production in the United States in centered in Texas, Oklahoma and Arizona, but it is also adapted to locations with more tropical climates, such as in Florida and Puerto Rico (Stephens, 1994). The crop is grown in India, Myanmar, Sri Lanka and Pakistan. Even in arid zones of the USA like Texas and Arizona it is being grown (Venkataratnam, 1973). Cluster bean has been grown in India since ancient time for vegetable, manure and fodder purposes. The presence of a number of wild relatives of cluster bean in Africa suggests that it was most probably originated in Africa (Gillette, 1958). It is possible that cluster bean was domesticated very early in the Africa and Arabia and made its way to Indo-Pakistan subcontinent. On the other hand, Whistler and Hymowitz (1979) mentioned that the name of cultigen in Arabic *hindia* suggests it to be an Indian origin. As it is well known that horses were the major trade between the Arabs and Indians, there is a possibility that Arabs have boarded their ships in large quantities

of fodder of cluster bean to feed their horses. The plants of *C. senegalensis* probably were cut and carried along the ship as fodder. Since the climatic conditions in the Indo-Pakistan subcontinent were favorable to *C. senegalensis*, seeds get germinated and became the basis of cluster bean. Hypotheses on the origin of the cluster bean appear quite speculative. Except for taxonomic study, no detailed molecular or genetical studies are accessible in the literatures to authenticate the claim or prove the hypotheses. Chavalier (1939) postulated that *C. senegalensis* probably extended up to Sindh where after domestication a few of its cultigens became cultivated in India whereas Vavilov (1951) suggested that India is the geographical centre of cluster bean variability. Dabas and Thomas (1986) indicated that the cluster bean perhaps has been domesticated in the western Rajasthan. Hymowitz (1972) believed that the African wild species *C. senegalensis* appeared to be the ancestor of West African *C. tetragonoloba*.

Guar is mainly grown in India, Pakistan, United States, and recently in China. It is native to the Indian subcontinent and crop is mainly grown in the dry habitats of Rajasthan, Haryana, Gujarat and Punjab (Anonymous, 2006; Whistler and Hymowitz, 1979; Strickland and Ford, 1984 and Sultan *et al.*, 2013). In addition to its major cultivation in India, guar is regarded as a cash crop in the southwest part of the United State, especially in Texas and Okladoma grown to limited extent in other parts of the world like Australia, Brazil and South Africa (Sultan *et al.*, 2013; Liu *et al.*, 2009 and Khare, 2004). Guar does not exist in a wild state and is believed to have originated from an African species imported to India as horse fodder by Arabian traders (Khare, 2004). It was turned into a gum-producing crop during the Second World War in the United States.

Classification

The name guar is believed to have arisen from the Sanskrit words *gau* and *ahaar*, which mean cow and fodder respectively. In the earlier literature, *C. tetragonoloba* was known as *Dolichos fabaeformis* or *C. psoralioides* (Stephens, 1998). It belongs to the tribe Galegae of the family fabaceae; which has three species, of which *C. tetragonaloba* is the only economically important one. The haploid chromosome number of guar is 7 (Patil, 2004).

Genus *Cyamopsis* and Species

Genus *Cyamopsis* is an old word genus belonging to the tribe Galegeae of *Paplionaceae*. There are four major species: (Singh *et al.*, 2014).

1. *Cyamopsis tetragonoloba*
2. *Cyamopsis senegalensis*
3. *Cyamopsis serrata*
4. *Cyamopsis dentata*

Cyamopsis was emphasized to be a separate genus with Africa as its probable center of origin (Gillett, 1958). Vavilov (1951) suggested India to be the center of variability for cluster bean. Trans-domestication process was proposed to explain the origin of cluster bean (Hymowitz, 1972) and it is reported that the cultivated cluster bean plant, *C. tetragonoloba*, developed from a drought tolerant wild African species *C.* senegalensis (Mudgil *et al.*, 2014).

Kingdom: Plantae

Division: Magnoliophyta

Class: Magnoliopsida

Order: Fabales

Family: Leguminosae

Tribe: Indigofereae

Genus: Cyamopsis

Species: *C. tetragonoloba*

Botanical Name: *Cyamopsis tetragonolobus* (L.)

Synonyms: *Cyamopsis psoralioides* L.

PartUsed: Seeds

Vernacular Name: Guar

The genus *Cyamopsis* distributed in tropical Africa and Asia as main center, with its largest diversity cultivars in India (Kirtikar and Basu, 1918 and Anonymous, 2006). It is widely cultivated in countries like India, Pakistan, United, State of America, Italy, Morocco, Germany and Spain and is thus considered as a new crop for western agricultural practices. Early taxonomy divided the genus into three species: *C. tetragonoloba; C. senegalensis; C. serrata* excluding *C. dentate* (Anonymous, 2006, Whistler and Hymowitz, 1979), the intermediate type between *C. senegalensis* and *C. serrata*. *C. senegalensis* is regarded as the ancestral form of guar, and has similar gum concentration, composition and viscosity characteristics (Strickland and Ford, 1984). *C. senegalensis* and *C. serrata* are quite different from *C. tetragonoloba* (L.) Taub. Both of the former species are shorter in height (about 30 cm) with smaller leaves and pods, compared with *C. tetragonoloba* (L.) Taub. Seeds of *C. senegalensis* and *C. serrata* are smaller, short and cylinder shaped when mature than *C. tetragonoloba* round shaped. Pod shattering was observed in *C. senegalensis* and *C. serrata* when mature, but not in *C. tetragonoloba*. These morphological differences can be utilized to differentiate the genetic relationships among guar cultivars. Evan if varieties are morphological distinct from each other, differences at the molecular level are not clear (Sultan *et al.*, 2013).

Ecological Adaptation

Guar is a hardy, drought-tolerant legume. It grows in a wide range of environments from the sub humid to semi-arid conditions in the tropics and subtropics with (300-)500-800 (-1500) mm of rainfall. The main production of guar for seed occurs where annual rainfall is less than 800 mm. In areas with higher rainfall, vegetative growth is greater, but seed quality is inferior, making guar more suitable as a green manure and fodder crop. Guar prefers a very hot climate. Mean monthly maxima in northern India may reach 35-40 °C, though in southern India extremes are lower. Optimum soil temperature for root development is 25-30 °C. Guar is cultivated up to 900 m altitude. It is highly susceptible to frost. The optimum temperature for germination is about 30 °C. At 20 °C, germination is retarded, at still lower temperatures the rate of germination is reduced. It can grow in most soils, but thrives in well-drained alluvial and sandy-loam soils of pH 7.0-8.0. Water logging is not tolerated. On heavy soils, guar should be grown on ridges to maintain root aeration. In an experiment using irrigation water with equal amounts of NaCl and $CaCl_2$, salinity levels up to 8.8 dS/m did not affect germination, early growth or grain yields.

Cyamopsis tetragonoloba (L.) can grow on a wide range of different soil types. Preferably in fertile, medium textured and sandy loam soils that are well drained because water logging decreases plant performance. In respect of soil acidity, guar grows best in moderate alkaline conditions (pH 7-8) and is tolerant of salinity. Thanks to its taproots which are inoculated with rhizobia nodules, it produces nitrogen rich biomass and improves soil quality (Undersander *et al.*, 1991).

Morphology

The Guar grows up right, reaching a maximum height of up to 5 ft in this arid zone of Rajasthan, Bikaner. It has a main stem with either basal branching or fine branching along with the stem. The "Taproots" of the guar plants can access soil moisture in low soil depth up to 1.5ft.

Additionally this legume develops "root nodules" with "nitrogen – fixing soil bacteria rhizobia" in the surface part of rooting system; the "Leaves" of Guar has an elongated oval shape having maximum 5cm in length in this area and of alternative position. The developing pods on plants axial are of bluish color containing 5 to 12 small oval seeds of 5mm length (TGW = 25 – 40g). The seeds of guar beans have a remarkable characteristic. Its kernel consists of a protein rich germ and a relatively large endosperm containing big amount of galactomanan which exhibits a great hydrogen bonding capacity having viscosifying effects in liquids.

Guar is an annual herb, which grows up to 60 cm high. Guar is a predominantly self pollinated crop of the Fabaceae (Bhosle and Kothekar, 2010).

Flower

Flowers are axillary, 6-30 flowered racemes. The structure of flowers is arranged in fives. The sepals use fused and hairy on the outside. The lower calyx teeth are longer than the upper ones. The corolla is butterfly shaped (flag, 2 wings, keel formed from 2 fused petals), small and reddish. There are 10 stamens (Anonymous, 2006).

Fruit

Developing from a carpel is an upright, 3.8-5 cm long, sparsely haired legume with 5-6 seeds, these have a very well developed, slimy endosperm (Anonymous, 2006, Bhosle and Kothekar, 2010).

Leaves

Alternate, trifoliate and leaflets are broad elliptical acuminate, dentate, pubescent on both surfaces. They measure 3.8- 7.5 cm long and 1.2-5 cm wide. The petiole is 2.5-3.8 cm long while stipules are 6-10 mm long (Whistler and Hymowitz, 1979, Strickland and Ford, 1984).

Root

The root and root tuber have symbiotic bacteria, which bonds nitrogen from the air (Khare, 2004).

Cyamopsis Collections in Gene Banks

The ICAR-National Bureau of Plant Genetic Resources, New Delhi, India and the Agricultural Research Institute, Lyallpur, Pakistan have representative collections of guar accessions. In the United States, the Texas Agricultural Experiment Station, Vernon, Texas and the Oklahoma State University, Stillwater, Oklahoma maintain collections. In addition to these actively used collections, the entire plant collection of 1300 accessions of 33 cultivars and local forms are stored at the National Seed Storage Laboratory, Fort Collins, Colorado. Germplasm is a basic tool for any crop improvement programme. Natural variation represents a huge and largely untapped resource, which has been subjected to selection over millions of years of evolution, with both basic and practical value as well as the potential to break yield barriers of agricultural plants (Johal *et al.,* 2008; Tanksley and McCouch, 1997; Zamir, 2001). The variation available in the germplasm is utilized as a source of useful genes to improve the cultivars. As the importance of guar was realizing 1950's germplasm collection was initiated. The collection initially began from Maharashtra for vegetable varieties. Later the Pant lntroduction Division of the ICAR-Indian Agricultural Research Institute (IARI), New Delhi continued the collection and maintenance work. There are about 4,901 accessions in ICAR-National Bureau of Plant Genetic Resources (NBPGR), New Delhi. These accessions have been catalogued based on the accession numbers and characterized for phenotypic traits like pubescence, days to 50 per cent flowering, days to 50 per cent maturity, plant height, branch number, total number of pods per plant, number of seeds per pod, seed yield per plant, seed color, gum content and disease resistance under field conditions (Dabas, 2001). These studies have identified the accessions that can be donors to traits like dwarfing, branching/unbranched, pod length, pods per plant, seed size, days to maturity, gum content and disease resistance. Classical approaches using the donor cultivars as sources have lead to the development of certain elite cultivars for cultivation in the previous decade. The development of elite cultivars and their widespread use in breeding programmes have reduced the utilization of available genetic resources. It has been estimated that for most crop species, less than 5 per cent of the biodiversity known to exist has been utilized in agriculture, particularly in the case of self pollinated crops (Tanksley and McCouch, 1997). Much of the diversity present in living systems is probably adaptive (Johal *et al.,* 2008).

The assessment of genetic diversity using quantitative traits has been of prime importance in many contexts particularly in differentiating well defined populations. The germplasm in a self pollinated crop can be considered as heterogeneous sets of groups, since each group being homozygous within it. Selecting the parents for breeding program in such crops is critical because, the success of such programme

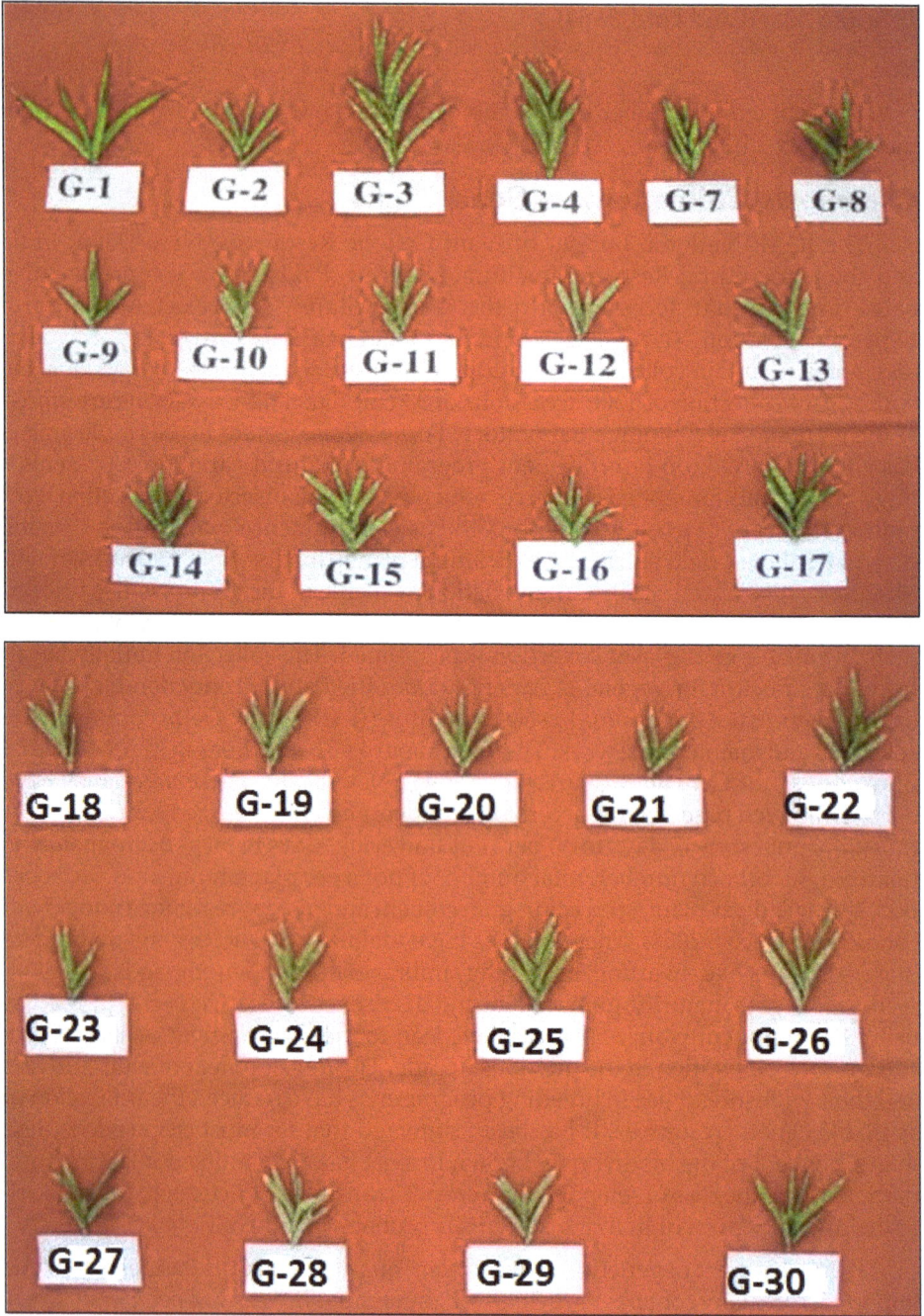

Plate 7.1: Variability in Cluster Characters of 30 Genotypes in Cluster Bean Observed under Study.

Plate 7.2: Variability in Pod Characters of 30 Genotypes in Cluster Bean Observed under Study.

depends upon the segregants of hybrid derivatives between the parents, particularly when the aim is to improve the quantitative characters like yield. To help the breeder in the process to identify the parents that nicks better, several methods of divergence analysis based on quantitative traits have been proposed to suit various objectives. Among them, Mahalanobis' generalized distance occupies a unique place and an efficient method to gauge the extent of diversity among genotypes, which quantify the difference among several quantitative traits. The ICAR-National Bureau of Plant Genetic Resources (ICAR-NBPGR), New Delhi and the 'Central Arid Zone Research Institute', Jodhpur, have evaluated 4869 germplasm lines and identified early maturing and high yielding genotypes (ex. IC 116804, IC 116868, IC 116869 HGI 1/P8-80 and others; Dabas *et al.,* 1982). Collection and evaluation of indigenous (731) and exotic lines (20) at Hisar center yielded lines maturing in <90 days1. Large numbers of germplasm accessions were put under medium term storage (Mishra *et al.,* 2009) and also for *ex situ* conservation. At S K Nagar, Gujarat several promising lines with enhanced disease resistance against bacterial leaf blight (GAUG 9406, GG 1, RGC 1027), *Alternaria* blight (GAUG 9406, GAUG 9005, GAUG 9003 and GC 1) and root rot (GAUG 9406, GG 1 and HGS 844) have been identified (Kumar, 2005). Some cluster bean lines have been released for grain (ex. Suvidha, Naveen, Sona, PLG-85, PLG-850, RGC-471) and vegetable (Pusa Navbahar, Pusa Sadabahar and Sharad Bahar) purpose (Kumar and Rodge, 2012). Further, two wild species *viz., C. serrata* and *C. senegalensis* have been introduced from USA. Efforts for transferring earliness trait from these two wild species into the cultivated species (*C. tetragonoloba*) using conventional methods were not successful (Kumar and Rodge, 2012).

Breeding Nature

Selection and breeding in guar in the United States aims at increased seed production and disease resistance, while in India cultivars have been selected for seed, vegetable, and multipurpose use. The American cultivars are derived from a very small number of introductions from India, leaving the genetic variability largely unutilized. An improved technique for controlled pollination of guar has been developed in India. Male sterility has been found, with pollen fertility probably being monogenically dominant over sterility. Recent research has concentrated heavily on increasing grain yield. This was justified by the strong and growing market for guar gum and the increasing number of its applications. Development of determined cultivars with a compressed flowering phase and improved tolerance of temporary water logging have been suggested to increase its adaptability. Despite the crop's good prospects, research on guar all but stopped in the United States around 1985.

Multiple approaches are being attempted in cluster bean to identify better lines suited for drought conditions: (i) germplasm collection from dry regions (ii) summer screening to identify tolerant genotypes (iii) crossing high yielders with drought tolerant lines and (iv) hybridization between two early lines. Cluster bean varieties maturing in about 80-90 days are now available (HG-365, RGC-936, RGC-563) (Dass *et al.,* 1973). Gum content has a positive correlation with seed yield (Lal and Gupta, 1977; Jhorar *et al.,* 1988) and negative correlation with seed weight. Gum content and endosperm showed negative correlation with protein content (Singh *et al.,* 1995). Both additive and non-additive gene effects determine the gum content in cluster bean (ICAR-AICRPAL, 2013). There is a need to isolate entries with high gum content (>32 per cent) and higher viscosity profile (4000 cP) (Kumar, 2005). ICAR-NBPGR has evaluated and identified lines with high gum content: 34 lines (>40 per cent), 98 lines (30-39.9 per cent) and 31 (19.2-29.9 per cent) (Kumar, 2005), similarly HGS 870 showed high gum content (31.78 per cent) and higher viscosity profile (5116 cP) (Gandhi *et al.*, 1978). Bacterial blight, alternaria leaf spot and powdery mildew are the important diseases that threaten guar production in several guar growing regions. Tolerant varieties with inbuilt resistance are the best option for combating these diseases thereby preventing heavy economic losses. Lodha (1984) reported that 'GAUG 63' had some tolerance to powdery mildew whereas HG75, G85, G102 and G225 were sources of bacterial leaf blight resistance (Kumar, 2005; Sharma *et al.*, 1999). Resistance to blight and leaf spot were reported to be dominant and recessive respectively (Singh *et al.*, 1995; Vig, 1965). Further studies indicated the presence of additive, dominance and epistatic interactions governing the resistance of these two diseases. Phenol and peroxidase enzymes have been reported to confer resistance against these diseases. Induced mutations through physical mutagens in cluster bean were used for the first time in India (Vig, 1969; Singh, 1972). Germination percentage, seedling survival and pollen fertility were greatly affected by increasing concentration of gamma rays (10-200 kR) (Chaudhary *et al.*, 1973). But, Chaudhary *et al.* (1973) reported an increase in yield, protein and gum contents in M_2 population irradiated with low doses (2-20kR) of gamma rays. High pods per plant and seed yield were observed in M_3 progenies of RGC 197, exposed to different doses (10-80 kR) of gamma rays (Yadav *et al.*, 2004). Literature indicates that lower doses of different mutagens single or in combination induce higher variability than compared to higher doses (Kumar *et al.*, 2013). An unsuccessful attempt was made to transfer earliness from wild cluster bean (*C. serrata*) into cultivated species (*C. tetragonoloba*) by Sandhu (1988). Several other approaches including stigma amputation, organic solvents and bud pollination could not overcome the stigmatic incompatibility barrier. Techniques like protoplast hybridization, ovary/embryo rescue, and or recombinant DNA technology could be used in transferring agronomically important genes from wild species. Markers (RAPDs, SSRs, AFLPs and ISSR) are powerful tools to analyze the genetic diversity and relationships. RAPD markers have been utilized in cluster bean to a large extent to analyze the genetic diversity of varieties or cultivars (Punia *et al.*, 2009 and Pathak *et al.*, 2010). Crossing leads to the hybridization of DNA from two plants belonging to the same genus with a different genetic makeup, but it becomes difficult with small flowers and results in less-consistent pollen production. In guar the flower is only 8 mm long and requires a magnifying lens for emasculation. Once the anthesis begins, ten anthers can be seen encircling the stigma. The pollen is viable from 2 hours before anthesis up to 11 hours after anthesis (Stafford, 1982). The flower morphology leads to self-pollination although an out crossing rate of 9 per cent has been observed (Gill, 2009).

Useful Cyamopsis/Cultivars

There are several varieties distinguished by height of plant size and shape of the pods. Two famous varieties are giant and dwarf. The former is commonly grown in Gujarat, while latter is popular in Punjab and Utter Pradesh. The different forms used for different purposes like as vegetable and as fodder. In Bombay three forms Pardeshi, Sotiaguvar and Deshi are famous (Anonymous, 2006 and Bhosle and Kothekar, 2010). Many cultivars of guar are available. In one study six guar cultivars were divided into two subgroups: subgroup I consisting of Kinman and subgroup II consisting of Esser, Lewis, Monument, Matador and Santa Cruz, indicating that Kinman is genetically less similar to the other five guar cultivars. Modern commercial lines are originated from brooks and mills. Kinman and Esser are sister lines developed from a cross between Brooks and Mills, and Lewis and Santa Cruz are from a cross between T64001 and PI 33870-B (Liu *et al.*, 2009).

A few number of cluster bean varieties have been developed worldwide for different agro-climatic conditions. Screening of various cluster bean germplasm collections till now have shown them to be susceptible or only tolerant.

Cyamopsis or cultivars of importance carrying some specific characters are described below:

Water Stress and Drought Tolerance

Crop productivity in arid and semi-arid regions, is often restricted due to several abiotic stresses but the amount of soil moisture available to plants during the growing season is a major limiting factor for crop yield (Boyer, 1982). The inadequate and erratic precipitation compels water stress in these regions. Cluster bean is a drought avoider crop (Kumar, 2005) and has higher capability to recover from water stress and provides reasonable seed yield and dry matter once the stress is relieved (Garg *et al.*, 1998). Sheoran *et al.* (1980) found that cumulative water stress delayed and reduced germination, decreased moisture uptake and adversely affected the hypocotyls and radical growth of cluster bean and resulted in accumulation of starch in the seedlings. The water stress imposed at flower initiation stage decreased relative water content, photosynthesis, starch and carbohydrate accumulations (Kuhad and Sheoran, 1986) and is most disadvantageous and critical to the growth of plant (Vyas *et al.*, 2001). Increasing intensities of water stress at the pre-flowering stage result in progressive and significant decline in the plant water potential, relative water content, total chlorophyll and soluble protein and nitrate reductase activities. While free proline and free amino acids showed an increasing trend with increasing water stresses. Even mild water stress at the pre-flowering stage may reduce plant growth and yield significantly (Vyas *et al.*, 1985). Whereas water stress imposed at vegetative and pod formation stages had negative effect on growth, photosynthesis, activity of various enzymes and seed yield of the plant (Vyas *et al.*, 2001). It was observed that early maturing or early flowering varieties/genotypes of cluster bean perform better compared to late maturing or late flowering varieties/genotypes under low rainfall conditions. Water stress is known to induce structural and physiological alterations in the root nodules and affect their nitrogen fixing ability (Pankhrust and Sprent, 1975). It significantly decreases shoot water potential, relative water content of leaves, net photosynthetic rate, total chlorophyll, starch and soluble proteins as well as nitrate reductase activity at various growth stages. Venkateswarlu *et al.* (1983) studied the effect of water stress on the nodulation and nitrogenase activities of cluster bean and found that increasing water stress significantly reduced the fresh weight of nodules but the number of nodules remained unchanged, while nitrogenase activities were adversely affected. The activity of nitrogen fixing bacteria in legumes reduces under moisture stress situations and the root nodules degrade after the peak flowering stage (Venkatesh and Basu, 2011). Bibi *et al.* (2014) assessed genetic associations of various seedling traits in cluster bean genotypes under water stress conditions and reported a significant correlation of shoot length, chlorophyll a/b, fresh and dry shoot weight with irrigation suggesting that the selection of drought resistance genotypes may be helpful to improve yield under water stress conditions. Water stress induced decline in plant water potential and leaf relative water content led to reduction in total chlorophyll, starch and soluble protein contents besides accumulation of free proline to various extents depending upon the genotype and growth stage at which drought was experienced (Shubhra *et al.*, 2004). Under water stress, leaf water potential, osmotic potential, chlorophyll and gum contents decreased and soluble sugar contents increased in cluster bean. Shubhra (2005) studied the impact of phosphorus application on leaf characteristics, nodule growth and plant nitrogen content in cluster bean and reported that phosphorus treatment may be effective to some extent for alleviating the effect of water deficit. The water stress reduces the availability and absorption of phosphorus and the phosphorus deficiency declines the leaf growth, photosynthetic rate (Brooks, 1986) and also reduces the uptake rate of nitrate and its assimilation by the nitrate reductase (Pilbeam *et al.*, 1993). Burman *et al.* (2009) explored the interactive effects of phosphorus nutrition and water stress intensities on water relations, photosynthesis, nitrogen metabolism and yield of cluster bean and suggested that beneficial effects of phosphorus application may be achieved only up to moderate levels of water stress in cluster bean. Similar positive effects of phosphorus application on grain yield and quality compared to control plants have been reported in cluster bean (Bhadoria *et al.*, 1997). Under water deficit conditions, application of potassium nutrition increases crop tolerance by utilizing the soil moisture, more efficiently and has substantial effect on enzyme activation, protein synthesis, photosynthesis, stomatal movement and the maintenance of the osmotic potential and turgor regulation of the cells (Lindhauer, 1987) and influences the physiological processes and yield of the crop (Lahiri and Kachar, 1985). The positive effects of potassium on water stress tolerance may be through promotion of root growth accompanied by a greater uptake of nutrients and water by plants (Rama Rao, 1986), reduction of transpirational water loss and through the regulation of the stomatal functioning under water stress conditions (Kant and Kafkafi, 2002), which is reflected as enhanced photosynthetic rate, plant growth and yield under stress conditions (Umar and Moinuddin, 2002) and maintenance of a high pH in stroma and prevention of the photo oxidative damage to chloroplasts (Cakmak, 1997). Maintenance of adequate potassium in soil improved plant water relation, photosynthesis and yield of cluster bean under simulated water stress at different developmental stage (Garg *et al.*, 2005). The stress tolerance of the plants can be improved with the exogenous use of stress alleviating chemicals (Farooq *et al.*, 2009). Kadian *et al.* (2014) studied the interactive effect of AM fungi and potassium on growth and yield under water stress and reported that AM colonization and potassium fertilizer can mitigate the deleterious effect of water stress on growth and yield in cluster bean.

Physiology and Abiotic Stresses

Cluster bean is a robust crop and copes with almost all the physiological and abiotic stresses; however, for good crop productivity these stresses should be properly addressed. The physiological aspects associated with seed maturation, seed coat color, germination, seedling/plant growth, seed yield and seed storage in cluster bean are taken in hand. Plant growth is greatly affected by environmental stresses, such as drought, high salinity and low temperature and pathogen infection during the life cycle. To encounter these stresses, plants have evolved mechanisms to increase their tolerance through physical, molecular and cellular adaptations. Abiotic stresses have become an integral part of crop production under changing climate scenario (Mittler and Blumwald, 2010). Among abiotic stresses drought, water stress and salinity are the main abiotic factors for yield reduction (Munns and Tester, 2008). Seed maturation refers to the morphological, physiological and functional changes that occur from the time of anthesis to the harvest of seeds (Khattra and Singh, 1995). Maturity is critical and the most important factor that determines the size, quality, planting value and storability of the seed (Jerlin *et al.*, 2001). Physiological maturation occurs commonly in seeds to gain the reproducing capacity of the younger generation and this normally coincides with the attainment of maximum dry weight when the flow of nutrients to the seed from the mother plant is ceased (Harrington, 1973). Khattra and Singh (1995) observed that accumulation of dry matter with loss of moisture is one of the characteristic features that could be expressed during seed development and maturation in any crop. The change in pod and seed color is considered as a promising visual index of seed maturation (Carlson, 1973). Based on physical, physiological, biochemical and visual indices, cluster bean seeds reach their physiological maturity at 18.2 per cent of seed moisture content (Renugadevi *et al.*, 2006).

Guar Meal

The sweet and tender young pods are consumed as a vegetable or snacks in north-western and southern India and the mature seeds can be eaten during food shortages. Young pods, fresh or dry forage are used as livestock feeds. The plant is also used as a green manure and cover crop. Guar yields up to 45 t/ha of green fodder, 6-9 t/ha of green pods and 0.7-3 t/ha of seeds (Undersander *et al.*, 1991; Wong *et al.*, 1997; Ecocrop, 2010; Ecoport, 2010). Guar meal is the main byproduct of guar gum production. It is a mixture of germs and hulls at an approximate ratio of 25 per cent germ to 75 per cent hull (Lee *et al.*, 2004). A protein-rich material containing about 40 per cent protein, it is used as a feed ingredient but may require processing to improve palatability and remove ant nutritional factors. The removal of the gum from seed enhances the protein contents (51 per cent) of guar meal in comparison to seed. In addition to the regular guar meal ("churi"), certain Indian manufacturers sell a high-protein guar meal ("korma"). Guar meal contains about 12 per cent gum residue (7 per cent in the germ fraction and 13 per cent in the hulls) (Lee *et al.*, 2005), which increases viscosity in the intestine, resulting in lower digestibilities and growth performance (Lee *et al.*, 2009). It is also good source of essential amino acids and useful protein supplement for chicks and laying hens.

Guar Gum

Guar gum is a Galactomannan (Galactose + Mannose) known as "guaran". It is a high molecular weight carbohydrate polymer. Guar gum is an extract of the guar bean, where it acts as a food and water store. The guar seeds are dehusked, milled and screened to obtain the guar gum. It is typically produced as a free flowing, pale, off-white colored, coarse to fine ground powder. Guar gum obtained from the endosperm of the seed of the legume plant *Cyamopsis tetragonoloba*. The crude gum is grayish white powder which contains 78-82 per cent galactomannan, 10-13 water, 4-5 per cent protein, 1.5-2 per cent crude fiber, 0.5-0.9 per cent ash and 0.5-0.75 per cent fat. Guar seeds contain 35-42 per cent gum. Guar gum is a potential raw material derived from endosperm used in paper industry, cosmetics, pharmaceutical, oil well drilling, explosives, ice-cream processed cheese products, dressings and sauces, beverages, baked goods, pastry icings, as meat binders, canned meat products, textile, mining industry, tobacco and other various industrial applications. (Undersander *et al.*, 1991; Saleem *et al.*, 2002).

Food Industries and Bakery Products

It is used as thicken and stabilize salad dressings, ice creams, backery products, meats, confectionaries, sausages and cheese due to its ability to bind water.

Paper Industry

Guar Gum provides better properties compared to substitutes. It gives denser surface to the paper used for printing. Guar Gum imparts improved erasive and writing properties, better bonding strength and increased hardness. Due to improved adhesion, it gives better breaking, mullen and folding strengths. It is added to pulp slurry before the formation of sheet on machine ensuring a regular distribution of pulp- fiber.

Textile Industries

Guar Gum gives excellent film forming and thickening properties when used for textile sizing, finishing and printing. It reduces warp breakage, reduces dusting while sizing and gives better efficiency in production.

Oil Field Applications

Industrial grade guar gum powder are use in oil well fracturing, oil well stimulation, mud drilling and industrial applications and preparations as a stabilizer, thickener and suspending agent. It is a natural, fast hydrating dispersible guar gum and is diesel slurriable. In the oil field industry, guar gum is used as a surfactant, synthetic polymer and deformer ideally suited for all rheological requirements of water-based and brine-based drilling fluids. High viscosity guar gum products are used as drilling aids in oil well drilling, geological drilling and water drilling. These products are used as viscosifiers to maintain drilling mud viscosities that enable drilling fluids to remove drill waste from deep holes. Guar gum products also reduce friction in the holes, and so minimizing power requirements. Some guar gum products act to minimize water loss should occur in broken geological formations.

Medicinal Uses

Cluster beans or Guar pods are rich in soluble dietary fibre and lowers blood cholesterol levels. Guar gum is a common ingredient in fibre-rich drinks marketed as health drinks and weight-loss drinks. Low on Calories, the cluster bean or guar contains Vitamin C, Vitamin K, Manganese, and Vitamin A, dietary fiber, folate, iron and potassium. The Vitamin K is important for maintaining strong bones and proper development of fetus. It is also good source of fiber, potassium and folate, each of these nutrients plays important role in cardio-protective role. It contains various vitamins namely vitamin A, B and K and minerals such as Potassium, Iron, Folate and Calcium in addition to other nutrients. This green vegetable and seeds of Gawal ki phali have immense medicinal value packed with numerous health benefits. Food grade guar gum contains 80 per cent guaran (a galactomannan composed of D-mannose and D galactose units) with an average molecular weight of 220 kDa. However, guar gum is not a uniform product and its viscosity may vary in proportion to the degree of galactomannan cross-linking but the crop has special importance because of gum content present in its seed. (Gupta *et al.*, 1978; Paroda and Arora, 1978; Anonymous, 1987; Patel *et al.*, 2002; Meena *et al.*, 2010).

Rich in Nutrients

Guar beans are rich in proteins, soluble fiber, Vitamin K, Vitamin C, Vitamin A, foliates, carbohydrates, phosphorous, iron, calcium, potassium with absolutely no cholesterol and fat content.

Good for Diabetics

Cluster beans contain glyconutrients (plant based saccharides or sugar) that help in controlling blood sugar levels in the body, combined with their low levels of glycaemic index (food's effect on glucose levels) means they can be safely eaten by diabetics without causing rapid fluctuations in blood sugar levels. Also, guar gum made from cluster beans by dissolving them in water is used as an remedy for diabetic patients by subordinating the amount of glucose in body.

Strengthens Bone Health

Gavar beans contain calcium that helps in strengthening the bones. Further, the presence of phosphorus in this vegetable aids in fortifying the bones and enhances bone health.

Cholesterol Control – Better for Heart

Gavar exerts heart healthy effects by lowering the levels of bad cholesterol (or LDL) in blood lowering the risks of a heart attack. Further, the presence of dietary fibre, potassium and folate in this vegetable prevents the heart from various cardiovascular complications.

Pharmaceuticals and Cosmetics

A high concentration of flavanoids and other phenolics compounds like kaempherol in guar seeds may expand its neutraceutical and pharmaceutical use. Leaves are used in asthma and to cure night blindness where as the pods and seeds are used to cure inflammation, sprains (Khare, 2004), arthritis (Katewa *et al*, 2004), as anti-oxidant, antibilious, laxatives and in polluting boiling. As per ayurveda the plant is used to reduce fire and can be used as cooling, digestive, tonic, galactogogue, useful in constipation, dyspepsia, anorexia, agalatia, hyetalopia and vitated condition of kapha and pitta. The Plant is also mentioned as aperitif and flatugenic (James, 2002).

Improved Blood Circulation

The presence of iron and phytochemicals in cluster beans increases body's hemoglobin production and allows blood to carry more oxygen throughout the body.

Manage Blood Pressure

The hypoglycemic (low blood sugar) and hypolipidemic (lowering the lipids) properties of cluster beans make them the best natural aid for hypertensive patients.

Recommended during Pregnancy

Iron and calcium present in cluster beans fills up the deficiencies of these minerals in the pregnant women. This vegetable is also loaded with high levels of folic acid that prevent the fetus from numerous birth defects and also various pregnancy related health issues. Further, Vitamin K contained in cluster beans is good for the bones and helps with better fetal development.

Acts as Digestive Aid

Cluster beans work as a good laxative (induces bowel movements and loosens the stool), stimulating bowel movement and improving your digestive system. They also help in flushing the unwanted toxins from the digestive system and prevent digestion-related problems such are irritable bowel syndrome, crohn's disease and colitis.

Calms the Brain

Hypoglycemic (low blood sugar) properties of cluster beans help in soothing the nerves. They are also found to reduce anxiety and tension and calm the person mentally.

Conclusion

Cluster bean is a photosensitive crop and requires specific climatic condition to grow; it tolerates high temperatures and dry conditions and is well adapted to arid and semi-arid climates, because of it is highly drought tolerating property, much of its area is concentrated in states like Rajasthan, Gujarat and Haryana. Moreover cluster bean cultivation has not been exploited for vegetable purpose. It is being an important legume and mainly utilized as a vegetable crop. Currently, it is gaining importance in industry as guar gum due to higher galactomannan content in the seed endosperm. Apart, the crop is rich in protein content (24-28 per cent) and thus it can be used as a good protein source. Bacterial blight, alternaria leaf spot and powdery mildew are the important diseases that threaten guar production in several guar growing regions. Tolerant varieties with inbuilt resistance are the best option for combating these diseases thereby preventing heavy economic losses; this is the best option to improve cluster bean breeding in India. The future of cluster bean depends on to develop good genotypes for the vagarious climatic condition is important. Cluster bean germplasm is genetically diverse and possesses potential variation for morphological and yield attributing characters and hence could be extensively evaluated for greater exploitation for use in breeding programs. Knowledge about the size and nature of diversity in the germplasm provides a platform for designing crop improvement programmes. Cluster bean hybrids with higher yield and productivity, development of variety suitable for summer season, resistance to diseases and early maturity will boost cluster bean productivity and availability of quality material. Guar gum and protein contents in guar seed are of prime importance from industrial point of view. Positive correlation was observed between seed yield and percentage gum content but there was a negative correlation between seed weight and gum percentage which may hamper breeding for bold seededness. Genetic analysis revealed that both additive and non additive gene effects were operating in the expression of this industrial component. Breeding for disease resistance like, bacterial blight (Xanthomonas oxonopodis), alternaria leaf spot and powdery mildew are the serious diseases of cluster bean. Inheritance of bacterial blight and alternaria leaf spot is not fully understood. It is observed that resistance for bacterial blight was a dominant trait whereas alternaria leaf spot is inherited by recessive trait additive, dominant and epistatic gene interaction played significant role towards contributing resistance in guar crop. Phenol and peroxidase enzyme played a significant role in imparting the resistance towards main diseases. Abiotic stresses have become an integral part of crop production under changing climate scenario in cluster bean. Among abiotic stresses drought, water stress and salinity are the main abiotic factors for yield reduction. Similarly significant reduction in root length, shoot length, dry weight, fresh weight, germination, protein content, catalase activity, tolerance index, vigour index, germination rate, germination relative index, mean daily germination of seeds were observed at increasing fluoride concentration in cluster bean. Breeding strategies need to exploit existing variation within the cluster bean germplasm for widening the genetic base. Eventually, the results of the present study can be used for varietal/genotype identification and parental selection, and will be helpful in augmentation of the cluster bean improvement programme. This study can be used in developing mapping population and for QTL mapping which can be used for marker assisted breeding and could be beneficial to the plant breeders, for enhancement of galactomannan content. In the scenario of changing climate these problems are likely to magnify. Therefore, abiotic and biotic stresses need to be tackled through breeding and for that systematic collection, conservation and utilization of biodiversity is necessity of the hour.

References

AICRPAL. 2013. Annual Progress Report 2003-04. (In) All-India Co-ordinated Research Project on Arid Legumes, Central Arid Zone Research Institute, Jodhpur, Rajasthan, 177.

Anderson, E. 1949. Endosperm mucilages of legumes. *Industrial and Engineering Chemistry,* **41** (12): 2887-2890.

Anonymous. 1987. Handbook of Agriculture, ICAR, New Delhi. pp. 905.

Anonymous. 2006. Wealth of India, Raw materials, First supplement series, Vol. II (Cl-Cy), National Institute of Science Communication and Information Resources, Council of Scientific and industrial Research, New Delhi India, 137-146.

Anonymous 2012. CRN India- analyzing the Indian stock market. *National Commodity and Derivatives Exchange Limited,* India, 1-3.

Bhadoria, R.B.S., Tomar, R.A.S. and Khan, H. 1997. Effect of phosphorus and sulphur on yield and quality of cluster bean (*Cyamopsis tetragonoloba*). *Indian J. Agron.,* **42**: 131–134.

Bhosle S.S. and Kothekar, V.S. 2010. Mutagenic efficiency and effectiveness in cluster bean (*Cyamopsis tetragonoloba* (L.) Taub.) *J. Phytol.,* **2** (6): 21–27.

Bibi, A., Shakir, A. and Sadaqat, H.A. 2014. Assessment of genetic association among seedling traits in guar (*Cyamopsis tetragonoloba* L.) genotypes under water stress conditions. *Int. J. Res. Stud. Biosci.,* **2** (8): 20–29.

Boyer, J.S. 1982. Plant Productivity and Environment. *Science,* **218**: 443-448.

Brooks, A. 1986. Effects of phosphorus nutrition on ribulose-1, 5-biphosphate carboxylase activation, photosynthetic quantum yield and amounts of some Calvin cycle metabolites in spinach leaves. *Aust. J. Plant Physiol.,* **13**: 221–237.

Burman, U., Garg, B.K. and Kathju, S. 2004. Interactive effects of thiourea and phosphorus on cluster bean under water stress. *Biol., Plant.* **48** (1): 61–65.

Cakmak, I. 1997. Role of potassium in protecting higher plants against photo-oxidative damage. (In) Johnston AE (Eds.) Food security in the WANA region, the essential need for balanced fertilization. International Potash Institute, Basel, Switzerland, pp 345–352.

Carlson, J.B. 1973. Morphology. (In) Caldwell B.E. (Ed.) Soybean improvement, production and uses. American Society of Agronomy, Wisconsin. pp. 17-66.

Chaudhary, M.S., Ram, H., Hooda, R.S. and Dhindsa, K. S. 1973. Effect of gamma irradiation on yield and quality of guar (*Cyamopsis tetragonoloba* (L.) Taub.). *Ann. Arid Zone,* **12**: 19-22.

Chevalier, A. 1939. Recherches sur les especes du genre *Cyamopsis,* plants fourrages pour les pays tropicaux et semiarids. *Rev. Bot. Appl.,* **19**: 242-249.

Dabas, B.S. and Thomas, T.A. 1986. Shattering guar and its significance. *Int. J. Trop. Agric.,* 2: 185–187.

Dabas, B.S., Mital, S.P. and Arunachalam, V. 1982. An evaluation of germplasm accessions in guar. *Indian J. Genet.*, **42:** 56-59.

Dabas, D.S., Mandal, S., Phogat, B. S., Bisht, I.S and Agrawal, R. C. 2001. Guar (*Cyamopsis tetragonoloba*) - A resume of research NBPGR, New Delhi: National Bureau of Plant Genetic Resources, **269:** 278-289.

Dass, S., Arora, N.D. and Singh, V.P. 1973. Heritability estimates and genetic advance for gum and protein content along with seed yield and its components in cluster bean [*Cyamposis tetragonoloba* (L.) Taub.]. *Haryana Agric. Univ. J. Res.*, **3:** 14-19.

Ecocrop. 2010. Ecocrop database. FAO. http: //ecocrop.fao.org/ecocrop/srven/ home. pp. 23-28.

Ecoport. 2010. Ecoport database. Ecoport. http: //www.ecoport.org. pp. 15-19.

Farooq, M. Wahid, A. and Kobayashi N. 2009. Plant drought stress: effects, mechanisms and management. *Agron. Sustain. Dev.*, **28:** 185–212.

Gandhi, S.K., Saini, M.L. and Jhorar, B.S. 1978. Screening of cluster bean genotypes for resistance to leaf spot caused by *Alternaria cyamopsis*. *Forage Res.*, **4:** 169-172.

Garg, B.K., Burman, U. and Kathju, S. 2005. Physiological aspects of drought tolerance in cluster bean and strategies for yield improvement under arid conditions. J. Arid Legume. **2** (1): 61–66.

Garg, B.K., Kathju, S. and Vyas, S.P. 1998. Influence of water deficit stress at various growth stages on some enzymes of nitrogen metabolism and yield in cluster bean genotypes. *Indian J. Plant Physiol.*, **3:** 214–218.

Gill, S.L. 2009. Evaluation of reciprocal hybrid crosses in guar. Texas Tech University, Texas, USA. pp. **113:** 4571-4577.

Gillett, J.B. 1958. *Indigofera* (*Microcharis*) in tropical Africa with the related genera *Cyamopsis* and *Rhynchotropis*. *Kew Bull. Add. Ser.*, **1:** 1-166.

Gupta, P.C., Sagar, V. and Pradhan, K. 1978. *Forage res.*, **4A:** 109-122.

Harrington, J.K. 1973. Biochemical basis of seed longevity. *Seed Sci. Technol.*, **1:** 453–461.

Hymowitz, T. 1972. The trans-domestication concept as applied to guar. *Economic Botany*, **26** (1): 49-60.

Hymowitz, T. 1979. Seed protein electrophoresis in taxonomic and evolutionary studies. *Theoretical and Applied Genetics*, **54:** 145-151.

Hymowitz, T. and Matlock, R. S. 1963. Guar in the United States. *Oklahoma Agricultural Experiment Station Technical Bulletin*, **611:** 1-34.

James, A. D. 2002. Handbook of medicinal herbs. CRC Press, Washington D.C. pp. 118-119.

Jerlin, R., Srimathi, P. and Vanangamudi, K. 2001. Seed physiological and maturity indices. (In) Vanangamudi K, Bharathi A, Natesan P *et al.* (Eds.) Recent techniques and participatory approaches on quality seed production.

Department of Seed Science and Technology, Tamil Nadu Agricultural University, Coimbatore. pp.123-126.

Jhorar, B.S., Solanki, K.R. and Jatasra, D.S. 1988. Combining ability analysis of kernel weight in cluster bean under different environments. *Indian J. Agric. Res.*, **22**: 188-192.

Johal, G. S., Balint-Kurti, P. and Weil, C. F. 2008. Mining and Harnessing Natural Variation: A Little Magic. *Crop Science*, **49**: 2066-2073.

Kadian, N., Yadav, K. and Aggarwal, A. 2014. Interactive effect of arbuscular mycorrhizal fungi and potassium on growth and yield in *Cyamopsis tetragonoloba* (L.) under water stress. *Researcher.* **6** (8): 86–91.

Kant, S. and Kafkafi, U. 2002. Potassium and abiotic stresses in plants. In: Pasricha NS, Bansal SK (eds.). Role of potassium in nutrient management for sustainable crop production in India. Potash Research Institute of India, Gurgaon, Haryana, pp. 233–251.

Katewa, S.S., Caudhary, B.L. and Jain, A. 2004. Folk herbal medicine from tribal areas of Rajasthan, India. *J. Ethano. Pharmacol.*, **92** (1): 41-46.

Khare, C.P. 2004. Indian Medicinal Plants. An Illustrated Dictionary New Delhi, 2nd ed. Springer, 189-190.

Khattra, S. and Singh, G. 1995. Indices of physiological maturity of seeds: a review. *Seed Res.*, **23** (1): 13–21.

Kirtikar, K.R. and Basu, B.D. 1918. Indian medicinal plant, Indian press., pp. 407.

Kuhad, M.S. and Sheoran, I.S. 1986. Physiological and biochemical changes in cluster bean genotypes under water stress. *Indian J. Plant Physiol.*, **29**: 46–52.

Kumar D 2005. Breeding for drought resistance. (In) Ashraf M and Harris PJC (Eds.) Abiotic stress: plant resistance through breeding and molecular approaches. Haworth Press, New York. pp. 145–175.

Kumar, D. 2005. Status and direction of arid legumes research in India. *Indian J. Agric. Sci.*, **75**: 375-391.

Kumar, D. and Rodge, A.B. 2012. Status, scope and strategies of arid legumes research in India: A Review. *J. Food Leg.*, **25**: 255-272.

Kumar, D. and Singh, N. B. 2002. Guar in India, Scientific publishers (India), Jodhpur. pp. 225.

Kumar, S., Joshi, U.N., Singh, V., Singh, J.V. and Saini, M.L. 2013. Characterization of released and elite genotypes of guar [*Cyamopsis tetragonoloba* (L.) Taub.] from India proves unrelated to geographical origin. *Genet. Resour. Crop Evol.*, **60**: 2017–2032.

Lahiri, A.N., Garg, B.K. and Kathju, S. 1987. Responses of cluster bean to soil salinity. Ann. Arid Zone. **26**: 33–42.

Lal, B.M. and Gupta, O.P. 1977. Studies on galactomannans in guar and some correlation for selecting genotypes rich in gum content. (In) Proceedings of first guar research workshop, CAZRI, Jodhpur, India. pp. 124-130.

Lee, J.T., Connor-Appleton, S., Haq, A.U., Bailey, C.A. and Cartwright, A.L. 2004. Quantitative measurement of negligible trypsin inhibitor activity and nutrient analysis of guar meal fractions. *J. Agric. Food Chem.*, **52** (21): 6492-6495.

Lee, J.T., Connor-Appleton, S., Bailey, C.A. and Cartwright, A.L. 2005. Effect of guar meal by- product with and without beta mannanase hemicell on broiler performance. *Poult. Sci.*, **84** (8): 1261-1267.

Lee, J.T., Bailey, C.A. and Cartwright, A.L. 2009. *In vitro* viscosity as a function of guar meal and beta mannanase content of feeds. *Int. J. Poult. Sci.*, **8** (8): 715-719.

Lindhauer, M.G. 1987 Solute concentrations in well watered and water stressed sunflower plants differing in K nutrition. *J. Plant Nutr.*, **10**: 1965–1973.

Liu, W., Hou, A., Ellen, B.P. and Dick, L.A. 2009. Genetic Relationship of guar commercial cultivars. *Chinese Agricultural Science Bulletin*, **25** (02): 133-138.

Lodha, S. 1984. Varietal resistance and evaluation of seed dressers against bacterial blight of guar (*C. tetragonoloba*). *Indian Phytopathol.*, **137**: 438-440.

Meena, A.K., Godara, S.L. and Gangopadhyay, S. 2010. Biochemical changes in constituents of cluster bean [*Cyamopsis tetragonoloba* (L.) Taub.] due to *Alternaria* blight. *Indian Phytopath.*, 64 (Supplementary Issue): 68-82.

Mishra, S.K., Singh, N. and Sharma, S. K. 2009. Status and utilization of genetic resources of arid legumes in India. (In) Perspective research activities of arid legumes in India. [D. Kumar and A. Henry (Eds.)], Indian Arid Legumes Society, CAZRI, Jodhpur, India. pp. 23-30.

Mittler, R. and Blumwald, E. 2010. Genetic engineering for modern agriculture: challenges and perspectives. *Annu. Rev. Plant Biol.*, **61**: 443–462.

Mudgil, D., Barak, S. and Khatkar, B.S. 2014. Guar gum: processing, properties and food applications – A Review. *J. Food Sci. Technol.*, **51 (3):** 409-418.

Munns, R. and Tester, M. 2008. Mechanisms of salinity tolerance. *Annu. Rev. Plant Biol.*, **59**: 651–681.

Pankhrust, C.R. and Sprent, J.I. 1975. Effects of water stress on the respiratory and nitrogen fixing activity of soybean root nodules. *J. Exp. Biol.*, **26**: 287–304.

Paroda, R.S. and Arora, S.K. 1978. Guar- its improvement and management. *The Indian Society of Forage Research, Hissar*: pp. 169.

Pathak, R., Singh, S.K., Singh, M. and Henry, A. 2010. Molecular assessment of genetic diversity in cluster bean (*Cyamopsis tetragonoloba*) genotypes. *J. Genet.* **89**: 243-246.

Patil, C.G. 2004. Nuclear DNA amount variation in *Cyamopsis* DC (Fabaceae). *Cytologia*. **69**: 59–62.

Patel, D.S., Patel, S.I. and Desai, A.G. 2002. Cluster bean varietal reaction to wilt disease caused by *Neocosmospora vasinfecta* E.F. Smith. *J. Mycol. Pl. Pathol.*, **32** (**1**): 120-121.

Pilbeam, D.J., Cakmak, I. and Marschner, H. 1993. Effect of withdrawal of phosphorus on nitrate assimilation and PEP carboxylase activity in tomato. *Plant Soil.* **154**: 111–117.

Punia, A., Yadav, R., Arora, P. and Chaudhary, A. 2009. Molecular and morphphysiological characterization of superior cluster bean (*Cyamopsis tetragonoloba*) varieties. *J. Crop Sci. Biotech.* **12**: 143-148.

Rama Rao, N. 1986. Potassium nutrition of pearl millet subjected to moisture stress. J. Potassium Res., **2**: 1–12.

Renugadevi, J., Natarajan, N. and Srimathi, P. 2006. Studies on seed development and maturation in cluster bean (*Cyamopsis tetragonoloba*). *Madras Agric. J.*, **93**: 195–200.

Saleem, M.I., Shah, S.A.H. and Akhtar, L. H. 2002. BR-99: A new guar cultivar released for general cultivation in Panjab province. *Asian Journal of Plant Sciences*, **1** (3): 266-268.

Sandhu, H.S. 1988. Interspecific hybridization studies in genus *Cyamopsis*. PhD Thesis: CCSHAU, Hissar, India. pp. 1- 187.

Sharma, B.S., Bhatnagar, K. and Cheema, H.A. 1999. Fungal management of stem blight disease of cowpea induced by *Macrophomina phaseolina*. *J. Mycol. Plant Pathol.*, **29**: 276.

Sheoran, I.S., Khan, M.I. and Garg, O.P. 1980. Differential behavior of guar genotypes to water stress during germination. *Guar Newsletters.* **1**: 3–4.

Shubhra, D.J. 2005. Impact of phosphorus application on leaf characteristics, nodule growth and plant nitrogen content under water deficit in cluster bean. *Forage Res.*, **31** (3): 212–214.

Shubhra, D.J. and Goswami, C.L. 2004. Influence of phosphorus application on water relations, biochemical parameters and gum content in cluster bean under water deficit. *Biol. Plant.*, **48** (3): 445–448.

Singh, A. 1972. Trisomics in *Cyamopsis psoraleoides*. *Can. J. Genet. Cytol.*, **14**: 200-204.

Singh, C., Singh, P. and Singh, R. 2014. Modern Techniques of Raising Field Crops. [C. Singh, P. Singh and R. Singh (2 nd. Eds.)], Oxford and IBH Publishing Co. Pvt. Ltd., New Delhi. pp. 1-583.

Singh, J.V., Lodhi, G.P. and Singh, V.P. 1995. Varietal improvement in guar. *Farm Dig.*, **5**: 23.

Smith, P.M. 1976. Evolution of Crop Plants (Ed. N.W. Simmonds), Longman, London, and New York. pp. 311-312.

Stafford, R.E. 1982. Yield stability of guar breeding lines and cultivars. *Crop Sci.*, **2**: 1009– 1011.

Stephens, J.M. 1994. Guar-[*Cyamopsis tetragonoloba* (L.) Taub.] Fact Sheet HS-608, a series of the Horticultural Science Department, Florida Cooperative Extension Service, Institute of Food and Agriculture Sciences, University of Florida, **60** (6): 956-964.

Stephens, J.M. 1998. Guar – *Cyamopsis tetragonoloba* (L.) Taub. HS608. Horticultural Sciences Department, Florida Cooperative Extension Service, Institute of Food and Agricultural Sciences, University of Florida. **60** (6): 956-964.

Strickland, R.W. and Ford, C.W. 1984. *Cyamopsis senegalensis*: potential new crop source of guaran. *J. Australia Institute Agri. Sci.*, pp. 47-49.

Sultan, M., Zakir, N., Rabbani, M.A., Shinwari, Z.A. and Masood, M.S. 2013. Genetic diversity of guar (*Cyamopsis tetragonoloba* L.) Landraces from Pakistan based on RAPD markers. *Pak. J. Bot.*, **45** (3): 865-870.

Tanksley, S. D. and McCouch, S. R. 1997. Seed banks and molecular maps: unlocking genetic potential from the wild. *Science*, **277** (5329): 1063-1066.

Umar, S. and Moinuddin 2002. Genotypic differences in yield and quality of groundnut as affected by potassium nutrition under erratic rainfall conditions. *J. Plant Nutr.*, **25**: 1549–1562.

Undersander, D.J., Putnam, D.H., Kaminski, A.R., Doll, J.D., Oblinger, E.S. and Gunsolus, J.L. 1991. Guar. University of WisconsinMadison, University of Minnesota (http://www.hort.purdue.edu/newcrop/afcm/guar.html). pp. 1-6.

Undersander D.J., Putnam D.H., Kaminski A.R., Kelling K.A., Doll J.D., Oplinger E.S. and Gunsolus J.L. 1991. Guar- Alternative field crops manual. University of Wisconsin Cooperative or Extension Service, Department of Agronomy, Madison, WI 53706. pp. 1-6.

Vavilov, N.I. 1951. The origin, variation, immunity and breeding of cultivated plants. Translated by Start K. *Chron. Bot.*, **13:** 1-366.

Venkataratnam, L. 1973. Beans in India, Directorate of Extension, Ministry of Agriculture, New Delhi. pp. 64.

Venkatesh, M.S. and Basu, P.S. 2011. Effect of foliar application of urea on growth, yield and quality of chickpea under rainfed conditions. *J. Food Legumes*, **24** (2): 110–112.

Venkateswarlu, B., Rao, A.V. and Lahiri, A.N. 1983. Effect of water stress on nodulation nitrogenase activity of guar (*Cyamopsis tetragonoloba* (L.) Taub.). *Proc. Indian Acad. Sci. (Plant Sci.)*. **92** (3): 297–301.

Vig, B.K. 1965. Effect of a reciprocal translocation on cytomorphology of guar. *Sci. Cult.*, **31:** 531-533.

Vig, B.K. 1969. Studies with 60Co radiated guar (*Cyamopsis tetragonoloba* (L.) Taub). *Ohio J. Sci.*, **69:** 18.

Vyas, S.P., Garg, B.K. and Kathju, S. 2001. Influence of potassium on water relation, photosynthesis, nitrogen metabolism and yield of cluster bean under soil moisture deficit stress. *Indian J. Plant Physiol.*, **6**: 30–37.

Vyas, S.P., Kathju, S. and Garg, B.K. 1985. Performance and metabolic alterations in *Sesamum indicum* L. under different intensities of water stress. *Ann. Bot.*, **56**: 323–333.

Whistler, R.L. and Hymowitz, T. 1979. Guar: Agronomy, Production, industrial use and nutrition. Purdue university press, Indiana. pp. 16-21.

Wong, L.J. and Parmar, C. 1997. *Cyamopsis tetragonoloba* (L.) Taub. Record from Proseabase. Farida Hanum, I and Van der Maesen, L.J.G. (Editors). PROSEA (Plant Resources of South East Asia) Foundation, Bogor, Indonesia. pp. 109-113.

Yadav, H.; Singh, S. and Haque, E. 2014. An Analysis of Performance of Guar Crop in India. Ministry of Agriculture, Government of India. pp. 1-98.

Yadav, S.L., Singh, V.V. and Ramkrishna, K. 2004. Evaluation of promising M_3 progenies in guar (*Cyamopsis tetragonoloba* (L.) Taub.). *Indian J. Genet. Plant Breeding*, **64:** 75-76.

Zamir, D. 2001. Improving plant breeding with exotic genetic libraries. *Nature Reviews Genetics*, **2** (12): 983 -989.

Chapter 8

Kale

Chander Parkash, S.S. Dey and Vijay Bhardwaj

ICAR IARI, Regional Station,
Katrain (Kullu Valley) – 175 129, H.P.
E-mail: cp1968@gmail.com

Kale (*Brassica oleracea* L.convar. *acephala* (DC.) Alef.) is a representative of cole group of vegetables, cool-season cooking green somewhat similar to collard and non heading cabbage that has gained recent widespread attention due to its health-promoting, sulfur-containing phytonutrients. Kale is grown for its large foliage which is mainly used as leafy vegetable. This leafy vegetable is helpful in reducing the chances of cancer if taken regularly. Among the vegetable crops kale is the richest source of carotenoids which is the precursor of vitamin-A. Besides, it also contains good amount of different minerals essential for human diet. There are several varieties of kale, all of which differ in taste, texture, and appearance.

Area and Distribution

Kale is native to the Mediterranean or to Asia Minor. It was introduced to America from Europe as early as the 17th century. The most important growing areas lie in mid and West Europe, North America, more rarely in tropical areas. Kale is mainly used as vegetable during the cold months. Its acreage is small and commercial cultivation is mainly done in Europe. Presently, the crop is also raised in commercial basis in USA. In India, the crop is mainly cultivated in large scale in Kashmir valley of Jammu and Kashmir where it is the main leafy vegetable. In Himachal Pradesh, the crop is cultivated only in small areas. In rest of the India the crop is not known widely.

It is mainly a winter vegetable and cultivated during *rabi* season. It can be grown in all types of soils with good moisture supply; however, medium-heavy soil is ideal. Kale is a very reliable cropper and can withstand relatively unfavourable conditions. It can tolerate freezing temperatures down to at least -6.7°C. It is a hardy plant, hence can be left in the field over the winters and the flavour actually improves when low temperatures prevail. It starts to peter out when temperatures start getting above 25°C on regular basis. It rarely suffers from pests and diseases.

Botany

Stem and Leaves

Kale is a form of cabbage (*Brassica oleracea* Acephala Group) in which the central leaves do not form a head. Kale (*Brassica oleracea* var. *acephala*) is a plant with a straight, non-branched stem 0.2 to 2 m tall, terminating in a loose rosette. For human consumption, it is mainly the curly leafed types that are grown. Curliness is caused by disproportionately rapid growth of leaf tissue along the margins. In some varieties the stem is not more than 30 cm long and is neither branched nor markedly thickened. At the apex of the stem is developed a rosette of generally oblong and sometimes red coloured leaves. Marrow stem kales form a long swollen stem and tree kales form stems over 2m in height.

There are also curly leafed forms with various coloured foliage the so called ornamental Kales. The collards and the tronchuda kales are rather low plants with rosettes of large leaves.

Flower

The first flower branches are generated in the leaf axil of the main flower stalk and the second in the leaf axil of the first ones. Plant growing under sufficient nutrient and good management practices can generate even the third and fourth branches. Due to their differences in branching habit, the stock plants of different ecotypes vary greatly in plant shape.

A healthy plant can generate 800-2000 flowers depending on the varieties and agronomic management. The flowering occurs on the main flower stalk first, followed by the first order of branches from top to the bottom and the second, third and fourth branches in that order. On one anthotaxy, flowers open orderly from the bottom to the top.

Kale has a complete flower with calyx, corolla, androecium, and gynoecium. Each calyx has four sepals, and the corolla is with 4 petals which are arranged in a cross. Two short and four long stamens (tetradynamous condition) are found inside the petals with each having an anther on the top. Maturing pollen grains are released by anther dehiscence.

Kale is a typical, cross-pollinated plant with strong self-incompatibility. Mainly insects mediate natural pollination. The percentage of hybrid seeds can be as high as 70 per cent when two varieties are planted together and pollinated naturally. Stigma receptivity and pollen viability are highest on the first day of flower opening. A stigma can be fertilized 6 days before or 2-3 days after flower opens. Pollen are viable 2 days before or 1 day after the flower opens, but the viability can be maintained for more than 7 days if pollen grains are stored in a desiccated state at room temperature and even longer at temperatures below 0°C. It takes about 36-48 h from pollination to fertilization at optimum temperature of 15-20°C. When the pollen grains land on the stigma, it takes 2-4 h for the pollen tube to begin elongation; 6-8 h later the pollen

Kale Cultivar 'KTK-64' under Testing in AICRP (VC) Trials.

Varieties availabe in Kale Crops.

tube penetrates the stigma tissue and fertilization is accomplished within 36-48 h thereafter. Pollen germination slows down below 10°C and the normal fertilization is affected at temperature higher than 30°C.

Fruit

Fruits are globular silique, 4-5 mm wide and sometime over 10 cm long, with two rows of seeds lying along the edges of the replum (false septum). A siliqua contains from ten to thirty seeds. When silique ripe, dehiscence takes place through the two valves breaking away from below upwards, leaving the seeds attached to the placentas.

Seed

Seed coat is featureless; position of the radical is indicated by a low ridge. The seed is globular to slightly oval, 2-3 mm in diameter, light brown at first, but becoming grayish- black to red- brown later on.

Origin and Domestication

Kale is originated from the wild cabbage (*Brassica oleracea* var. *sylvestris*) in the Mediterranean region in Asia and to have been brought to Europe around 600 B.C. by groups of Celtic wanderers. Curly kale was an important crop during ancient Roman times and a popular vegetable eaten by peasants in the Middle Ages. Until the end of the Middle Ages, kale was the common green vegetable in all of Europe. English settlers brought kale to the United States in the 17th century.

During the Roman era large number of cole crops has been mentioned and some curly forms resembled to the kale. The meaning of the word *cauli* during the middle age is kale as mentioned by various authors. During this time all the cole crops spread over Europe but were mainly cultivated for medicinal purposes. During the sixteenth century numerous herbals came into picture and one of them was kale. But seed production of all cole crops was confined to Europe until the recent past. Thus, it may be concluded that kale is one of the oldest known cole crops and cultivated in the past mainly for medicinal purposes. «Kale» is a Scottish word derived from *coles* or *caulis*, terms used by the Greeks and Romans in referring to the whole cabbagelike group of plants. The first mention of the kales (colewarts) in America was in 1669; but because of their popularity in European gardens it is probable that they were introduced somewhat earlier.

Both ornamental and dinosaur kale are much more recent varieties. Dinosaur kale was discovered in Italy in the late 19th century. Ornamental kale, originally a decorative garden plant, was first cultivated commercially as in the 1980s in California. Ornamental kale is now better known by the name salad savoy.

Today, one may differentiate between varieties according to the low, intermediate or high length of the stem, with varying leaf types. The leaf colours range from light green through green, dark green and violet-green to violet-brown. Russian kale was introduced into Canada (and then into the U.S.) by Russian traders in the 19th century. These leafy nonheading cabbages bear the Latin name *Brassica oleracea* variety *acephala*, the last term meaning «without a head.» They have many names in many languages, as a result of their great antiquity and widespread use. All varieties of collards appear rather similar, but the kales show interesting diversity: tall and short; highly curled and plain leaved; blue-green, yellow-green, and red; erect and flat-growing; in various combinations and gradations of these characters.

Horticultural Types

A large number of cultivars are available worldwide. The cultivars are mainly classified on the basis of stem length and grouped into dwarf and medium-tall types. Besides, stem length the cultivars are also classified on the basis of leaf colour and curliness. There is another form called Chinese kale (*Brassica alboglabra*) which is a typical tropical and hot weather vegetable crop and does not resemble to more popular curly kale. In India, both smooth leaf and curly leaf kale are cultivated.

There are several varieties of kale, commonly divided into groups *viz.*, curly kale (*Brassica oleracea* var. *acephala* sub var. *laciniata* L.), smooth leaved kale (*Brassica oleracea* var. *acephala* sub var. *plana* Peterm), thousand head kale (*Brassica oleracea* var. *acephala* sub var. *millecapitata* Lev.), tree kale (*Brassica oleracea* var. *acephala* sub var. *palmifolia* DC) and marrow stem kale (*Brassica oleracea* var. *acephala* sub var. *medullosa* Thell), all of which differ in taste, texture and flavour.

In India, there is no identified variety of kale so far at the national level. Local cultivars like Khanyari Green, GM Dari, Siberian Kale and many others are cultivated in the Kashmir valley. However, two lines, developed at ICAR-IARI, Regional Station, Katrain and two lines developed at SKUAS and T, Shalimar, Jammu and Kashmir have recently been entered in ICAR-AICRP (VC) trials for multi-location testing for the first time.

Nutritional Value

Kale has gained widespread attention due to its health promoting, sulphur containing phytochemicals. Therefore, The American Cancer Society has recommended that Americans should increase intake of kale. It is the organo-sulphur compounds in this food that have been main subject of phytonutrient research and these include glucosinolates and methyl cysteine sulfoxides. From over 100 different glucosinolates present in plants, 10-15 are present in kale, which sufficiently appears to lessen the occurrence of a wide variety of cancers, including breast and ovarian cancers. Sulforaphane, a chemical found in kale, not only helps in detoxifying cancer causing chemicals and inducing colon cells to commit suicide, but in a recent study it shows to help stop proliferation of breast cancer cells, even in the later stages of their growth.

Kale is a rich source of vitamins A, C and B_6, proteins, minerals, calcium, fibres, antioxidants and anticarcinogenic compounds. Vitamin A and C, magnesium, potassium, proteins and caloric value of kale is highest among all the Brassica vegetables (Table 8.1).

Among the different leafy vegetables, kale contained maximum amount of β-carotene (a precursor of vitamin A) as presented in Table 8.2.

Table 8.1: Nutritive Value of Kale

Nutrients	100g Chopped Kale
Proteins	3.3 g
Fats	0.7 g
Carbohydrate	10 g
Fibre	2.0 g
Ca	135 mg
Mg	1.7 mg
P	56 mg
K	447 mg
Mn	34 mg
Ascorbic acid	120 g
Vitamin A	9620 IU

Table 8.2: Comparative Amount of β-carotene in different Leafy Vegetables

Common Name	Botanical Name	Variety	β-carotene Content (mg/100g of sample)
Kale	Brassica oleracea var. acephala	K-64	0.4316
Lettuce	Lactuca sativa	Great Lakes	0.2407
Leaf mustard	Brassica juncea	Pusa Sag-1	0.2432
Palak (beet)	Beta vulgaris	Pusa Harit	0.1858
Spinach	Spinacea oleracea	Virginia Savoy	0.2382

Cultural Practices

Soil and Climatic Requirements

Kale grows more rarely in tropical areas as it prefers cooler climates. Kale is the most robust cabbage type - indeed the hardiness of kale is unmatched by any other vegetable. Kale will also tolerate nearly all soils if drainage is sufficient. Well-drained loamy soil with relatively high in organic matter are suitable for kale. Cover crops may be turned under to maintain organic matter. The optimum pH for proper growth and development is 5.5-6.5. If the pH is too high, manganese is frequently unavailable which results in a chlorotic condition of the leaves. If the pH is too low, an application of lime is recommended to correct it as per necessity.

Kale is cultivated as a rabi season crop like other crops of the cole group. However, in the areas with mild weather it can be raised twice a year. In hilly areas of mid Himalaya this crop can be cultivated almost throughout the year for vegetable purpose. However, seed of this crop is not available widely. For seed production the crop has specific chilling temperature requirements (4-7°C for 6-8 weeks) which can only be met in the specific temperate regions. There is a need to popularize this highly nutritious leafy vegetable which can be cultivated through out India.

Kale prefers cooler climates, thus it can be grown round the year in the temperate regions and during winters in the tropical areas. It can be seeded directly or transplanted during October to December in the plains and main season for sowing in hills is from March through September. For best results, planting should be adjusted in such a manner that harvesting is done in coolest months.

Sowing and Transplanting Requirement

Generally raised bed nursery is preferred for raising the seedlings successfully. Flat bed nursery is not suitable when there are chances of rainfall. Sufficient amount of organic manures in the nursery is needed for raising healthy seedlings. Around 400-500 g seeds are enough to produce seedlings for planting in an area of one hectare.

Planting distances are conditional on the time of planting, cultivars, soil and location. Dwarf cultivars may be spaced at 45 cm, for late planting, these are planted much closer at 30 cm. The medium tall types are generally planted at a distance of 60 cm. If sown directly thinning should be done 7-10 days after germination.

Manure and Fertilizer Requirements

Basal dose of 20-25 tones/ha of farmyard manure is recommended at the time of land preparation. Application of vermi-compost and other organic manure like poultry manure are recommended if available. A dose of 150 kg N and 175 kg each of P and K per ha should be applied. The N is given in equal split doses at planting time, 30-40 days after planting and 15-20 days prior to first harvesting. After transplanting sufficient amount of organic manures and nitrogenous fertilizers ensure luxurious growth of leaves and more yield. Application of sufficient nitrogen after each leaf cutting is needed for quick vegetative growth and getting more yield.

Irrigation

Like other cole crops, irrigation should be done based on the rainfall, weather condition and type of soil. Irrigation should be done along with nitrogenous fertilizers after harvesting of leaves for quick rejuvenation. Subsequent irrigations are given at 15-20 days interval. If there is no rainfall irrigation should also be done immediately after planting. Maintaining uniform soil moisture for tender growth and maximum use of soil nutrients is must. As much as 12-14 inches of water may be needed. Soil type does not affect the amount of total water needed, but determines frequency of water application. Lighter soils need more frequent water applications, but less water applied per application.

Intercultural Operations

Hoeing and weeding should be done regularly to keep the crop weed free. Usually 2-3 hoeing and weeding are sufficient. Once the leaves cover the soil, there is no need of hoeing. A shallow hoeing is recommended to avoid root injury.

Harvesting

Harvesting of kale starts from November and continues till the end of January. It should be done at appropriate vegetative stage to obtain best quality harvest. The whole rosette of dwarf varieties is harvested at one time. Harvesting is mainly done for tender leaves. The leaves can be removed 3-4 times at an interval of 20-25 days as the crop produces leaves rapidly after removal of leaves. The number of harvestings depends upon the type of cultivars. Generally, 2-3 harvestings are recommended for dwarf cultivars while for relatively long stem type 3-5 harvestings can be practiced. The leaves along with petioles are harvested and bundled properly for marketing. Kale can be harvested in three ways: whole plant, bunched leaves, or stripped leaves. Stripped kale is pre-packaged for fresh market. In all methods, yellow or damaged leaves must be removed before packing. Whole plant harvest is generally not practiced.

Yield

The autumn sown crops usually yield more, compared to early sown crops. On an average it yields 100-250 q/ha. The yield of long-stemmed varieties is more than the dwarf varieties.

Post Harvest Management

Leafy greens such as collards, kale, rape, Swiss chard, and beet greens are handled like spinach. Because of their perishability, they should be held as close to 0°C as possible. At this temperature, they can be held for 10 to 14 days. Relative humidity of at least 95 per cent is desirable to prevent wilting. Air circulation should be adequate to remove heat of respiration, but rapid air circulation will speed transpiration and wilting. Satisfactory pre-cooling is accomplished by vacuum cooling or hydro-cooling. These leafy greens are commonly shipped with package and top ice to maintain freshness. Research has shown that kale packed in polyethylene-lined crates and protected by crushed ice keeps in excellent condition for 3 weeks at 0°C but only 1 week at 5°C and three days at 10°C. Vitamin content and quality are retained better when wilting is prevented.

Diseases and Pests

The occurrence of disease and pests are less in kale as compared to other cole crops. If precautions are taken at right time they can be managed effectively and do not become major problem in cultivation of kale.

Diseases

Damping Off

It is a primary disease in nursery and generally occurs in two stages. In pre emergence stage young seedlings are killed before they reach the surface of the soil after the germination of seeds. Post emergence occurance is characterized by topping over of the infected seedlings. The disease is caused by the soil borne pathogens like, *Phytopthora*, *Pythium*, *Fusarium* and *Rhizoctonia*.

There should be proper sanitation to prevent the disease in nursery. During nursery preparation sufficient amount of organic manure should be applied. Proper drainage and air circulation in nursery are vital for minimizing the occurrence of the disease. The soil in the nursery can be sterilized by drenching of 0.3 per cent solution of Captan/Captaf/Thiram @ 5 litres/m^2 or 0.1 per cent of formaldehyde and covering the soil for 2-3 days with polyethylene sheets.

Downey Mildew

It is caused by fungal pathogen *Perenospora parasitica*. In India it occurs during winter and early summer in northern plains and hilly regions, respectively. It can attack the crop at any stage of development. Small chlorotic, irregular, translucent, light green lesions appear on leaf lamina in the initial stage with pronounced downey growth of fungus on lower surface during high humid conditions. Lesions dry out and become necrotic leading to defoliation and death of the infected plants.

Seed treatment with metalyxal compound @ 2g/kg of seeds is effective. One protective spray of Ridomil MZ 72@ 0.025 per cent before the onset of favourable condition followed by Dithane M 45@ 0.2 per cent at 10 days interval is recommended for minimizing disease occurrence.

Black Rot

It is caused by the bacterium *Xanthomonas campestris pv. campestris*. It causes heavy damage in subtropical and warm temperate areas of world. The spread of the disease is limited at temperature below 5°C and above 35°C. The typical disease symptom is occurrence of V-shaped chlorosis on the margin of the leaves. In severe cases blackening of veins and internal leaf chlorosis is also common. The disease is transmitted through seeds.

Hot water seed treatment (52±2°C) for 30 minutes is highly effective and only satisfactory method to control the disease. The spraying of streptocyclin @ 100-200 ppm is also recommended to prevent the spread of the disease.

Insects and Pests

Diamond Black Moth (*Plutella xylostella*)

The caterpillars severely damage leaves in various stages of growth. They also damage the growing tips thus causing heavy loss.

Conventional pesticides are not effective against DBM. Spraying of cypermethrin 10 EC @ 1g/litre is recommended. Spraying of cartap hydrochloride followed by neem seed kernel extract (5:1000 w/v) was also found effective.

Cabbage Butterfly (*Pieris brassicae*)

The caterpillars eat up leaves from the margin onward leaving primary vein intact. They also damage the seed crop to a great extent.

Spraying of Malathion @ 0.2 per cent is found effective to manage the insect.

Aphids (*Bravicoryne brassicae, Myzus persicae*)

They cause severe damage in seedling to seed setting stage. They suck the sap of leaves, stem, flowers and pods of the seed crop.

Spray of 0.2 per cent Malathion or Thiodan at 10-15 days interval are recommended for effective control of aphids.

References

Annonymous. 2004. Production recommendations for vegetables. Sher-e-Kashmir University of Agricultural Saciences and Technology of Kashmir, Shalimar Campus, J&K.

Chander Parkash, 2009. Identification of self-incompatible lines in kale. International Conference on Horticulture (ICH-2009) held at Dr. Prem Nath Agricultural Science Foundation, Bangalore from November 9-12, 2009. pp. 79-80.

Chander Parkash, Dey S.S., Dhiman M.R. and Bhatia,R 2010. Kale: Nutritionally important vegetable. *Intensive Agriculture* 49 (4): 18-20.

Chander Parkash, Sharma S.R., Dhiman M.R. and Seema 2008. Cultivation of under utilized vegetable crops. Bulletin. IARI, Regional Station, Katrain (Kullu Valley), H.P.

Chander Parkash, Seema, Dhiman, M.R. and Sharma S.R. 2008. Genetic variability in Kale. National Seminar on «Sustainable Horticultural Research in India; Perspective priorities and preparedness» held at Babasaheb Bhimrao Ambedkar University, Lucknow (UP) from 14-15th April, 2008. pp.33-34.

D'Antuono, L.F. and Neri R. 2003. Traditional crop revised: yield and quality of palm-tree kale, grown as a mechanized processing crop, as a function of cutting height. Acta Hort., 598: 123-127.

Kohlmeier, L. and Su, L. 1997. Cruciferous vegetable consumption and colorectal cancer risk: Meta-analysis of the epidemiological evidence. FASEB J., 11: 369.

Korus, A. and Kmiecik, W. 2007. Concept of carotenoids and cholorophyll pigments in kale (*Brassica oleracea* L. var. *acephala*) depending on the cultivar and the harvest date, Electronic J. of Polish Agric. Univ.Vol. 10 (1): http: //www.ejpau. pl/volume10/issur1/art-28. html.

Chapter 9

Onion

M.A. Vaddoria[1] and Ganesh Kulkarni[2]

[1]Vegetable Research Station,
[2-]Department of Genetics and Plant Breeding,
JAU Junagadh, Gujarat
E-mail: mavaddoria@jau.in

The Biological Diversity Act, 2002 has defined three types of Biodiversity namely: Ecosystem diversity, Species diversity and Genetic diversity. The species diversity includes flora and fauna in wild as well as cultivated plants and domestic animals breeds. In general, the variety of life on Earth and its biological diversity is commonly referred to as biodiversity. The number of species of plants, animals and microorganisms, the enormous diversity of genes in these species; the different ecosystems on the planet such as deserts, rainforests and coral reefs are all part of a biologically diverse Earth. Biological diversity means the variability among all living organisms from all sources including inter alias, terrestrial, marine and other aquatic ecosystems and biological diversity within a species and of ecosystems. Biodiversity is the degree of variety in nature and not nature itself. Indian cosmology estimates 84 lakh species of living organisms in the entire universe but the biologists have described only 15 lakh species and estimates to have a total of around 50 lakhs.

Biodiversity of India

Out of the 1.4 million known species of living organisms only about 2,50,000 are higher plants and 1.03 million are animal. According to another estimate, worldwide there are 2,70,000 known species of vascular plants. India is the seventh largest country of the world with an area of about 32,67,500sqkms (Negi, 1993). India ranks sixth among the 12 megabiodiversity centers of the world, and is home for an unusually large number of endemic species. It supports 15,000 species of flowering plants 5,000 of them exclusively providing shelter to 317 species of mammals (Rathore and Jasrai 2013). India is unique, not so much of its numerical species but for the range of biodiversity attributable to a variety of biogeographically and physico-environmental situation. Nature has endowed India with a rich biological diversity, which includes over 40,000 species of plants and 75,000 species of animals. India has about 12 per cent of the global plant wealth amongst which there are nearly 3,000 tree species. However, nearly a third of the total plant species of India are endemic (Negi, 1993). Indian flora is extremely varies in extent, composition and endemism. In India there are over 30,000 species of higher plants belonging to 174 natural orders. There are over 600 species of Pteridophytes including ferns. Of the higher plants, there are 11,124 species of dicots with 1,831 genera. The family Orchidaceae is the largest family of flowering plants, contributing nearly 1,700 plant species.

Nomenclature of Onion

The onion takes its name from the city built by Onias (B. C. 173) near the Gulf of Suez. The onion has been domesticated a very long time and is one of the earliest of cultivated plants. Drawings of it are found on the Egyptian monuments. Its indigenous form, however, is not well understood. Under long continued cultivation and selection the bulbs have developed into large shapely organs. It has for centuries found favor with the Egyptians and Israelites and is now cultivated and popular in almost every country of the world.

Onion (*Allium cepa*) is a popular vegetable grown for its pungent bulbs and flavorful leaves. Onion has been recognized as an important functional food since ancient times. The oldest known 'recipe book', a 4000-year-old Mesopotamian clay tablet, features the use of onions, garlic and leeks as flavouring (Bottero 1985). The medicinal and health benefits of onion are due to the presence of flavonoids, anthocyanins, fructo-oligosaccharides and organosulphur compounds (Goldman 2011). Early historic records refer to it frequently as an article of food, also as a preventive of thirst while on the march or traveling in the desert.

In olden times the growing of the crop was confined chiefly to the alluvial river valleys; but by improvement the different varieties have been adapted to a diversity of conditions. It is only within the last quarter of a century that rapid growth and development of the industry has taken place in the world.

Origin

It is widely grown throughout the world. *Alliums* are native to Asia and the Middle East and have been cultivated for over five thousand years.The place of its origin is unknown; but it occupied a vast area in Western Asia during a very early epoch, extending, perhaps, from Mediterranean basin,Palestine to India and central Asia.

The center of origin of onion, *Allium cepa* is remains debatable, the middle Asiatic countries in the region of Iran and Pakistan are considered to be the primary centres of origin of onion. The near east Asiatic and Mediterranean regions are considered to be the secondary centres of origin. The next most important centre is located in western North America.

Vavilov (1926) proposed southwest Asian gene center as primary center of domestication and variability of onion. Based on ecotypes and wild forms, Vavilov and Burkinich (1929) confirmed that Afghanistan and adjacent countries are the genetic center of origin of the cultivated forms of onion and garlic. More than 600 species of *Allium* were reported to be distributed in Afghanistan, Turkey, Iran and central Asia comprising Turkmen SSR, Uzbek SSR, Tadzhik SSR, Kirgiz SSR and Kazakh SSR, and Mongolia (Kotlinska*et al.*, 1990). The secondary center of origin in the Mediterranean gene center represents the area from which onion with large bulbs was selected. From central Asia, the supposed onion ancestor probably migrated first towards Mesopotamia, where onion is mentioned in Sumerian literature (2500 B.C.), then to Egypt (1600 B.C.), India and South East Asia. From Egypt *A. cepa*was introduced into Mediterranean area and from there to all Roman Empire.

Botany

Allium is the onion genus, with 600-920 species, making it one of the largest plant genera in the world.As of March 2014, the World Checklist of Selected Plant Families accepts 920 species. (Peterson *et al.*, 1988.)

The genus *Allium* belongs to family *Alliaceae*.The detail classification of *Allium* is given in Table 9.1.

Table 9.1: Classification of *Allium cepa* L.

Kingdom	Plantae	Synonyms and Common Names
Sub-kingdom	Tracheobionta	*Scientific Name: Allium cepa* L.
Super division	Spermatophyta	*Synonyms:*
Division	Liliopodia	*Allium ascalonicum* L.
Subclass	Liliales	*Allium esculentum*Salisb.
Order	Liliaceae	*Allimporrumcepa*Rehb.
Genus	*Allium* L.	*Cepa rotunda* Dod. (URL-2)
Species	*Allium cepa* L	

The onion belongs to the *Allium cepa*, a widely variable species forming a part of the botanical family which includes the lilies and the several forms of asparagus and smilax. It is generally a biennial, although some forms, such as the tree onion and multipliers, are perennial. The latter are used for bunching purposes. Usually, however, it is grown for bulbs as an annual. The bulbs are variable in colour, being yellow, red, white and the intermediate shades of these colours. Now and then a bulb does not develop, and the neck, or stalk just above the bulb, remains relatively thick. Such onions are termed scullions, and they may be regarded as run down or reverted forms. The seed stalks are slender and tall. The seeds are angular in shape and black, and are borne in a dense and compact cluster at the end of the seed stalk.

Allium cepa is cultivated mainly as a biennial, but some types are treated as perennials. It is propagated by seeds, bulbs or sets (small bulbs). Bulbs have a reduced disc-like rhizome at the base. Scapes are up to 1.8 m tall and gradually tapering from an expanded lower part. The leaves have rather short sheaths and differ in size and are near circular in cross-section but somewhat flattened on the adaxial side. The umbel is subglobose, dense, many-flowered (50 to several hundred) and with a short persistent spathe. Pedicels are equal and much longer than the white and star-like flowers with spreading tepals. Stamens are somewhat exerted, and the inner ones bear short teeth on both sides of the broadened base. The fruit is a capsule approximately 5 mm long (Plate 9.1). The wide variation in bulb characteristics indicates intensive selection. Bulb weight may be up to l kg in some southern European cultivars, and the shape covers a wide range from globose to bottle-like and to flattened-disciform. The colour of the membranous skins may be white, silvery, buff, yellowish, bronze, rose red, purple or violet. The colour of the fleshy scales can vary from white to bluish-red. There is also much variation in flavour, the keeping ability of the bulbs and the ability to produce daughter bulbs in the flrst season. Great variability in ecophysiological growth pattern has developed. There exist varieties adapted to bulbing in a wide range of photoperiodic and temperature conditions. Similarly, adaptation exists for bolting and flowering in a broad range of climates, but non-bolting strains are found in many shallots (Hanelt, 1986a) Organs not selected for by humans, *e.g.* the flower and the capsule, have been very little affected by domestication and exhibit no striking variations

Umble of Onion

Individual Flower of Onion

Onion Seeds

Plate 9.1: Onion Flower and Seeds.

Biodiversity in Onion

Onion and Types

It is a large genus and contains several major agricultural crops including bulb onion (*Allium cepa*), shallot (*A. cepa* syn. *A. ascalonicum*), Japanese bunching or Welsh onion (*A. fistulosum*), chive (*A. schoenoprasum*), garlic (*A. sativum*) and leek (*A. ampeloprasum* syn. *A. porrum*). All these common/domesticatedAlliums have a basic chromosome number of 8 andmost are diploids (*A. cepa*, 2n=2x=16) (Collum Mc, 1976).

Vvedensky (1944) classified the cultivated Allium species into four sections:*Cepa* (bulb onion), *Phyllodolon* (Japanese bunching onion), *Porrum* (garlic and leek) and *Rhizirideum* (chive). Alater classification based on morphological criteria, crossability and karyotype (Traub, 1968), also divided them among four sections (*Allium,Cepa, Fistulosa* and*Rhizirideum*)with further divisions into sub-sections. However, difficulties arose, because of (i) few morphological characters upon which the classification was based and (ii)strong barriers to crossing separate, but morphologically similar species.

Evolution of the genus has been accompanied by ecological diversification. The majority of species grow in open, sunny, rather dry sites in arid and moderately humid climates. However, *Allium* species particularly *Allium cepa* L have adapted too many other ecological niches. Different types of forests, European subalpine pastures and moist subalpine and alpine grasslands of the Himalayan and Central Asian high mountains all contain some Allium species, and gravelly places along river-banks do as well. Even saline and alkaline environments are tolerated by some taxa.

Allium species from these diverse habitats exhibit a parallel diversity in their rhythms of growth (phenology). Spring, summer and autumn-flowering taxa exist. There are short- and long-living perennials, species with one or several annual cycles of leaf formation, and even continuously leaving ones. Species may show summer or winter dormancy. For many species annual growth is limited to a very short period in spring and early summer when the cycle from leaf sprouting to seed maturation is completed in 2 or 3 months. The cultivated Allium crop species are listed in Table 9.2.

Description and Variability in Important *Allium* Species

1. *Allium altaicum*

It is native to Asiatic Russia,Mongolia, Kazakhstan and Northern China. It produces narrowly egg-shaped bulbs up to 4 cm (1 $^{1}/_{2}$ inches) in diameter. Scape is round in cross-section, up to 100 cm (39 inches) tall. Leaves are round, up to 50 cm (20 inches) long. Flowers are pale yellow, up to 20 mm ($^{13}/_{16}$ inch) across. Ovary is egg-shaped; stamens longer than the tepals.

Table 9.2: Cultivated *Allium* Species and their Areas of Cultivation

Sl.No.	Botanical Names of the Crop Groups	Other Names Used in the Literature	Area of Cultivation	English Names
1	A. altaicum Pall.	A. microbulbum Prokh.	South Siberia	Altai onion
2	A. ampeloprasum L.			
3	Pearl-onion group	A. ampeloprasum var. sectivum Lued.	Atlantic and temperate Europe	Pearl onion
4	A. cepa L. Common onion group	A. cepa ssp. cepa/var. cepa	Worldwide	Common onion
		A. cepa ssp. australe Kazakova		
5	Ever-ready onions	A. cepa var. perutile Stearn	Great Britain	Ever-ready onion
6	Aggregatum group	A. ascalonicumauct. hort.	Nearly worldwide	Shallot,
		A. cepa var. aggregatum		potato onion
		G. Don, var. ascalonicum Backer, ssp. orientalis Kazakova		multiplier onion
7	A. fistulosum L.		East Asia, temperate Europe and America	Japanese bunching onion, Welsh onion
8	A. obliquum L.		West Siberia, East Europe	Oblique onion
9	A. oschaninii O. Fedtsch		France, Italy	French shallot
10	East Asian group	A. aobanum Araki, A. wakegi	China, Japan, South East Asia	Wakegi onion
11	Eurasian group	A. cepa var. viviparum (Metzg.) Alef.	North America, Europe,	Top onion
		A. cepa var. proliferum (Moench) Alef	North-East Asia	Tree onion, Egyptian onion, Catawissa onion
12	A. victorialis L.	A. microdictyon Prokh.	Caucasus, Japan, Korea	Long-root onion

Source: Fritsch and Friesen (2002).

2. *Allium ampeloprasum*

This is commonly known as wild leek or broadleaf wild leek. Its native range is southern Europe to western Asia. It has been differentiated into three cultivated vegetables, namely leek, elephant garlic and kurrat. Wild populations produce bulbs up to 3 cm across. Scapes are round in cross-section, each up to 180 cm tall, bearing an umbel of as many as 500 flowers. Flowers are urn-shaped, up to 6 mm across; tepals white, pink or red; anthers yellow or purple; pollen yellow. (Gleasonand Cronquist, 1991).

3. *Allium fistulosum*

Commonly known as bunching onion, long green onion, Japanese bunching onion, scallion, spring onion, Welsh onion, is a species of perennial plant. The species is very similar in taste and odor to the related *Allium cepa*, and hybrids between the two (tree onions) exist. The Welsh onion does not develop bulbs and possesses hollow leaves (fistulosum means «hollow») and scapes. In addition to culinary uses, it is also grown as an ornamental plant (Fritsch and Friesen, 2002).

4. *Allium obliquum*

Common name lop-sided onion or twisted-leaf onion, is a Eurasian species of wild onion with a range extending from Romania to Mongolia. It is also widely cultivated as an ornamental. Allium obliquum produces an egg-shaped bulb up to 3 cm long. Scape is up to 100 cm tall, round in cross-section. Leaves are flat, shorter than the scape, up to 20 mm across. Umbels are spherical, with many yellow flowers crowded together (Lazkov and Turdumatov,2010).

5. *Allium victorialis*

Commonly known as victory onion, Alpine leek, and Alpine broad-leaf allium, is a broad-leaved Eurasian species of wild onion. It is a perennial of the Amaryllis family that occurs widely in mountainous regions of Europe and parts of Asia (Caucasus and Himalayas). Some authors consider certain East Asian and Alaskan populations as constituting subspecies platyphyllum within the species *Allium victorialis*. (Rabinowitch and Currah, 2002).The photographs of important allium species are given in Plate 9.2.

Intraspecific Classification

The great variability within the species has led to different proposals for intraspecific groupings, whose historical development has been discussed in detail by Hanelt (1990). Kazakova (1978) presented the most recent version of a classical system which held shallots apart at species level and recognized three subspecies, eight varieties and 17 cultivar groups (named '**conculta**') based on quantitative characters. This rather cumbersome classification of *A. cepa* involves statistical methods. The characteristics used are affected strongly by environment and need to be tested in a range of climates. Also, in modern breeding, many 'classical' cultivar groups have been crossed and the boundaries between the different taxa are becoming blurred, making it difficult to place material within the scheme. The broadly accepted concept of the species *A. cepa* used here includes races with many lateral bulbs and/or shoots, which rarely bolt, and which are partly seed-sterile, namely shallots and potato onions. Other morphological and karyological characters, isozyme and molecular-marker patterns are almost identical to those of *A. cepa* (Klaas, 1998). Here a simple informal classification will be applied, similar to that of Jones and Mann (1963), accepting two large and one small horticultural groups. The advantages of fiexibility and the lack of nomenclature constraints have been discussed in detail elsewhere (Hanelt, 1986b). This approach is convenient for both breeders and horticulturists.

Plate 9.2: Photographs of Cultivated Allium Species.

Allium altaicum

Allium ampeloprasum

Allium fistulosum

Allium obliquum

Allium victorialis

Grouping of Onion

The grouping in onion genotypes is carried out by many scientists as the genepool of onion is more than 20000 in all over the world.The genotypes were sub divided into Common group, Aggregatum group and Ever-ready onion groups. (Fritsch and Friesen, 2002)

1. Common Onion Group

This group is economically most important *Allium* crop. It includes open-pollinated traditional and modern cultivars, hybrids and local races, cultivated in the world. The bulbs are large and normally single, and plants reproduce from seeds or from seed-grown sets. The majority of cultivars grown for dry bulbs belong to this group. Maximumdiversity exists in North India,Pakistan, former Soviet Union, Europe,Middle East and in Mediterranean area (Astley *et al.*, 1982).

2. Aggregatum Group

The group is of minor economic importance. The bulbs are smaller than in common onions, formin an aggregated clusters. Reproduction is exclusively vegetative through daughter bulbs.The cultivationis mainly in Europe, America and Asia for dry bulbs. In tropical areas, these are used as substitutes toonion.Shallots are the most important subgroup of the Aggregatum group and the only ones grown commercially to any extent. They produce aggregations of many small, narrowly ovoid to pear-shaped bulbs, which often have red-brown (coppery) skins. The plants have narrow leaves and short scapes (Rabinowitch and Kamenetsky,2002).

3. Ever-ready Onion Group

This third group of *A. cepa* may be distinguished from the other two by its prolific vegetative growth and by the lack of a dormant period. Bulbs or leaves can be gathered at all times of the year. It is used mainly as a salad onion and was commonly cultivated in British gardens in the mid-20th century. Detailed descriptions were given by Jones and Mann (1963).

Status of Genetic Resources/Germplasmin in the world as well as in India

The availability of diverse crop germplasm is a prerequisite for the success of breeding programmes (Glaszmann*et al.*, 2010). Onion has been grown from ancient times in different regions, and growers have selected desirable populations according to their needs. As a result, a wide range of landraces and cultivars harbouring huge genetic variation for various traits are available (Brewster, 2008).

Astley *et al.* (1982) were the first that made an overview of global edible Allium genetic resources. They identified major Allium collections worldwide, and presented the numbers of accessions per species per collection. Furthermore a draft Allium descriptor list was included in their report together with a list of collecting priorities. In total ca. 9000 accessions were reported to be present worldwide and the number of onion accessions was by ca. 7000 by far the largest. The collection of local/modern cultivars and landraces of Allium cepa (dry bulb onions and shallots) was considered as an important future collection priority as modern F1 hybrids were thought to quickly replace old landraces.

In the late 1980s, the threat of genetic erosion due to the introduction of F_1 hybrids led to the collection of onion landraces and modern cultivars in various countries (Fritsch and Friesen, 2002). These germplasm collections represent a wide range of onion gene pools and are available in various gene banks (Table 9.3). Developing core collections representing this diversity and evaluating a common set of lines under different set of environmental conditions would enhance the utility of bulb onion germplasm.

Table 9.3: List of Allium Gene Banks in the World as well as in India

Name of Institute	Web page	Number of Accessions	Country
AVRDC: The world vegetable center	https://avrdc.org	1100	Taiwan
ICAR-Directorate of Onion and Garlic Research	http://www.dogr.res.in	2050	India
European CooperativeProgramme for Plant Genetic Resources	http://eurisco.ipk-gatersleben.de	14400	Germany
N.I. Vavilov Research Institute	http://www.vir.nw.ru	1888	Russia
National Institute of Agrobiological Sciences	http://www.nias.affrc.go.jp	1352	Japan
Royal Botanical Gardens	http://www.kew.org	1100	UK
US Department of Agriculture	http://www.usda.gov	1304	USA
Warwick crop center	http://www.warwick.ac.uk	1755	UK

Source: Khosa *et al.*, 2016.

The onion genepools; number of accessions based on the IPGRI (www.ipgri.cgiar.org)

Genepools	No. of Accessions
Primary	
A. cepa	12740
A. vavilovii	20
A. galanthum	34
A. roylei	4
Secondary	
A. fistulosum	951
A. altaicum	121
Tertiary	
A. pskemense	21
A. oschaninii	41

Source: Kik, 2008.

Genetic Resources/Germplasm Status in India

Status of Research on Onion in India

Systematic research programmesin onion were started in 1960 at Pimpalgaon,Baswant, Nashik and later on at ICAR-Indian Agricultural Research Institute (ICAR-IARI), New Delhi and ICAR- Indian Institute of Horticultural Research (ICAR-IIHR), Bengaluru. National Horticultural Research and Development Foundation (NHRDF), Nashik established by National Agricultural Co-operative Marketing Federation of India Ltd. (NAFED) is carrying out research and development activities on export oriented crops, especially onion. Development of multiplier onion varieties was done by Tamil Nadu Agricultural University (TNAU),Coimbatore. Prior to this, research on collection and maintenance of land races and standardization of agro-techniques was attempted by State Agricultural Departments.With the concept of coordinated projects and Agricultural Universities, the work on onion research was strengthened, in terms of varietal development for different seasons and standardization of production techniques in early nineties.

The R&D in onion got impetus with the establishment of National Research Center on Onion and Garlic at Nashik in 1994. This center was shifted to present location at Rajgurunagar in 1998 and upgraded to Directorate with the addition of All India Network Research Project on Onion and Garlic in 2008.Besides concentrating on genetic improvement and biotechnology of onion,the ICAR-Directorate of Onion and Garlic Research (ICAR-DOGR) is also working on development of agrotechnologies including post-harvest management practices.This work is also being supplemented by NHRDF and some universities. At present different state agricultural universities, ICAR institutes across the country and private companies are working on different R&D aspects to improve and sustain production and productivity of onion. The status of work conducted in India in areas of onion improvement, is presented below.

Crop Improvement

A large numbers of landraces including some wild species are available in India particularly in North-eastern states. As per reports from Singh and Rana (1994), ICAR-National Bureau of Plant Genetic Resources (ICAR-NBPGR) has conducted extensive plant exploration in different allium-growing states/regions inIndia. Kale *et al.* (1994) undertook a detailed survey of traditional and non-traditional onion-growing areas of the state of Maharashtra, and India in general, and collected148 red-skin and 33 white-skin types of onion,evaluated and identified some lines on the basis of maximum average bulb weight, high TSS and centerness. As per Singh and Rana (1994), some of the cultivated Indian accessions have been identified to be resistant/tolerant to purple blotch (*Alternaria* species), *Stemphylium* blight and garlic mosaic virus. However, sources of resistance to many diseases and pests such as neckrot (*Botrytisallii* Munn.), basal rot (*Fusarium* species), black mould (*Aspergillus niger* Tieghem) are yet to be identified. Many farmers in various parts of the country are growing old land races of onion. For example, Pune Fursungi, a red coloured land race is being cultivated in Nashik and Pune areas of Maharashtra in late *kharif* and *rabi* seasons.The Junagadh, Saurashtra and Mehsana areas of Gujarat are dominated by Pili Patti, which is commonly grown in *rabi* season. Bellary Red, another red onion land race is prevalent in Karnataka and land race Sukhsagar is being cultivated in West Bengal. K.P. onion dominates in Andhra Pradesh, whereas Nirmal Local occupies large area in Madhya Pradesh. Further, the multiplier type of onion has been a unique feature in Tamil Nadu. This variability is being maintained in national germplasm collection of onion at ICAR-DOGR, which is the National Active Germplasm site for onion. The present status of collection of germplasm at ICAR-DOGR is given in Table 9.4 (Jai Gopal, 2015). At present nearly eighty varieties have been developed by different public as well as private institutes.

Table 9.4: Status of Onion Germplasm Collection at ICAR-DOGR

Sl.No.	Category	No. of Accessions
1	Dark Red	274
2	Light Red	429
3	White	450
4	Yellow	50
5	Exotic onion	237
6	Wild species	12
	i. Allium altaicum.	
	ii. A. ampeloprasum	
	iii.A. cepaxA. cornutum (PRAN)	
	iv. A. cepax A. fistulosum	
	v. A. chinense	
	vi. A. fistulosum.	
	vii. A. flavum	
	viii. A. galanthum	
	ix. A. guttatum	
	x. A. hookeri	
	*xi. A. schoenoprasum*var.*schoenoprasum*	
	xii. A. tuberosum	

Wild *Alliums* harbour a range of useful traits but crossing barriers restrict their use (Chuda and Adamus 2009). Interspecific hybridization in genus *Allium* allows researchers to transfer useful traits from closely related alliums into bulb onion, and introgressions from *Allium roylei, Allium galanthum* and *Allium fistulosum* have opened new avenues for improvement (Scholten*et al.*, 2007, Vu *et al.*, 2012). These species can be utilized to develop onion cultivars resistant against various diseases along with desirable agronomic traits, such as earliness and better rooting system. Further, *Allium cepa, A. fistulosum* and *A. roylei* species have been used to develop Alien Monosomic Addition Lines (AMAL), which are useful for increasing genetic variability in cultivated alliums and for comparative genetic studies (Shigyo*et al.*, 1996, Vu *et al.*, 2012). However, the use of wild alliums to improve the genetics of commercial varieties is a very long-term approach that can take up to 20 years, as was the case when downy mildew resistance from *A. roylei* was introgressed into bulb onion (Scholten *et al.*, 2007).

Basic research in breeding for resistance, processing qualities and export worthy varieties are lacking.Thrust in these areas can help to improve onion productivity and export.

1. Biennial nature, high cross-pollination and sharp inbreeding depression in onion are still challenges for breeders with conventional approaches. There is thus an opportunity to use biotechnology particularly molecular approaches and functional genomics to overcome these problems.

2. Due to poor maintenance of breeders' stock, many varieties are out of production chain or could not even make entry into the chain. Farmers find easy and economical to produce their own seed of onion but due to ignorance of out-crossing they are not able to maintain purity. Due to supply of spurious seed by many seed merchants, the spread of good varieties has been hampered. Thus there exists opportunity to produce and distribute good quality seed of true-to type varieties and capture the market of onion seed.Seed multiplying agencies working in public sector need to be sensitized in this regard.

3. Thrust is required to increase the storage capacity in the country. Infrastructure facilities need to be created in a way that about 30-40 per cent produce is stored in the cold storages to significantly reduce the post-harvest losses.

References

Astley, D., Innes, N.L. and Van der Meer, Q.P. 1982. Genetic Resources of *Allium* species, IBPGR secretariat, Rome, Italy p.38.

Bottero, J. 1985. The cuisine of ancient Mesopotamia. Biblic. Archaeol. 481: 30 – 47.

Brewster, J. L. 2008 Crop Production Science in Horticulture 15. Onion and Other Vegetable Alliums 2ndEdition. CAB Internatinal.

Chuda, A., and Adamus,A.2009. Aspects of interspecific hybridization within edible Alliaceae. Acta Physiol. Plant 31, 223–227.

Collum Mc G. D. 1976. Onion and allies *Allium* (*Liliaceae*). (In)N.W. Simmonds (Eds). Evolution of crop plants., Longman, London, pg. 186-190.

Fritsch, R.M. and Friesen, N. 2002. Evolution, domestication and taxonomy. pp.5-30. (In) H.D. Rabinowitch and L. Currah (Eds.), *Allium* Crop Science: Recent Advances, CABI Publ., Wallingford, UK.

Glaszmann, J. C., B. Kilian, H. D. Upadhyaya, and Varshney,R.K. 2010. Accessing genetic diversity for crop improvement. Curr. Opin.Plant Biol. 13, 1–7.

Gleason, H. A. and Cronquist,A.J. 1991. Manual of the Vascular Plants of Northeastern United States and Adjacent Canada, New York, Botanical Garden, Bronx.

Goldman, I. L. 2011 Molecular breeding of healthy vegetables. EMBO Rep. 12, 96–102.

Hanelt,P 1986a. Formal and informal classification of the infra-specific variability of cultivated plants- advantages and limitation. (In) Styles, B.T. (Eds.) Intraspecific classification of wild and cultivated Plants. Clarendone Press, Oxford pp. 139-156

Hanelt, P. 1986b. Formal and informal classifications of the intraspecific variability of cultivated plants –advantages and limitations. (In) Styles, B.T. (Ed.) *Intraspecific Classification of Wild and Cultivated Plants.* Clarendon Press, Oxford, pp. 139–156.

Hanelt, P. 1990. Taxonomy, evolution and history. (In)Rabinowitch,H.D. and Wrewster,J.L. (Eds) *Onion and allied crops,* Vol.1 *Botany, Physiology and Genetics.* CRC Press, Boca Raton. Florida, USA. pp.1-26.

Jai Gopal 2015. Onion Research in India: Status and Challenges *Progressive Horticulture,* 47 (1): 1-15

Jones, H.A. and Mann, L.K. 1963.*Onions and Their Allies: Botany, Cultivation and Utilization.* Leonard Hill, London and Interscience, New York, 285 pp.

Kale, P.N.; Warade, S.D. and Midmore, D.J.1994. International symposium on alliums in the tropics, Bangkok, Thailand. 15-1993. *Acta Hort.,* **358**: 153-156.

Kazakova, A. A. 1978. *Luk*Kul'turnaja Flora SSSR, X, Kolos, Leningrad, USSR, pp. 264

Khosa, J. S., John Mc Collum, Ajmers, Dhatt and Richard Mac Knight 2016. Review:

Enhancing onion breeding using molecular tools. Plant Breeding, 135: 9–20.

Kik, C. 2008. *Allium* genetic resources with particular reference to onion. Proc. XXVII IHC - Cultiv. Utiliz. Asian, Sub-Trop., Underutilized Hort. Crops. Eds.-in-Chief: Dae-Geun Oh and ChieriKubotaActaHortic770 : 135-138.

Klaas, M. 1998 Applications and impact of molecular markers on evolutionary and diversity studies in the genus *Allium.* Plant Breeding 117, 297–308.

Kotlinska T.P.; Havranek, M.; Navratill, L.; Gerasimova,A.; Pimakhov and Neikov, S. 1990. Collecting onion,garlic and wild Species of Allium in central Asia, USSR.Plant Genetic Resources Newsletter,83184: 31-32.

Lazkov, G.A. and Turdumatov,N.K. 2010. New and rare species of the genus *Allium* (*Alliaceae*) for the flora of Kyrgyzstan. BotanicheskiiZhurnal. Moscow and Leningrad 95: 1637-1639

Negi, S. S.1993. Biodiversity and its Conservation in India. Indus Publising Co., pp 1-12

Peterson, PM, CR Annable and Rieseberg,L.H. 1988. Systematic relationships and nomenclatural changes in the 'Allium douglasii' complex. Systematic Botany 13: 207-214.

Rabinowitch, H.D. and Kamenetsky, R. 2002. Shallots. (In) H.D. Rabinowitch and L. Currah (Eds.), Allium Crop Science – Recent Advances. CABI Publishing, Wellingford, UK.P.409-430.

Rabinowitch, Haim D. and Currah, Lesley 2002. *Allium Crop Science: Recent Advances* (preview). CABI. p. 26. ISBN 978-0851-99510-6.

Rathore, Aparna and Yogesh T. Jasrai 2013. Biodiversity: Importance and Climate Change Impacts. International Journal of Scientific and Research Publications, 3 (3): 1-5

Scholten, O. E., A. W. V. Heusden, L. I. Khrustaleva, K. Burger-Meijer,R. A. Mank, R. G. C. Antonise, J. L. Harrewijn, W. Van Haecke, E.H. Oost, R. J. Peters, and Kik, C. 2007. The long and winding roadleading to the successful introgression of downy mildew resistanceinto onion. Euphytica 156, 345–353.

Shigyo, M., Y. Tashiro, S. Isshiki, and Miyazaki, S. 1996. Establishment of a series of alien monosomic addition lines of Japanese bunching onion (*Allium fistulosum*L.) with extra chromosomes from shallot (*A. cepa* L. *Aggregatum* group). Genes Genet. Syst. 71, 363–371.

Singh, B.P. and Rana, R.S. 1994. Collection and conservation of *Allium* genetic resources: an Indian perspective. *Acta Hort.*, 358: 181–190.

Traub, H.P. 1968. The subgenera sections and subsections of *Allium* L. Plant Life, 24: 147–163.

Vavilov N.I. 1926 *Origin and geography of cultivated Plants.* English translation by D Love (1992) Cambridge Univ. Press, Cambridge,U.K.

Vavilov N.I. and Burkinich D.D. 1929 ZemeledeleheskiiAfganistan.Tr.Po.Prikl. Botanike,GenetikeIselekcii V/R.T.Z. pp156-158

Vu, H. Q., Y. Yoshimatsu, L. I. Khrustaleva, N. Yamauchi, and Shigyo.M,2012 Alien genes introgression and development of alien monosomic addition lines from a threatened species, *Allium roylei* Stearn, to*Allium cepa* L. Theor. Appl. Genet. 124, 1241–1257.

Vvedensky, A.I. 1944. The genus *Allium* in the USSR, Herbertia, 11, 65 -218

Chapter 10

Ridge Gourd

D.K. Singh[1] and M. Sarkar[2]

[1]*Department of Vegetable Science,*
G.B.Pant University of Agriculture and Technology,
Pantnagar – 263 145, Uttarkhand
E-mail: dks1233@gmail.com
[2]*Hill Millet Research Station*
Navasari Agricultural University, Waghai, Navsari – 394 730, Gujarat

"Cucurbit" is a term coined by Liberty Hyde Bailey for cultivated species of the family, Cucurbitaceae. The family Cucurbitaceae comprises largest group of summer vegetables. All together there are 2 well defined subfamilies, 8 tribes, about 118 genera and 825 species. Out of these, approximately 20 species belonging to 9 genera are under cultivation (Jeffery, 1990). In India, cucurbits occupy an area of 0.43 million hectare having annual production of 4.25 million tones giving productivity 10.52 tones per hectare (FAO, 2006).

Ridge gourd [*Luffa acutangula* (Roxb.) L.], 2n=2x=26, is one of the important cucurbitaceous vegetable crops. It is frost sensitive, predominantly tendril bearing vines which are found in subtropical and tropical regions around the globe. This vegetable is annual and largely monoecious, bearing ridged fruits. Immature fruits are used as vegetable and mature fruits are used for fibre. Plant produces long cylindrical fruits, 10-40 cm in length and 6-10 cm in diameter. The fruit has about ten distinct longitudinal acutely angled ribs. There are existence of variation in fruit shape like small fruited cluster type, long fruited type and dwarf to long vine types. It is commercially grown in several countries. Immature tender fruits of non bitter varieties are of major economic importance and are used as cooked vegetables. Like sponge gourd the fibrous spongy network of mature dried fruits is used in several ways. **(Choudhury and Thakur, 1965).** This vegetable is popular both as spring-summer and rainy season crop known as ribbed gourd or angled gourd or silky gourd or angled loofah or vegetable gourd. The vernacular names of ridge gourd are bhol, turai, tori, parteek, jhinga, janhi, gisoda, turiya, beera kaya, heeray kayi, peechinga, dodaka, ghosavala, ghosale or peerkankai. It is seed propagated vegetable and grows rapidly in warm condition. Ridge gourd along with other gourds occupies an area of 1.5 million hectares with the annual production of 18.9 million tones in the world. In India, the production of gourds is about 3.50 million tones from an area of 0.36 million hectares **(FAO, 2006)**.

Taxonomy

Family: Cucurbitaceae

Subfamily: Cucurbitoideae

Tribe: Benincaseae

Subtribe: Luffinae

Genus: *Luffa*

Species: *acutangula*

Origin and History

Ridge gourd is an essentially old world genus, consisting of two cultivated (*Luffa acutangula* and *Luffa cylindrica*) and two wild species (*Luffa graveolens* and *Luffa echinata*). This crop has a long history of cultivation in the tropical countries of Asia and Africa. The name "Luffa" or "Loofah" is of Arabic origin. Its fruit characteristics have been described in Egyptian writing and in early Chinese literature. Sanskrit name "Koshataki" indicates its early cultivation in India. This gourd has originated from Asian subtropical areas, probably from India **(Kalloo, 1993)**. *Luffa acutangula* contains three varieties: var. *acutangula* (called angular loofah, Chinese okra, ridged gourd), which is grown in South Eastern Asia and other tropical areas; var. *amara*, a wild or feral form confined to India; and var. *forskalii*, which is probably a feral form **(Heiser and Schilling, 1990)**.

It is generally monoecious in nature but hermaphrodite, andromonoecious, trimonoecious and gynoecious flowering behaviour has also been reported **(Swarup, 2006)**. The advanced sex form of dioecism is only found in the wild primitive species *Luffa echinata*. Chromosome counting and relative morphology of the wild and cultivated species and chromosome coupling in interspecific hybrids implies that *Luffa graveolens* is the prime species that has given rise to the two cultivated monoecious species (*Luffa acutangula* and *Luffa cylindrica*). **Singh and Bhandari (1963)** raised a hermaphrodite variety of *Luffa acutangula* to that of a specific status *Luffa hermaphrodita*, which has not been accepted because of its easy crossability and fully fertile hybrids with *Luffa acutangula*.

Distribution of Diversity

By and large ridge gourd is grown throughout India as suitable agro-climates are available but the diversity is found in the North Eastern region, the Northern plain including Tarai region, Central region and the Western and Eastern Peninsular region **(Arora and Nayar, 1984)**.

Nutritional and Medicinal Value

This vegetable is low in saturated fat and cholesterol, high in dietary fible, vitamin C, riboflavin, Zn, thiamin, Fe, Mg and Mn. The nutritional value makes it suitable for maintaining optimum health, weight lose. It has excellent cooling properties. In addition to culinary properties, it has therapeutic properties and it is used for fibre extraction. It is easily digestible vegetable. It contains a gelatinous compound called luffein. It has lot of medicinal use. The oil from the seeds is known to cure coetaneous complaints, the roots have laxative effects and the juice from the leaves is used to cure granular conjunctivitis of the eye, adrenal type diabetes and hemorrhoids. The juice prepared from ridge gourd is a natural remedy for jaundice. It helps to purify blood and lowers the blood sugar level, also good for stomach and skin. It is having abortifacient, antitumor, ribosome inactivating and immunomodulatory activities. The fruit is demulcent, diuretic and nutritive. Seeds possess purgative and emetic properties. The pounded leaves are applied locally to aplenitis, haemorrhoides and leprosy. The leaf juice prevents the lids adhering at night from excessive meibomian secretion **(Rahman *et al.*, 2008)**.

Cytology

Luffa acutangula has chromosome number, 2n=2x=26, and all species of genus *Luffa* are cross compatible. Though the *Luffa* species can be crossed among themselves, they have remained isolated in nature. The isolation of species may be genic and not chromosomal. Cytological investigations have been conducted by **Dutt and Roy (1971 and 1990)** among the two cultivated species *L. acutangula* and *L. cylindrica* and two wild species *L. graveolens* and *L. echinata*. Chromosome count in all the species were found to be same (2n=26). Comparative morphology of the wild and cultivated species and chromosome pairing in interspecific hybrids suggests *L. graveolens* as the prime species which has given rise to the two cultivated monoecious species *L. acutangula* and *L. cylindrica*. The sex expression *i.e.*, monoecism and dioecism have, however, evolved independently because the advanced sex form of dioecism to be found only in wild primitive species *L. echinata*. In general both way crosses between *L. acutangula* and *L. cylindrica* are successful in the sense that the F_1 hybrid plants were obtained **(Thakur and Choudhury, 1967)**. High pollen sterility existed in the interspecific hybrids and seed setting was poor. Fertility was restored when F_1 hybrid was crossed to either of the parents. Cytological studies by **Pathak and Singh (1949)** of the interspecific hybrids and later **Trivedi and Rao (1976)** found that each chromosome of haploid complement of the two species are sufficiently homologous but non-homologous segments are present in normally pairing chromosomes of F_1 hybrids between two species. Without artificial pollination, however, the two species do not hybridize, in view of the sterility barriers, which are effective in preventing exchange of genes between the two species. Amphidiploids have been produced between *L. acutangula* and *L. graveolens* having 2n=52 chromosomes. Some degree of pollen sterility was recorded which may be due to the formation of univalents and multivalent **(Dutt and Roy, 1976)**. Autotetraploidy was induced in *L. acutangula* (4x = 52); however, such autotetraploids have no economic value **(Roy and Dutt, 1972)**.

Genetics

Genetics of few traits have been worked out in ridge gourd (Table 10.1).

Table 10.1: Genetics of different Traits in Ridge Gourd

Trait	Number of Gene	Type of Gene Action
Colour of node	Monogenic (NN)	Blue colour dominant to green
Seed surface	Monogenic (PP)	Pitted surface dominant to smooth
Sex expression	Digenic	Monoecious condition dominant to hermaphrodite
Fruit bearing habit	Monogenic	Non-clustering habit of fruit dominant over clustering
Corolla colour	Monogenic (YY)	Orange-yellow dominant over lemon-yellow
Fruit surface	Monogenic (RR)	Ridged fruit surface dominant over smooth
Fruit taste	Monogenic (BB)	Bitterness dominant over non-bitterness. The gene Ba is suppressor gene
Type of stamen	Monogenic (FF)	Partially fused stamens dominant over free

Source: Peter, 2008.

The genetics of flowering habit are controlled by two independent suppressor genes *A* and *G*, *A* suppressing the maleness and *G* femaleness. The genotypes of various flowering habits, *i.e.* monoecious and trimonoecious (*A-G-*), andromonoecious (*aaG-*), gynoecious (*A-gg*) and hermaphrodite (*aagg*), have been studied **(Choudhury and Thakur, 1965)**. **Singh *et al.* (1948)** studied monoecious, hermaphrodite and andromonoecious sex forms and in all of which true breeding was found on selfing. This indicated homozygosity for above forms. **Singh (1958)** identified four sex forms in ridge gourd. **Singh *et al.* (1948), Richharia (1948b),** and **Roy and Mishra (1967)** studied inheritance of sex in *Luffa*. Again they reported that principally two genes are involved in production of various sex forms which are depicted in Table 10.2.

Table 10.2: Sex Forms in *Luffa*

Sex Forms	Genotypes	Behaviour on Selfing
Monoecious	AAGG	Monoecious
Gynoecious	A-gg	Unstable and do not breed true
Androecious	aaGG	-
Andromonoecious	a'a'GG	Andromonoecious
Gynomonoecious	AAg'g'	-
Hermaphrodite	aagg	Hermaphrodite
	a'a'gg	Hermaphrodite
	aag'g'	Hermaphrodite
	a'a'g'g'	Hermaphrodite

Different gene combinations giving rise to intermediate sex forms are unstable and rarely found in nature. **Richharia (1948a)** analysed a cross between a monoecious selection of *Luffa acutangula* (Jhingha) and a hermaphrodite selection (Satputia). F_1 was monoecious and F_2 segregated into three forms *i.e.,* female, monoecious and hermaphrodite. Based on the data it has been suggested that sex inheritance in the mentioned cross depends on two factor pairs AA and BB. The A is epistatic to B which prevents expression of femaleness. Thus plants of constitution aaBB or aaBb are females, AABB, AaBB, AaBb are monoecious and aabb are hermaphrodite. Scientific reports of **ICAR-IARI (1963)** contained following information on the inheritance of sex in ridge gourd.

Inheritance of sex in ridge gourd		
P monoecious →	**X**	hermaphrodite
	↓	
F_1 →	monoecious	
	↓	
F_2 monoecious :	Androecious : gynoecious	: monoecious
9	3 3	1

The above finding indicates action of two pairs of genes in *L. acutangula* in sex expression.

The corolla colour, the type of androecium, the fruit surface and the seed surface are monogenically controlled. The bitterness is controlled by *Bi* gene **(Thakur and Choudhury, 1966)**. Yield has a positive correlation with the number of fruits per plant. Fruit length and days to flowering exhibit high heritability and genetic advance **(Kalloo, 1993)**.

Floral Biology and Botany

Ridge gourd is monoecious with annual vines. Tendrils are branched. Leaves are five to seven lobed, nearly glabrous. Flowers are yellow and showy. There are five petals. The staminate flowers are in racemes while the pistillate flowers are solitary and short or long pedunculate. Anthers are free. Pistil has three placentae with many ovules. The stigmas are three and bilobate. Sepals are 5-partite, gladular. Corolla obovate, petals are 5, 2.0 to 2.5 cm in length and yellow in colour. Stamens are usually 3, synandrous or syngenesious. Carpels are 3, syncarpous, overy inferior, filiform with 10 longitudinal ridges on which swollen glands are borne. ovules are many with one short style and trifid stigma **(Hazra and Som, 2009)**. Generally, the plants have a fairly long tap root with lateral roots, confined to top layer of about 60 cm and adapted to grow in river-beds to utilize sub-terranean moisture and also have xerophytic habit. The stems are branched (3 to 8), prostate or climbing, the branches cover large areas; nodes usually root when they touch the soil. The fruit is oblong or cylindrical with ridges. Fruit has about ten distinct longitudinal acutely angled ribs. The rind becomes dry on maturity **(Ram, 1997)**. Fruits are many seeded pepo, club shaped, variable in size, crowned by large sepals, stipule persistent. Seeds are numerous, black and pitted, average 100 seed weight is 9.5 gram. The key characters of this genus are fleshy fruits and flowers are in racemes. Other related species are *L. operculata* (L.) Cogn. grown in tropical America, *L. umbellata*, *L. pentalldra*, *L. giganta*, *L. scabra* and *L. narylandica* **(Peter, 2008)**.

Breeding Behaviour

Ridge gourd is highly cross-pollinated in nature due to predominant monoecious sex form. It is entomophillous and major pollinator is bee. Although 100 per cent natural cross pollination occur in monoecious and gynoecious sex form, yet reasonable extent of self fertilization happens because of geitenogamy due to foraging of insect pollinators in both pistillate and staminate flowers of the same vine. It does not suffer much from inbreeding depression. Anthesis of both staminate and pistillate flowers generally starts in the afternoon hours from 3.00 to 8.00 p.m. Anther dehiscence in the staminate flower generally coincides with the anthesis. Pollen fertility is maximum on the day of anthesis which may remain viable upto 1 to 3 days after anthesis in ambient condition depending on the season. Stigma receptivity is at its maximum at the time of anthesis although the stigma becomes receptive 6 hours earlier to 48 hours after anthesis. Best time of hybridization is between 2.00 to 6.00 p.m. on the day of anthesis. Bagging the staminate and pistillate flowers prior to opening followed by hand pollination and subsequent bagging for two more days is the commercial method of hybrid seed production in ridge gourd. In monoecious sex form, harvesting the fully mature staminate flower buds in the morning hours that will open in the afternoon hours on that day and keeping them covered with thin moist cloth for their use in cross-pollination in the afternoon hours is the conventional practice commercially **(Hazra and Som, 2009)**.

Breeding Objectives

☆ Plant growth: vigorous and well branched

☆ Earliness: first pistillate flower at a lower node number

☆ High female to male sex ratio

☆ Fruit quality:

a. Shape and size — uniformly cylindrical, medium-long fruits. Fruits with bulged blossom end and thick neck are not preferred. Extra long fruits often break and get damaged on transportation.

b. Colour — fruit colour to be selected depends on consumer's preference. As usually green or dark green colour is preferred.

c. Non bitter fruit

d. Tender and non fibrous fruit at marketable stage

☆ Yield: early and high productivity

☆ Biotic and abiotic resistance

Breeding Methods Adopted

☆ Single plant selection

☆ Mass selection

☆ Pedigree method

☆ Bulk population method

☆ Recurrent selection

☆ Back crossing

☆ Interspecific hybridization

☆ Heterosis

☆ Polyploidy breeding method

Works on Crop Improvement Exploring Biodiversity

Rich genetic diversity in wild and cultivated species of *Luffa* has been augmented as natural habitat in North-Eastern region of India. *L. acutangula* var. *amara* occurs in Peninsular India and *L. echinata* in the Western Himalaya and upper gangetic plains. Another important species, *L. graveolens* occurs in Bihar, Sikkim and Tamil Nadu. The Indian Institute of Vegetable Research (IIVR) is a National active germplasm site for the systematic management and utilization of germplasm wealth by collection, evaluation, maintenance and distribution of germplasm **(Rai et al., 2008)**. IIVR collections of ridge gourd are 68 in number **(iivr.org.in)**. The ICAR-National Bureau of Plant Genetic Resources (ICAR-NBPGR), New Delhi is the prime organisation for plant genetic resource management at national level. **Chandra 1995** reported that ICAR-NBPGR has collected 216 numbers of accessions in ridge gourd since 1986. These collections were evaluated systematically for 50 agro-botanical and economic characters and found to vary in vine length, sex type, fruit shape, number of fruiting node, fruiting potential, marketable fruit yield and span of fruiting period *etc*. Some promising lines in ridge gourd and *L. hermaphrodita* have been identified. Gynoecious lines are available and can be used for production of hybrids. *L. cylindrica* as female is easily crossable with *L. acutangula* as male. According to **Dhillion et al., 2001** NBPGR has collected 511 numbers of accessions in ridge gourd among which some lines *viz.*, NIC 20402 is found for early fruiting; IC 92779, IC 12136, NIC 957, NIC 10216, NIC 10224 and NIC 20213 are found high yielder; NIC 10222, NIC 10232, NIC 10288, NIC 10213 and NIC 10215 are having cluster fruit bearing. **Stoner, 1970** reported the ridge gourd variety, Pusa Nasdar is resistant to Red pumpkin beetle. At Govind Ballab Pant University of Agriculture and Technology total 104 numbers of accessions in ridge gourd are available among which PRG-6, PRG-7, PRG-48, PRG-54, PRG-88 and PRG-92 are found important in respect to different horticultural traits **(Ram et al., 1997)**. The Genetic Resource and Seed Unit (GSRU) of A.V.R.D.C.-The World Vegetable Centre have collected 306 numbers of diverse germplasm till 2006 **(GRSU, 2006)**. The evaluation of indigenous and exotic germplasm and their hybridization resulted in the selection of some superior varieties in ridge gourd (Table 10.3).

In China a ridge gourd F_1 hybrid, Fen Keng has been developed from the cross, KR 91-1-1 X 2-1-1. It is early maturing, vigorous, produces dark green fruits, 62 cm long with an average yield of 23.7 tones/ha **(Bose, 2002)**.

Out of nine species of *Luffa* found worldwide, seven are native to India. These comprise two major crops [*L. acutangula* (L.)Roxb. And *L. aegyptiaca* Mill.], one minor crop (*L. hermaphrodita* Singh and Bhandari) and four wild taxa (*L. graveolens* Roxb., L.

Table 10.3: Improved Cultivars of Ridge Gourd Identified/Released in India

Cultivar	Parentage	Source
Pusa Nasdar	Selection from a local cultivar of Neemuch, Madhya Pradesh	IARI, New Delhi
Pusa Sadabahar	Selection from local land races	ICAR-IARI, New Delhi
CO-1	Selection from a local type	TNAU, Coimbatore
CO-2	Selection from a local cultivar	TNAU, Coimbatore
PKM-1	Induced mutation	TNAU, Coimbatore
Haritham	Selection from local cultivar	KAU, Kerala
Punjab Sadabahar	Selection from a local cultivar	PAU, Ludhiana
Satputia	Selection from a local variety of Bihar	RAU, Sabour (Bihar)
Arka Sumeet	Pedigree selection of IIHR-54 x IIHR 24	IIHR, Bengaluru
Arka Sujat	Pedigree selection of IIHR-54 x IIHR 18	IIHR, Bengaluru
IIHR-8	Selection from local land races	IIHR, Bengaluru
Swarna Manjari	Pure line selection	HARP Ranchi
Swarna Uphar	Pedigree selection	HARP Ranchi
Konkan Harita	Selection from local cultivar	KKV, Dapoli, Maharashtra
Pant Tori-1	Pureline selection from indigenous germplasm	G.B.P.U.A and T., Pantnagar
Hisar Kalitorai	Selected from local germplasm	HAU, Hisar, Haryana

Source: Swarup, 2006 and Kalloo *et al.*, 2006.

echinata Roxb., *L. tuberosa* Roxb. and *L. umbellata* M. Roem.) **(Chakravarty, 1982).** *L. acutangula* (Ridge gourd) and *L. aegyptiaca* (Sponge gourd) are widely cultivated in the plains and low hills of India **(Chandra, 1995)** and a lesser known local cultivar 'Satputia' (Sometime classified as a separate species, *L. hermaphrodita*) is commonly grown in the Indo-gangetic plains of Bihar and Uttar Pradesh **(Sirohi *et al.*, 2005).** In a research carried out by **Prakash *et al.*, 2013** regarding morphological variability in cultivated and wild species of *Luffa* (Cucurbitaceae) from India, it was found that characters such as size, lobbing pattern and surface texture of leaf, flower colour, flowers if borne single or in cluster, fruit shape, pericarp texture, seed shape, coat colour and extension on seed coat were distinct among the cultivated and the wild species and therefore found more reliable for taxonomic delineation. The researchers have collected 70 germplasms of different species of *Luffa viz.*, *L. acutangula*, *L. aegyptiaca*, *L. acutangula* var. *amara*, *L. hermaphrodita*, *L. graveolens* and *L. echinata* from 10 different states of India and reported morphological diversity among them (Table 10.4). **Filipowicz *et al.*, 2014** reported that over a period of about ten years, C. Heiser, E. Schilling and the Indian geneticist B. Dutt collaborated on *Luffa*, using biosystematics approaches such as experimental crossing to infer reproductive barriers, chromosome counts and measurement of vegetative and reproductive characters from living plants. The material from their cultivation experiments was carefully documented and deposited in the herbarium of the Indiana University in Bloomington. The results led to the recognition of three species in the New World: *L. quinquefida* occurring from Mexico (Gulf of California) to Nicaragua, *L. operculata*

Ridge Gourd | 235

Table 10.4: Morphological Diversity among different *Luffa* Species found in India

Sl.No.	Characters	L. aegyptiaca	L. acutangula	L. hermaphrodita	L. graveolens	L. echinata
1	Stem (cross section)	Angular	Angular	Angular	Sub-angular	Sub-angular
2	Tendril (Number)	Three to four	Three to four	Three to five	Two	Two
3	Cotyledonary leaf shape	Elliptic-oblong	Elliptic	Elliptic	Elliptic	Elliptic
4	Leaf margin	Entire-sinuate	Dentate	Dentate	Denticulate	Denticulate
5	Leaf shape	Ovate, Orbicular	Ovate, Orbicular, reniform	Ovate, Orbicular, reniform	Sub-orbicular, reniform	Sub-orbicular, reniform
6	Leaf lobbing	Shallow-deeply five-seven lobbed	Shallow-deeply five lobbed	Shallow-deeply five lobbed	Lobes are not prominent	Shallowly five lobed
7	Leaf texture	Densely scabrid on upper surface, puberulent on lower surface	Both surfaces puberulent	Both surfaces puberulent	Upper surface sparsely scabrid, lower smooth or puberulous on nerves	Sparsely puberulent on both surfaces, scabrous
8	Leaf colour	Dark green with patchy areas on dorsal surface	Uniformly light green	Light green	Dark green	Dark green
9	Sex form	Unisexual and monoecious	Unisexual and monoecious	Bisexual	Unisexual and monoecious	Unisexual and monoecious
10	Flower colour	Bright yellow	Light yellow	Light yellow	Bright yellow	White
11	Fruit shape	Elongated, cylindrical	Elongated-cylindrical and spindle (in amara type)	Elongated-cylindrical	Cylindrical-oblong	Spherical-round
12	Fruit skin texture	Smooth	Rugose with deeply ribbed surface	Smooth	Tuberculate hairs, rough	Echinate, long bristles
13	Fruit born singly or in cluster	Single	Single	Cluster	Single	Single

Sl.No.	Characters	L. aegyptiaca	L. acutangula	L. hermaphrodita	L. graveolens	L. echinata
14	Seed shape	Oblong-elongated	Ovate	Ovate	Oblong-ovate	Oblong-ovate
15	Seed coat colour	Ivory, black, greyish-black	Greyish black-black	Greyish black-black	Grey	Grey
16	Seed coat extension	Wing	Beak (notched)	Beak (notched)	Absent	Absent
17	Seed coat surface	Smooth, non-shiny	Smooth, non-shiny	Smooth, shiny	Rough, non-shiny	Rough, non-shiny

from Panama to southern Brazil and *L. astorii* in coastal Venezuela, Ecuador and Peru. Linaeus originally described *L. acutangula* as "habitat of Tataria, China", but subsequent collection revealed the presence of that species also in Northern India. Further distinct populations occur in Yemen, where they have been interpreted as feral introduced *L. acutangula* and in Australia, where they were described as *L. cylindrica* var. *leiocarpa* (a synonym of *L. sylvestris*). These researchers have concluded the presence of eight monophyletic species of *Luffa* (Figure 10.1), three in New World, four in Asia and one in Australia with their correct name and natural distribution (Table 10.5).

Table 10.5: Species Diversity in *Luffa* with their Natural Distribution Worldwide

Species	Natural Distribution (Country)
Luffa acutangula	India, Yemen
Luffa saccata	Australia
Luffa graveolens	India, Nepal, Bangladesh, Myanmar
Luffa aegyptiaca	Australia, Laos, Indonesia, Samoa, Papua New Guinea, India, Vietnam, China
Luffa echinata	Egypt, India, Sudan
Luffa astorii	Per, Ecuador, Galapagos,
Luffa quinquefida	Mexico, Nicaragua, USA
Luffa operculata	Panama, Guyana, Senegal, Peru, Brazil

In numerous attempts over three years of crossing between *L. astorii* and *L. operculata*, **Heiser et. al. (1988)** obtained a single viable seed that gave rise to a plant with abnormally developed flowers that produced no anthers. No fruits were secured when it was pollinated by the parental species. Hybrids between *L. astorii* and *L. quinquefida*, and between *L. operculata* and *L. quinquefida* were fairly readily obtained with the latter species as the pistillate parent. Few seeds could be grown to maturity, and pollen viability and seed set in the F_1 plants were low. They found *L. astorii* differs morphologically and physiologically from *L. operculata* and *L. quinquefida*. The corollas in *L. astorii* are deep yellow with lobes of the male flowers over 12 mm long, while *L. operculata* and *L. quinquefida* have smaller male flowers with pale or deep yellow corolla respectively. *L. astorii* flowers in early July, *L. operculata* in late August or September and *L quinquefida* in mid-June. *L. saccata* differs from *L. graveolens* in the much longer pedicels of its male flowers **(Telford et al., 2011).**

The hermaphrodite sex form in *L. hermaphrodita* with compound racemes which is a primitive character is believed to have been derived from the monoecious species *L. graveolens*. High genetic variability with low genetic erosion is reported in *L. hermaphrodita* **(Chandra, 1995).** Generally, the cultivars bear seven fruits in each inflorescence and this appears to be a constant feature. Based on this trait, it is locally called 'Satputria or 'Satputia' meaning seven children **(Ali and Pandey, 2005-06).** To salvage and conserve the current spectrum of landrace diversity in different agri-horticultural crops from Adilabad district in Andhra Pradesh, India, **Pandravada *et al.*, 2014** have undertaken some biodiversity surveys. A *Luffa* species

Figure 10.1: Fruits and Flowers of *Luffa*. The scale bar corresponds to 1 cm.

A. *Luffa acutangula*, India, Rajasthan. B. *Luffa acutangula* var. *amara*, mature dehiscent fruit, India, Rajasthan. C. *Luffa aegyptiaca* immature fruit, Australia, Northern Territory. D. *Luffa graveolens* immature fruit, India, Uttar Pradesh. E. *Luffa astorii* immature fruit, Peru, Lambayeque. F. *Luffa aegyptiaca* dried fruit with the fibers that are used as a vegetable sponge, Africa, Tanzania G. *Luffa operculata* dried dehiscent fruit, Colombia, Rio Magdalana. H. *Luffa astorii* flower, Peru, Lambayeque. I. *Luffa saccata* male flower, Australia, Northern Territory. J. *Luffa echinata* male flower, Botanical Garden, Germany, Munich. K. *Luffa quinquefida* male flower. L. *Luffa quinquefida* leaf and immature fruit, both from Mexico, Sonora. M. *Luffa astorii* leaf, Peru, Lambayeque.

having small, elliptical and cluster bearing fruits was seen (Figure 10.2) being cultivated in the backyards of the tribal farmers in 10 mandals of Adilabad district.

Figure 10.2: *Luffa hermaphrodita*: **(Left) flowering in bunches; (right) fruiting in typical clusters and (inset) seeds found in Adilabad district, Andhra Pradesh, India**

Close observation of the plants and fruit samples *in-situ* and characterization for different qualitative and quantitative trait; specially the nature of flowering and fruiting, during field evaluation at NBPGR Regional Station, Rajendranagar, Hyderabad confirmed the plant species as *L. hermaphrodita*, popularly known as Satputia which is traditionally occur in the Indo-Gangetic plains and the cultivation of this species in Adilabad distric is the first report of its distribution in South India. In this district this cultivar is under cultivation since a long time and referred to as 'gutti beera (clustered ridge gourd) in Telugu language denoting its cluster bearing nature and 'todka' in gondi language.

Another wild species *L. echinata* is a spreading herb, wild bifid bristly or smooth tendrils and extremely bitter in taste, grows widely in Pakistan, India, Bangladesh and Northern Tropical Africa. In India, it is mainly found in Gujarat, Bihar, Rajasthan and Madhya Pradesh. The flowering occurs in September and October. It is known by various names *viz.*, English: Bristly Luffa; Sanskrit: Koshataki; and in Hindi: Bindaal, Bidali, Kukurlata and Ghagerbel. Being tremendously medicinally important it is well known ethanomedicinal plant used in Ayurveda. Different plant parts like root, stem, leaf and fruit are used in curing jaundice, fever, enlarged liver or spleen, diabetes, hemorrhoid, leprosy, rheumatism *etc.* **(Modi and Kumar, 2014).** A complete mature and dried fruit (Figure 10.3–10.6) of this plant more particularly the mature sponge of the fruit is used for curing dog bite victim by squeezing the sponge after soaking it in glass of water for 5-10 minutes. The squeezed material of the fruit is extremely bitter and given to victim of dog bite in the morning with empty stomach and one glass is enough to cure the patient without any side effect **(Yadav and Kumar, 2013).**

Dried fruits of *Luffa echinata* which is used in to treat dog bite victims

Figure 10.3

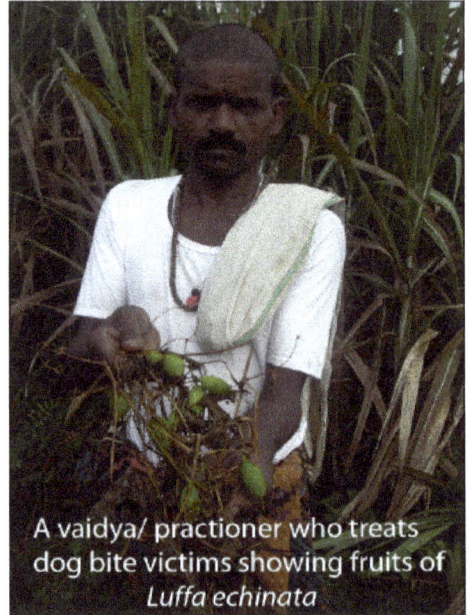
A vaidya/ practioner who treats dog bite victims showing fruits of *Luffa echinata*

Figure 10.4

Green Plants of *Luffa echinata* with flowers

Figure 10.5

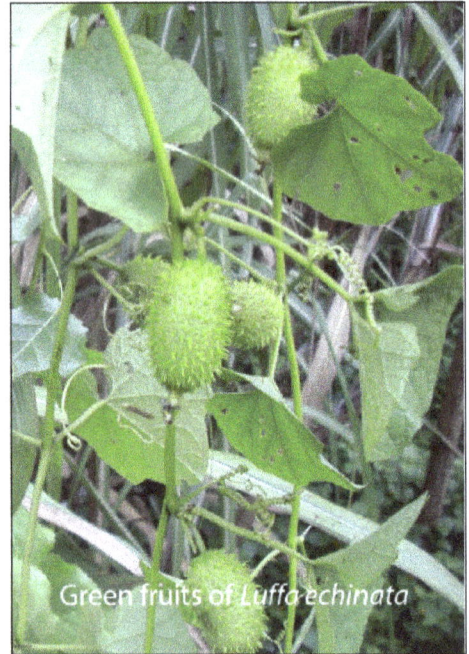
Green fruits of *Luffa echinata*

Figure 10.6

In a study by **Pradeepkumar** *et al.,* **2010** in Trichur, Kerala, an off type was detected in a population of ridge gourd which was characterized by the production of rudimentary male flowers in racemes (Figure 10.7) in contrast to the bright yellow flowers in male fertile plants.

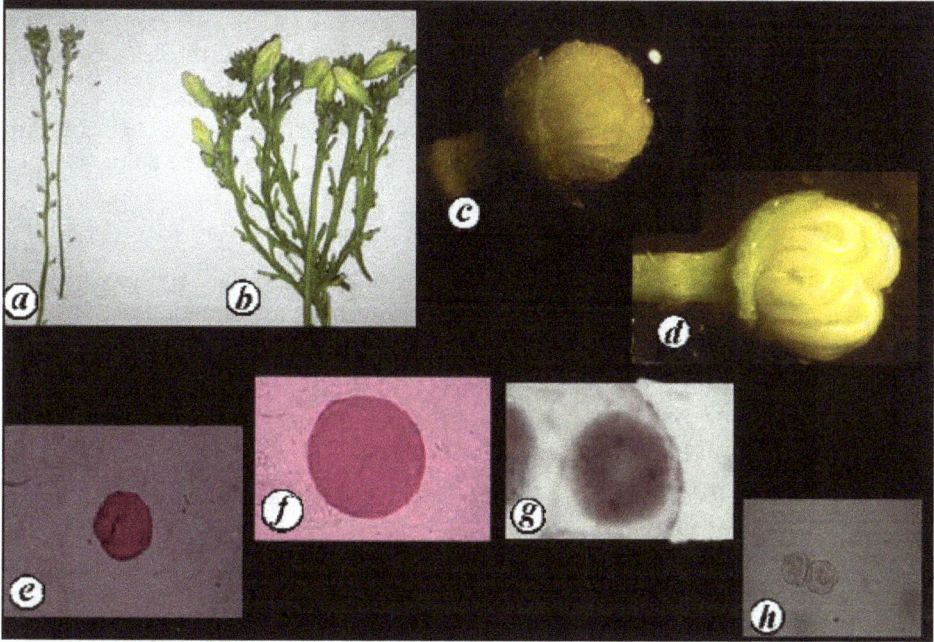

Figure 10.7

a: Sterile male flowers in raceme; b: Male fertile and male sterile flowers; c: Anther of male sterile plant; d: Anther of male fertile plant; e: Pollen of male sterile plant; f: Pollen of male fertile plant; g: Meiosis of pollen mother cells (PMCs) in male sterile plant; h: Microspore of male sterile plant.

On comparison of the anthers of the suspected male sterile line with those of monoecious ridge gourd variety named Haritham, a striking difference was noticed with respect to the morphology of anther lobes. In the male sterile line it was flat, highly pubescent and with few shrunken pollen grains whereas in cv. Haritham, it was plump and filled with abundant large fertile pollen grains. It was sporogenous form of male sterility. Female flowers of the male sterile line were crossed with normal monoecious variety Haritham and all the F_1 population observed, was male sterile in sex expression characterized by shrunken and sterile microspores, indicating the heritability and dominant nature of that male sterile character. Like muskmelon, male sterility can be exploited in heterosis breeding programme and development of F_1 hybrids in ridge gourd. Therefore these biodiversity can effectively be proven as tool for crop improvement.

References

Ali A M and Pandey A K 2005-06 Systematic Studies on the Family Cucurbitaceae of Eastern Bihar, India. Cucurbit Genetics Cooperative Report, **28-29**: 66-69.

Arora R K and Nayar E R 1984 Wild relatives of crolp plants in India, NBPGR Sci Monograph No. 7. pp. 90. National Bureau of Plant Genetic Resources, New Delhi – 110012.

Bose T K, Kabir J and Maity T K 2002 Vegetable Crops: Volume I. Calcutta, Naya Prakash. pp. 505.

Chakravarty H L 1982 Fascicles of Flora of India, Fascicle II, Cucurbitaceae. Botanical Survey of India, Howrah, India. pp.73.

Chandra U 1995 Distribution, domestication and genetic diversity of *Luffa* gourd in Indian subcontinent. Indian Journal of Plant Genetic Resources **8** (2): 189–196.

Choudhury B and Thakur M R 1965 Inheritance of sex forms in *Luffa*. Indian Jurnal of Genetics and PlantBreeding, **25:** 188.

Dhillon B S, Varaprasad K S, Srinivasan K, Singh M, Archak S, Srivastava U and Sharma G D 2001 (Eds) National Bureau of Plant Genetic Resources: A Compendium of Achievements. National Bureau of Plant Genetic Resources, New Delhi. pp 329.

Dutt B and Roy R P 1971 Cytogenetic investigations in *Cucurbitaceae* I. Interspecific hybridization in *Luffa*. Genetica, **42:** 139-156.

Dutt B and Roy R P 1976 Cytogenetic studies in an experimental amphidiploid in *Luffa*. Caryologia, **29:** 16-25.

Dutt B and Roy R P 1990 Cytogenetics of the Old World specis of *Luffa*. In: Bates. D.M., Robinson, R.W. and Jeffrey, C. (Eds) Biology and Utilization of the *Cucurbitaceae*. Cornell University Press. Ithaca. New York. pp. 134-140.

GRSU, AVRDC-The World Vegetable Centre 2006 URL: http: //203.64.245.173/avgris/search.asp

Hazra P and Som M G 2009 Vegetable Seed Production and Hybrid Technology. Kalyani Publishers, New Delhi-110002. pp. 156.

Heiser C B Jr and Schilling E E 1990 The genus *Luffa*: a problem in phytogeography. In: Bates. D.M., Robinson, R.W. and Jeffrey, C. (Eds) Biology and Utilization of the *Cucurbitaceae*. Cornell University Press. Ithaca. New York. Pp. 120-133.

Heiser C B, Schilling E E and Dutt B 1988 The American species of *Luffa* (Cucurbitaceae). Systematic Botany, **13:** 138–145.

IARI New Delhi 1963 Scientific Reports. Indian Agricultural Research Institute (ICAR), New Delhi. pp. 209.

iivr.org.in/germplasm-collection

Jeffery C 1990 Systematics of the cucurbitaceae: an overview. (In) Bates, D.M., Robinson, R.W. and Jeffery, C. (Eds) Biology and Utilization of the Cucurbitaceae. Ithaca, New York, Cornell University Press. pp. 3-9.

Kalloo G 1993 Loofah-Luffa spp. (*In*) Kalloo, G. and Bergh, B.O. (Eds). Genetic Improvement of Vegetable Crops. Pergamon Press. pp. 265-266.

Kalloo G, Rai M, Kumar R, Prassana HC, Singh M, Kumar S, Singh B, Ram D, Pandey S, Lal H, Rai S, Pandey KK and Sathpathy S 2006. New Vegetable Varieties from IIVR Varanasi. Indian Horticulture **51** (3): 16-22.

Modi A and Kumar V 2014 *Luffa echinata* Roxb.-A review on its ethanomedicinal, phytochemical and pharmacological perspective. Asian Pacific Journal of Tropical Disease, **4** (Suppl 1): S7-S12.

Pandravada S R, Sivaraj N, Jairam R, Sunil N, Begum H, Reddy M T, Chakrabarty S K, Bisht I S and Bansal K C 2014 *Luffa hermaphrodita*: First Report of its Distribution and Cultivation in Adilabad, Andhra Pradesh, South India. Asian Agri-History, **18** (2): 123-132.

Pathak G N and Singh S N 1949 Studies in the genus *Luffa*. I. Cytogenetic investigations in the interspecific hybrid L. cylindrica X L. acutangula. Indian Journal of Genetics and Plant Breeding, **9**: 18-26.

Peter K V and Pradeepkumar T 2008 Genetics and Breeding of Vegetable Crops. Indian Council of Agricultural Research, New Delhi-110012. pp. 192.

Philipowicz N, Schaefer H and Renner S S 2014 Revisiting *Luffa* (Cucurbitaceae) 25 Years After C. Heiser: Species Boundaries and Application of Names Tested with Plastid and Nuclear DNA Sequences. Systematic Botany, **39** (1): 205-215.

Pradeepkumar T, Hegde V C, Sujatha R and George T E 2010 Characterization and maintenance of novel source of male sterility in ridge gourd *Luffa acutangula* (L.) Roxb. Current Science, **99** (25): 1326.

Prakash K, Pandey A, Radhamani J and Bisht J S 2013 Morphological variability in cultivated and wild species of *Luffa* (Cucurbitaceae) from India. Genetic Resources and Crop Evolution, **60**: 2319-2329.

Prakash K, Pati K, Arya L, Pandey A and Verma, M 2014 Population structure and diversity in cultivated and wild *Luffa* species. Biochemical Systematics and Ecology, **56**: 165-170.

Rahman A H M N, Anisuzzaman M, Ahmed F, Rafiul-Islam A K M and Naderuzzaman A T M 2008 Study of nutritive value and medicinal uses of cultivated cucurbits. Journal of Applied and Scientific Research, **4** (5): 555-558.

Rai M, Pandey S and Kumar S 2008 Cucurbit research in India: a retrospect, p. 285-293. (In) M. Pitrat (Ed). Cucurbitaceae 2008, Proc. IXth EUCARPIA Mtg. on Genetics and Breeding of Cucurbitaceae, 21-24 May 2008. INRA, Avignon, France.

Ram H H 1997 Vegetable Breeding. Principles and Practices. Kalyani Publishers, Ludhiana-141008. pp. 423.

Ram H H, Singh D K, Upadhyay R, Kushawa, M L, Singh A and Tewari D 1997 Promising landraces and breeding lines of cucurbits for marginal lands in India. (In) E I Bassam, R K Behl and B Prochnow (Eds.) Sustainable Agriculture for food, Energy and Industry: Strategies towards Achievement-Proc. of the International Conference held in Braunchweig, Germany, June, pp 143-149.

Richharia R H 1948a Sex condition in interspecific cross *Luffa aegyptiaca* X *Luffa acutangula*. Current Science, **7**: 275-276.

Richharia R H 1948b Sex inheritance in *Luffa acutangula*. Current Science, **17** (12): 359.

Roy R P and Dutt B 1972 Cytomorphological studies in induced polyploids of *Luffa acutangula* Roxb. Nucleus, **15**: 17-21.

Roy R P and Mishra A R 1967 Cytogenetic investigations in Cucurbitaceae. Ph.D. thesis, University of Patna, Bihar, India.

Singh D and Bhandari M M 1963 The identity of an imperfectly known hermaphrodite *Luffa*, with a note on related species. Bailey, **11**: 132-141.

Singh H B, Ramanujam S and Pal B P 1948 Inheritance of sex forms in *Luffa acutangula*. Nature, **161**: 775.

Singh S N 1958 Studies in sex expression and sex ratio in *Luffa* species. Indian Journal of Horticulture, **15**: 66-75.

Sirohi P S, Munsi A D, Kumar G and Behera T K 2005 Cucurbits, pp.34-58. (In) B.S. Dhillon, R.K. Tyagi, S. Saxena and G.J. Randhawa (Ed.). Plant Genetic Resources: Horticultural Crops. Narosa Publishing House Pvt. Ltd., New Delhi, India.

Stoner A K 1970 Breeding for insect resistance in vegetables. Hortscience, **5** (2): 76-79.

Swarup V 2006 Vegetable Science and Technology in India. Ludhiana, Kalyani Publishers. pp. 426-431.

Telford I R H, Schaefer H, Greuter W and Renner S S 2011 A new Australian species of *Luffa* (Cucurbitaceae) and typification of two Australian Cucumis names, all based on specimens collected by Ferdinand Mueller in 1856. PhytoKeys, **5**: 21–29.

Thakur M R and Choudhury B 1966 Inheritance of some qualitative characters in *Luffa* species Indian Journal of Genetics and Plant Breeding, **26**: 79-86.

Thakur M R and Choudhury B 1967 Interspecific hybridization in the genus *Luffa*. India Journal of Horticulture, **24**: 87-94.

Trivedi R N and Rao R 1976 Interspecific hybridization and amphidiploid studies in the genus *Luffa*. Genet lber, **28**: 83-106.

Yadav U and Kumar M 2013 *Luffa echinata*: A valuable medicinal plant for the victims of dog bite. Octa Journal of Environmental Research, **1** (1): 1-4.

Chapter 11

Large Cardamom

S. *Sreekrishna Bhat*

Indian Cardamom Research Institute, Spices Board,
Myladumpara, Idukki – 685 553, Kerala
E-mail: sk9bhat@yahoo.co.in

Large cardamom (*Amomum subulatum* Roxb.) is one of the oldest spices used by man. India is the largest producer of large cardamom in the world with 54 per cent share, followed by Nepal (33 per cent) and Bhutan (13 per cent). It is a member of Zingiberaceae family under the order Zingiberales (formerly known as Scitaminae) is one of the main cash crop cultivated in the sub-Himalayan mountains at an altitude ranging from 800 – 2000 m. Generally the cardamom is grown under agro-forestry system in the Himalayas either under the Himalayan alder (*Alnus nepalensis*) or mixed forest tree species. This crop thrives well in 6°C to 25°C with well distributed annual rainfall of 200-350 cm. It is well adapted to the hilly forest ecosystem where the fertility status is high due to natural nutrient recycling. Apart from Sikkim and Darjeeling district of West Bengal, large cardamom is also cultivated to a limited extent in some of the other North Eastern states. (Gupta, 2000). It is one of the main cash crops cultivated in Sikkim, Darjeeling district of West Bengal, Nagaland, Uttarakhand, Manipur, Arunachal Pradesh and some other parts of the North Eastern India covering an area of 26400 ha with an annual production of 5000 metric tones (Anon., 2016). Its local name is Alainchi. *A. subulatum* is indigenous to moist deciduous and semi-evergreen forests of sub-Himalayan tracts. The presence of several wild relatives and the tremendous variability within the cultivated species support the view of its origin in Sikkim (Subba, 1984, Rao *et al.*, 1993, Singh and Singh, 1996). The chromosome no. 2n = 54 (Das *et al.*,1998). Roxburgh (1820a) was first to describe this plant in his 'Plants of the Coast of Coromandel' and in 'Flora Indica' (1820b). The genus *Amomum* Roxb. is the second largest genus of the family Zingiberaceae with 150 species (Tripathi and Prakash, 1999).

The family is represented here by 19 genera and 88 species out of 22 genera and about 170 species reported from India. (Ved Prakash and Mehrotra,1996). This family has provided important aromatic spices that are widely used. The important spices of this family are *Zingiber* (ginger), *Curcuma* (turmeric), *Alpinia* (galangal), *Kaemferia*, all representing rhizomatous spices and *Elettaria* (small cardamom), *Amomum* and *Aframomum* (large cardamoms) representing seed spices (Anon. 1977). Earlier explorers like Baker J.G. (1890) have listed eight species of Amomums from eastern tropical Himalayan region (North Eastern States). *viz., A. aromaticum* Roxb., *A. corynostachium* Wall., *A. costatum* Benth., *A. dealbatum* Roxb., *A.kingii* Baker, *A. pauciflorum* Baker, *A.linguiforme* Benth. and *A.subulatum* Roxburgh. However, later workers Ved Prakash and Mehrotra (1996) listed only six species of Amomums, out of which the occurrence of *A.fulviceps* in North East India is to be confirmed. The North-Eastern India has largest concentration of Zingiberaceous flora (next to Malaysia) especially in number of genera.

Under the best conditions, the crops will start bearing fruits in 24 months. Flowering starts from March–May and harvesting begins from September–October, and may extend up to November in higher altitudes. In India, it was used as early as the 6th century BC in Ayurvedic preparations, as mentioned by Susrata (Sharma *et al*, 2000). It is also useful in treatment of flatulence, loss of appetite, gastric troubles, congestion and liver complaints (Jafri *et al.*, 2001). In China large cardamom is used in two of the following preparation. (i). Hsiang Sha Yangwe Pien - Tablets for gastro-intestinal troubles such as indigestion, vomiting and diarrhea. (ii). Xiang Sha Liu Jun Wan - Pills for the some stomach complaints and also used as an expectorant. (Gupta, 1983)

Generally the cardamom is grown under agro-forestry system in the Himalayas either under the Himalayan alder (*Alnus nepalensis*) or mixed forest tree species. This has proved to be a sustainable land use practice supporting multiple functions and ecosystems. This system helps in the management of slope land and helps conserving soil and water, maintain soil fertility by organic recycling. The cardamom based agro forestry system is a good traditional practice that would meet the demands of both economy activity and conservation of forest resources. Besides its economic importance it plays very important role in the conservation of agro ecosystem.

Etymology

The specific epithet *'subulatum'* is derived from a Latin word 'subula' means "awl", referring to the awl-shaped tips of bract and calyx projecting above the inflorescence

Vernacular Names

Bengali – *Bara Elachi;* English – *Large Cardamom, Greater Cardamom, Nepal Cardamom, Bengal Cardamom*; Hindi – *Bada Elaichi*; Malayalam – *Perelam*; Sanskrit – *Brihadaela;* Tamil – *Periyelam*; Telugu – *Peddayelakai*.

The area and production in Sikkim and West Bengal are given in Table 11.1.

The aboriginal inhabitants of Sikkim-The Lepchas-were believed to be the first to collect large cardamom capsules from natural forests primarily for the purpose of medicine and as an aromatic edible wild fruit. While those cardamom forests eventually converted into ownership and the crop was domesticated in the process

Table 11.1: Area, Production and Wholesale price of Large Cardamom (2015-16)

Sikkim				West Bengal				Price/kg (Rs). (2015-16)
Total Area (ha)	Yielding Area (ha)	Production (tones)	Productivity (kg/ha)	Total Area (ha)	Yielding Area (ha)	Production (tones)	Productivity (kg/ha)	
23,082	17,520	4,435	253	3,305	2,829	865	306	1,409

Source: Spices Board's Annual Report-2016.

(Sharma *et al*, 2008). It is the most important revenue-earning crop of Sikkim having the largest area and the highest production in the country. It is used as a spice and in several ayurvedic preparations. It is used as an ingredient as well as a flavouring agent with masala and curry powders; in flavoring sweet dishes, cakes and pastries; as a masticatory and for medicinal purposes. It is also used in industrial sector for flavouring toothpastes, sweets, soft drinks, toffees, flavoured milk and alcoholic drinks. The ripe fruits are eaten raw by people of Sikkim and Darjeeling and are considered a delicacy (Gyatso *et al*.,1980). Number of capsules per spike ranges from 10 to 35 depending on the cultivars. The fruit is a maroon colored echinated capsule. The seeds are white when immature and become dark brown to black when mature. Seeds are embedded in a mucilaginous coat. The number of seeds per capsule varies from 30 to 80 in different cultivars of large cardamom (Biswas *et al.*, 1986). It contains 2-3 per cent essential oils and possesses medicinal properties like carminative, stomachic, diuretic, cardiac stimulant (Nigam *et al.*, 1961, Verma *et al.*, 2010). The fruit on average comprises 70 per cent seeds and 30 per cent skin (Pruthi, 1993). Seeds contains 8.6 per cent moisture, 5 per cent total ash value, 1.5 per cent ash insoluble in acid, 3.5 per cent water soluble ash value, 4.88 per cent alcohol extract, 4 per cent non-volatile ether extract and 91.4 per cent of total solid (Shukla *et al.*, 2010).

Large Cardamom Gene Bank

Explorations for collection of germplasm of Large cardamom was carried out from the year 1986 by Indian Cardamom Research Institute (ICRI), Spices Board, Regional Station, Gangtok at various tracts of Sikkim (East, West, North and South Sikkim) Darjeeling district of West Bengal and Arunachal Pradesh for the collection of germplasm and exploitation of desirable genotypes for crop improvement programme and a gene bank at ICRI farm, Pangthang (East Sikkim) (2160 m MSL) and Kabi farm (North Sikkim) (1630 m amsl) consisting of 301 accessions has been established. Based on the evaluation of the high yielding accessions, two selections ICRI Sikkim -1 and ICRI Sikkim -2, have been released for cultivation in Sikkim and Darjeeling district of West Bengal. Both the selections are from cultivar Sawney. At Indian Cardamom Research Institute (ICRI), Spices Board, Gangtok, Large cardamom descriptor has been prepared and received IC nos. from ICAR-NBPGR, New Delhi for all the accessions collected.

Botanical Description

Large cardamom is a rhizomatous perennial herb having subterranean rhizomes with leafy shoot. The stem is a pseudo stem with 9 to 11 leaf sheath. The leafy shoots are known as tiller which varies from 30 to 80 in a single plant. The colour of the tiller varies from green to maroon. The height of the tiller ranges from 1.0 to 2.5 m depending upon the cultivar and girth 7 – 7.5 cm, swollen at base, green to pink. Leaves simple, alternate, distichous, 9 – 11, sessile to petiolate, produced mainly towards the upper 2/3 of shoot, lower 1/3 covered by sheaths; lamina, oblong-lanceolate, 30 – 81 X 7.5 – 10.5 cm, thick, drooping, dark green on upper side, pale beneath, glabrous; margin entire, wavy, hyaline, slightly revolute, glabrous; apex long-acuminate, up to 4.5 cm long. Ligule 4 – 6 mm long, deeply emarginate, apex

Figure 11.1: Large cardamom (*Amomum subulatum* Roxb.).

nearly rounded. Inflorescence - spike 6 – 12 X 3 – 6 cm, 1 – 3 per leafy shoot, radical, clavate, compact, 1 or 2-flowered, a little elevated above the soil, elongate during flowering, with 20 to 40 flowers 4.7 – 5.1 cm long, yellow, borne singly from each bract and bracteole and flowers are zygomorphic, bisexual, pollinated by bumble bees. Calyx cylindric, tubular, 2.5 – 3.5 X 0.6 – 0.9 cm, 3-clefted, longer than corolla tube, reaches ¾ of the flower, pale green-yellow at tip, white towards base; Corolla tube 1.8 – 2 cm long, 4 mm at mouth. Labellum 2.4 – 2.6 X 1.2 – 1.3 cm, oblong, longer than corolla lobes, dark yellow. Lateral staminodes subulate, 4 – 9 mm long, Stamen 1, perfect, 2.2 – 2.6 cm long, nearly equal to the lip; filament slightly concave, 6 – 9 x 2 – 3 mm, creamy white; anther 2-celled; thecae oblong, 0.9 – 1.1 x 0.1 cm, dehiscing throughout the length. Ovary inferior, barrel-shaped, 4 – 5 x 3 – 4 mm, pink, pale towards tip, sparsely hairy, 3-loculed; ovules many, style filiform, 3.7 – 3.9 cm long, sparsely hairy, white; stigma subglobose, 1 mm across, minutely red-spotted, situated at the tip of the anther thecae, Inflorescence 13 –19 cm long; peduncle 6 – 11 cm long, elongate during fruiting. Fruit a capsule, 14 – 20 per spike, conical, 2.5 – 3 cm long, 5.5 – 6 cm in girth, fresh weight *c.*4.39 g, dry weight 0.64 g, maroon-coloured; 10 – 13-winged, wingless towards base, irregularly lobed, glabrous; bract, bracteoles and calyx are persistent in fruit; calyx 3 – 3.7 cm long. Seeds 40 – 50, 2.5 m, black, bold, glabrous, sweet (Table 11.2) (Thomas *et al.*, 2009).

Plant Propagation

Propagation of large cardamom is done through seeds and rhizomes. Conventionally, large cardamom is propagated vegetatively though suckers having

Table 11.2: Comparison of Characters between Wild *Amomum subulatum* Roxb and its Cultivars

Sl.No.	Characters	Wild	'Dzongu' Golsey	'Green Golsey'	'Sawney'	'Seremna'	'Varlangey'
1	Rhizome colour	Dark pink	Pale pink	Pale pink	Dark pink	Dark pink	Pale pink
2	Leafy shoot - height (cm)	150 - 285	100 - 145	155 - 200	150 - 190	150 - 200	200 - 250
3	Leafy shoot - girth (cm)	7 - 7.5	3.7 - 4	5 - 6	4.5 - 5.3	5 - 5.7	5.5 - 6.3
4	Tiller colour	Green-pink	Green	Green	Maroon	Maroon	Maroon
5	Tiller nature	Robust	Not robust	Robust	Robust	Robust	Robust
6	Sheath width (cm)	2.2 - 2.8	2.3	4	2.8 - 3	2.7 - 3	3.5
7	Lamina nature	Slightly drooping	Erect	Erect	Slightly drooping	Well drooping	Slightly drooping
8	Lamina length (cm)	30 - 81	22 - 45	35 - 55	40 - 60	40 - 60	60 - 80
9	Lamina width (cm)	7.5 - 10.5	6.5 - 8	4.5 - 10	6 - 10	7 - 10	6 - 10
10	Lamina shape	Oblong-lanceolate	Elliptic- lanceolate	Lanceolate	Lanceolate	Lanceolate to elliptic- lanceolate	Narrowly lanceolate
11	Lamina - length of tip (cm)	4.5	3.2	3.1	3.5	4.5	5
12	Lamina margin	Wavy	Straight	Wavy	Wavy	Wavy	Wavy
13	Petiole length (cm)	0 - 2.5	0 - 0.3	0 - 0.5	0 - 0.5	Usually absent	0 - 0.4
14	Ligule size (mm)	4 - 6	3 - 5	4 - 5	5 - 7	4 - 5	5
15	Ligule colour	Light pink emarginate	Pink emarginate	Pink emarginate	Pink emarginate	Pink emarginate	Pink emarginate
16	No. of spikes/Tiller	1 - 3	1 or 2	2	2 or 3	1 - 3	2 - 4
17	Spike length (cm)	6 - 12	6 - 8	9 - 12	7 - 11	6 - 10	8 - 12
18	Spike width (cm)	3 - 6	2.5 - 3	3 - 3.5	3 - 3.5	2.5 - 4	3 - 4
19	Peduncle length (cm)	3.5 - 6	1.5 - 2	2.5	2 - 2.5	2 - 2.5	2
20	Bract length (cm)	3 - 5	3.2 - 5	3.8 - 4.5	3 - 4	3.2 - 4	2.2 - 4.1
21	Bract width (cm)	2 - 3	1.5 - 2.3	1.2 - 2	1.2 - 1.9	1.1 - 2.2	1.5 - 2.1
22	Bracteole length (cm)	2.7 - 3.1	2.5 - 3.3	2.5 - 3.1	2.5 - 3.1	2.9 - 3.4	2.3 - 3.6
23	Bracteole width (cm)	2 - 2.4	1.2 - 1.5	1 - 1.9	1.3 - 1.4	1.3 - 1.8	1.5 - 2

Sl.No.	Characters	Wild	'Dzongu' Golsey	'Green Golsey'	'Sawney'	'Seremna'	'Varlangey'
24	Calyx length (cm)	2.5 - 3.5	3.9 - 4.6	4.5 - 4.8	4 - 4.5	4.2 - 5.1	3.8 - 4.5
25	Calyx beak (cm)	0.5 - 1	1 - 1.2	1 - 1.5	0.8 - 1.7	1.2 - 1.7	0.5 - 2
26	Flower length (cm)	4.7 - 5.1	5 - 5.5	5 - 6	6 - 6.7	5 - 7	5.5 - 6.5
27	Corolla tubelength (cm)	1.8 - 2	2 - 2.5	2.1 - 2.5	2.5 - 2.7	2.5 - 2.8	2.2 - 2.8
28	Labellum length (cm)	2.4 - 2.6	2.8 - 3	2.8 - 3.4	2.8 - 3	3.4 - 3.6	2.3 - 3.4
29	Labellum shape (apex)	Nearly truncate	Round	Round	Round	Round	Round
30	Bands on labellum	Bands not splitting, parallel	Bands splitting, not parallel	Bands splitting, not parallel	Bands splitting, not parallel	Bands splitting, not parallel	Bands splitting, not parallel
31	No. of veins	9 - 12 pairs	7 - 9 pairs	7 or 8 pairs	8 or 9 pairs	8 or 9 pairs	6 or 7 pairs
32	Colour of midrib	Dark yellow	Dark yellow	Dark yellow	Pale yellow	Dark yellow	Dark yellow
33	Staminodes (mm)	4 - 9	4 - 6	reduced	2 - 3/reduced	2 - 3/reduced	2 - 4/reduced
34	Stamen size (cm)	2.2 - 2.6	2.3 - 2.6	1.9 - 2	2.2 - 2.5	2.8 - 3	1.8 - 2.3
35	Filament length (cm)	0.6 - 0.9	0.6 - 1.1	0.6 - 1	0.8 - 1	0.7 - 1	0.5 - 0.8
36	Ovary size (mm)	4 - 5	3	4 - 5	4 - 6	4	4 - 6
37	Style length (cm)	3.7 - 3.9	3.9 - 4.5	4 - 4.8	4.7 - 4.9	4.8	3.4 - 3.8
38	Capsules/Spike	14 - 20	7 - 8	6	13 - 14	8 - 10	20
39	Capsule length (cm)	2.5 - 3.0	2.4 - 3	2.5 - 3	2.5 - 3	2.5 - 3	3 - 4
40	Capsule diameter (cm)	2.5	2.7	2	2.22	2	2 - 2.75
41	Capsule colour	Maroon	Dark maroon	Light pink and upper part green	Maroon	Dark maroon	Dark maroon
42	Capsule shape	Conical	Oblong- round	Oblong-round	Conical	Spherical	Conical
43	Capsule fresh weight (g)	4.39	7	5.05	5.70	5.15	8
44	Capsule dry weight (g)	0.64	1.50	1	1.40	1.20	1.60
45	Seeds/Capsule	40 - 50	55 - 70	35 - 50	35 - 50	65 - 75	55 - 70
46	Husk thickness (mm)	1 - 4	1 - 3	4 - 5	1	1	2 - 3
47	Flowering period	April - May	March - May	May - August	March - May	April - June	May - July

Table 11.3: Comparison of Characters between different Cultivars of Large Cardamom

Sl.No.	Characters	Ramsey	Ramla	Sawney	Varlangey	Seremna	Dzongu Golsey
1	Plant growth	Vigorous	Vigorous like Ramsey	Vigorous like Ramsey	Vigorous like Ramsey	Less vigorous	Less vigorous
2.	Suitability	High altitude	Medium to High altitude	Medium to high altitudes	Medium to high altitudes	Low altitude	Low altitude
3	Plant type/Morphology	Robust growth	Robust growth	Robust growth	Robust growth	Less robust	Less robust
4	Plant height (M)	1.5 - 2	1.5 - 2	1.5 - 2	2 - 2.5	1.5 - 2	1 – 1.5
5	Tiller colour	Maroon	Maroon	Maroon	Maroon/Greenish tiller with maroon spots	Green	Green
6	Leaf shape	Narrow	Broad and long	Broad and ovate	Narrow leaves with wavy margins	Drooping leaves	Narrow and erect
7	Flower blooming period	May 2nd fortnight	May	High Alt: May 2nd fortnight Mid alt: April-May	High alt: June-July Mid Alt: May	March-April	March
8	Spikes/tiller	2	2	2-3	3-4	2-3	2
9	Capsules/spike	20-25	15-20	20-25	20-25	15-20	10-15
10	Capsule Size	Small	Medium	Bold	Bold	Med.round	Bold
	(LXB) cm	2 x 1.8	2.1 x 1.9	2.5 x 2	3 x 2.7	2.2 x 2.1	2.7 x 2.6
11	Seeds/capsule	25 - 35	30 - 40	35 - 50	50 - 70	65 - 70	60 - 70
12	Month of Harvest (i) High altitude (ii) Medium altitude	Oct.-Nov.	Oct-Nov.	(i) Nov. (ii) Sept-Oct.	(i) Oct-Nov. (ii) Nov-Dec.	Sept.-Oct.	Aug.-Sept.
13	Pest and disease	Susceptible to viral disease	Susceptible to viral disease	(i) Susceptible to viral diseases (ii)Susceptible to Alternaria leaf spot	(i) Susceptible to viral dis. (ii)Susceptible to Alternaria leaf spot	Susceptible to Alternaria leaf spot	Tolerant to chirkey and susceptible to foorkey

a portion of rhizome with one mature and two new tillers. This method ensures high productivity if collected from high yielding elite garden free from *Chirke* and *Foorke* diseases. Micro propagation is a method of rapid multiplication of elite/high yielding clones.

Floral Biology and Pollination

Large cardamom is essentially cross-pollinated crop, insect pollination is the rule. The flower morphology is adapted for such a mode of pollination. The flower blooming period was March-April in low altitudes and May-July in higher altitudes. Some minute variation in floral morphology was observed among the cultivars (Thomas *et al.*, 2009). Each spike consists of about 20-40 flowers, which open in the acropetal sequence over a period of about 15-25 days. Flower opening starts at early morning hours, *i.e.* 3-4 a.m., anthers dehisce almost simultaneously whereas stigma receptivity starts an hour later and last for 24 hours. The stigma was found to be receptive even after 36 hours from the time of flower opening during rain-free days. 3 to 4 months was required for fruit formation and seed development.

Large cardamom is essentially cross-pollinated crop, due to its heterostylic (pin type) nature of flowers, though they are self fertile insect pollination is the rule. The flower morphology is adapted for such a mode of pollination. Bumble bees (*Bombus* sp.) are the main pollinators. Bumble bees are effective pollinators due to its compatible size with the flowers and is having brush-like hairy structures on its dorsal thorax which helps in carrying pollen mass and depositing it on stigmatic surface while entering the flowers. The highest foraging activity of bumble bees is seen during 6-7 a.m., but foraging activity less in cloudy and rainy days. (Gupta and John,1987).

Harvesting and Curing

The optimum time of harvest of capsule is obtained when the top most capsule becomes mature and when the colour of the seeds becomes dark brown to black. It takes 6-8 months for the harvest from pollination. The mature spikes are harvested, and then capsules are separated and dried. The traditional system of curing is bhatti system where the cardamom capsules came in direct contact with smoke which turns the capsules to dark brown or black colour with a smoky smell. Improved curing techniques (flue pipe system of curing) are being introduced where cardamom is dried by indirect heating at 45-50°C, which helps to retain the characteristic violet colour, aroma and high oil content of the capsule. It contains 1.95 to 3.32 per cent of essential oil having characteristic aroma and possesses medicinal properties. It is reported as an official drug in Ayurvedic Pharmacopoeia due to its curative as well as preventive properties for various ailments. The major constituent of large cardamom essential oil is 1,8-cineole. The monoterpene hydrocarbon content is in the range of 5 to 17 per cent of which lamonene, sabeinene, and pinenes are significant components. The terpinols comprise approximately 5 to 7 per cent of the oil. Due to the presence of these compounds, it has pharmacognostic properties such as analgesic, antimicrobial, cardiac stimulant, carminative, diuretic, stomachic (Bisht *et al.*, 2011).

Genetic Diversity

Large cardamom seedling population has large variability due to cross pollination and hence very rich in its genetic diversity. It grows from about 800m altitude to about 2000m altitude in the eastern Himalayas. Because of the wide range of altitudes it inhabits, there are different cultivars adapted to different altitudes and agro-climatic conditions.

There are mainly six cultivars of large cardamom *viz., Ramsey, Ramla, Sawney, Golsey, Varlangey (Bharlangey)* and *Seremna or Lephrakey* (Gyatso *et al.,* 1980 and Upadhyaya and Ghosh, 1983). Another cultivar Bebo is also getting importance and is spreading to more areas in Arunachal Pradesh. Seven wild species such as *A. linguiforme, A. kingii, A. aromaticum, A. corynostachyum, A. delbatum, A. costatum* and *A. plauciflorum* are naturally occurring in the region. (Ghanashyam Sharma *et al.,* 2008).

There are distinct cultivars suited to different altitudes and diverse agro-climatic situations, hence increasing the scope of introduction and area expansion of suitable cultivars in the NE states. Cultivars suited for high altitudes >1515m amsl) are Ramsey, Varlangey and Ramla. *Sawney* is suited for mid (975 – 1515 m amsl) altitudes and cultivars Golsey and Seremna are suited for low (<975m MSL) altitude areas. The characteristic features of important cultivars are as follows.

Ramsey

The name Ramsey was derived from two Bhutia words – 'Ram' meaning mother and 'sey' for gold (yellow). This cultivar is well suited for higher altitudes, on steep slopes. Grown up clumps of 8-10 years age group possesses 30-60 tillers. The tillers colour is maroonish green to maroon. Second half of May is the peak flowering season. Capsules are small, the average being 2.27 cm in length with 2.5 cm diameter, with 20-25 capsules in a spike, each containing 20-30 seeds. The harvest is during October-November. Peak bearing of capsules is noticed in alternate years. This cultivar is more susceptible to viral diseases like foorkey and chirke especially if planted at lower altitudes. It occupies a major area under large cardamom in Sikkim and Darjeeling district of West Bengal. Two subtypes of this cultivar viz.,Kopringe and Garadey from Darjeeling district having stripes on leaf sheath, are reported to be tolerant to chirke virus (Karibasappa *et al.,* 1987).

Ramla

The plants are tall and vigorous like *Ramsey* and have capsule characters like *Dzongu Golsey;* the colour of tiller is maroon like *Ramsey* and Sawney. The leaves of Ramla are very broad compared to all other cultivars. Cultivation is restricted to a few mid-high altitude plantations in north Sikkim. The capsules are medium sized, dark pink with 25-35 seeds per capsule. They are susceptible to *foorkey* but are moderately tolerant to *chirke* disease.

Varlangey

This cultivar grows medium and high altitude areas in South Regu (East Sikkim) and at high altitudes at Gortak (Kalimpong sub-division in Darjeeling district of

West Bengal). Its yield performance is exceptionally high at higher altitude areas *i.e.*1500 m and above. It is a robust type and total tillers may range from 60 to 80 in a clump of 4-5 years age. Plants 200 – 250 cm high; leaves 60 – 80 cm long. Colour of tillers is like that in Ramsey *i.e.* maroonish-green to maroon towards collar zone; girth of tillers is more than that of Ramsey. Each productive tiller on an average produces almost three spikes with an average of 20 capsules/spike. Size of capsules is bigger, capsules 3 – 4 cm long and bold with 55-70 seeds. Harvest begins in the last week of October. This cultivar is also susceptible to foorkey and chirke diseases.

Etymology: The Bhutia word 'Varlangey' means 'bold' which indicates the large size of capsule.

Sawney

This cultivar got the name from Sawan in Nepali, corresponds to August by which month this becomes ready for harvest at low and mid altitudes. This cultivar is widely adaptable, especially suited for mid and high altitudes *i.e.* around 975-1500 m. It is robust in nature and consists of 40-80 tillers in each clump of 4-5 years of age. Plants 150 – 190 cm high; leaves 40 – 55 cm long. Colour of tillers maroon. Each productive tiller on an average produces two spikes. Average length and diameter of a spike is 6 cm and 11 cm respectively. Flowers are longer (6.23 mm) and yellow in colour with pink veins. Capsules 2.5 – 3 cm long, seeds 35 – 50 per capsule. Second half of May is the peak flowering time (Rao *et al.*, 1993).

Etymology: 'Sawan' is the name of a month in Nepali calendar, which comes in the month of August and indicates the harvesting season of the cultivar

Harvest begins in September-October and may extend up to November in high-altitude areas. Large cardamom cultivars are susceptible to both chirke and foorkey viral diseases. Cultivars such as Red sawney and Green Sawney derived their names from capsule colour. Mongney, a strain found in south and west districts of Sikkim, is a non-robust type with its small round capsules resembling mostly that of Ramsey.

Golsey (Dzongu Golsey)

The name has derived from Hindi and Bhutia words; 'Gol' means round and 'sey' means gold. This cultivar is suitable to low altitude areas below 975MSL especially in Dzongu area in North Sikkim. Plants are not robust like other cultivars, and consist of 20-50 straight tillers with erect leaves. Alternate, prominent veins are extended to the edges of leaves (Biswas *et al*, 1986). Plants 100 – 145 cm high; leaf margin straight; ligule pubescent outside. Unlike *Ramsey* and *Sawney,* tillers are green in colour. Each productive tiller on an average produces two spikes. Flowers are bright yellow. On an average each spike is 5.3 cm long with 9.5 cm diameter and with seven capsules. Capsules are big and bold, completely dark maroon in colour and contain about 60-62 seeds. This cultivar becomes ready for harvest in August-September. *Golsey* is tolerant to *chirke* and susceptible to *foorkey* and leaf streak diseases. The cultivar is known for its consistent performance though not a heavy yielder.

Etymology: The Hindi word 'Gol' means 'round' and indicates the shape of the capsule, Bhutia word 'sey'means 'yellow', which indicates the flower colour.

Green indicates the characteristic green colour of bract, leafy shoot and upper part of the fruit.

Seremna

This cultivar is grown in a small pocket at Hee-Gaon in West Sikkim at low altitude and is known for its high yield potential. Plant features are almost similar to Dzongu Golsey but the leaves are mostly dropping, hence named as Sharmney. Total tillers range from 30 to 60 and is not robust in nature. On an average 2-3 spikes emerge from each productive tiller with an average of 10.5 capsules per spike, each having 65-70 seeds. This cultivar is having narrow adaptability as it is not performing well in other low altitude areas.

Etymology: The Nepali word *'Seremna'* means drooping, which indicates the drooping nature of leaves.

Babo

This cultivar is grown in Siang district of Arunachal Pradesh. Plant is medium to tall in height (2.5 m) with 10-12 leaves per tiller. The plant has unique features of rhizome and tillering. The rhizome rises above the ground level with roots penetrating deep into the soil and the young tillers are covered under thick leafy sheath. It is supposed to be tolerant to foorkey disease. The spikes have relatively long peduncle (10-15 cm) and the capsules are bold, red or brown or light brown; seeds contain low level of essential oil (2 percent v/w). Flowers are yellowish red. Spikes bear 10-12 capsules. Capsules mature in October and weigh 1.3-1.4g with 80-85 seeds. The capsule is edible and closely resemble the commercial large cardamom. An inflorescence contains about 12 capsules (Upadhyaya and Gosh, 1983).

Area of Diversity of Large Cardamom Genotypes

The maximum diversity of large cardamom was noted in North Sikkim up to an altitude of 1829 m above mean sea level with the cultivars of Ramsey, Sawney and Dzongu Golsey. Seto Golsey and Mongney were observed in some areas of Singhik and Kabi. The concentration of Dzongu Golsey were observed in pasingdang, Gnansangdong, Lingtem and other areas of Dzongu area of North Sikkim. Ramsey and Sawney are found in Tingda, Kabi, Phensong, Chawang, Labi, Jalikhola, Phodong, Mangan, Singhik, Manul, Nega and other areas of North Sikkim.

In East Sikkim, the diversity and concentration of large cardamom were noted from Ranipool, Pakyong, Assam Lingzey, Chungecenty, Goucharan, Sang, Nazitam, Martam, Rongli, Rigu, Dickling, Rakdong, Tumin, Tintek, Pangthang, Penlong, Lindak, Samdong and other areas with the cultivars of Sawney, Ramsey, Varlangey and Golsey. The area of diversity in South and West Sikkim for large cardamom was noted in different areas of Soreng, Hee, Bermiok, Kaluk, Sreebadam, Gezing, Hee-Pechrek, Ben, Ravangla, Lekship, Kewsing and some other areas with the cultivars of Ramsey, Ramla, Sawney, Dzongu Golsey, Seremna (Laphrakey) and Green Golsey. (Thomas *et al.*, 2009). Study revealed that the cultivar Ramsey, Ramla, Varlangey is suitable for high altitude, Sawney is suitable from medium to high altitude and Dzongu Golsey, Seremna and Green Golsey is suitable for low altitude areas of Sikkim. Distinct characters of six cultivars are presented in Table 3.

Role of Ecosystem

Large cardamom is shade loving plant. It helps in maintain biological diversity. Due its shade loving nature, large cardamom is playing an important role in reducing deforestation. It also helps in the regulation of green house gas and thus controls pollution. Large cardamom agro forestry practice support conservation of tree biodiversity in the region. Biodiversity is an important indicator for the sustainability, and biological diversified systems such as cardamom based traditional agro forestry have a greater capacity for adaptability ecological resilience and show more sustenance. Such agro forestry help in soil and water conservation by proper land-use, protecting from deterioration of soil quality. Such system help in increasing of organic matter and nutrient levels in soil and help in controlling of soil erosion and soil conservation by improving the fertility levels. Such system also plays an important role in the conservation of water and thus helps in providing quality water for local consumption. The large cardamom agro forestry practice also acts as a habitat for pollinators and biological agents of pests and disease organisms. It supports birds and other wildlife which influences the ecological structure and functioning of ecosystem. As large cardamom is a shade loving plant, therefore it plays an important role to check soil erosion and control landslides. Large cardamom agro forestry practice is the best landscape management for biodiversity conservation and excellent slop management, soil fertility maintenance and resilience to extreme conditions (Sharma *et al*, 2008).

Role in Economic Growth

Large cardamom has helped to alleviate many small household from poverty. Sikkim produces 80 percent of India's large cardamom, which enjoys a high value market in Pakistan, Singapore and the Middle-East. It is also exported to UAE, Iran, USA, Afghanistan, UK, Malaysia, South Africa, Japan and Argentina. The major domestic markets in India are Amritsar, Kolkata, Delhi, Mumbai, Guwahati and Kanpur (John, 1984 and Sharma *et al*, 2008).

Firewood, timber, fodder, food and medicines are other byproduct from the system. The spikes after peeling of capsule and fruit-stalk of large cardamom can be dried, powdered and used as a base for manufacture of agarbathi. Even the husk of capsule of large cardamom is dried, powdered and used in agarbathi making. The large cardamom seeds with silver coating sold as 'supari' (Chandrasekhar, 1987).

Indigenous Traditional Knowledge (ITK) in Biodiversity Conservation

Indigenous Traditional Knowledge (ITK) form an integral part of biodiversity conservation. The knowledge residing with traditional farmers and seed custodians is crucial in guiding strategies to preserve and sustainable use of domesticated, semi-domesticated and wild plant species. This ITK has been acquired by the farming community from generation to generation is valuable for crop improvement as that of Plant Genetic diversity. Under climate change conditions for the development of climate resilient improved varieties, the ITK plays greater role in identifying the gene pool in the form of traditional varieties and helps in building sustainable livelihood systems based on local knowledge and local biodiversity.

Traditional knowledge systems in large cardamom farming: biophysical and management diversity in Indian mountainous regions Conversion of forests to other forms of land management has been the general trend in mountainous areas. In the changing land management from forests to other forms, cardamom agro forestry in the Himalayas is a good traditional practice that would meet the demands of both economic activity and conservation. Cardamom has been a major agriculture cash crop and an export agricultural commodity in Southeast Asian countries in the last couple of decades. This is purely the adaptive approach through indigenous traditional ecological knowledge (ITEK) of the communities to devise an integrated natural resource management that will increase agricultural production in a sustainable manner. (Mukherjee and Chakraborty,2010).

Agroforestry Practices

In Sikkim and Darjeeling, agro-forestry is an integral part of the farming system, where trees are integrated extensively with crop and livestock production. Traditional agro forestry systems in the Eastern Himalaya show the way to reconciling short-term food and livelihood needs with long term environmental conservation and enhancement. The combination of trees, grasses, herbs and shrubs along with large cardamom plantation arrest the flow of water, reduce the risk of soil erosion and water pollution hazards.

(a) Cardamom Based Agroforestry System

Alaninchi bari

Large cardamom plantations under the mixed tree species, support fuel, fodder and timber apart from economic return from cardamom

Cardamom-based agro forestry is a purely traditional land use adaptation in the fragile, inaccessible, vulnerable and marginal mountain slopes of the Sikkim Himalaya first initiated by the indigenous communities of Sikkim. This traditional adaptive management system has been a potential livelihood support to the small holders, a means to biodiversity conservation, environmental services and ecological health and social and economic well being of the people (Sharma *et al.*,2007). The large cardamom based agro forestry system is observed to accelerate the nutrient cycling, increases the soil fertility and productivity, reduces soil erosion, conserves biodiversity, conserves water and soil, serves as carbon sink, improves the living standards of the communities by increasing the farm incomes and also provides aesthetic values for the mountain societies (Buckingham, 2004). Large cardamom with shade trees on hill slopes unsuitable for crop production is ecologically sustainable. Some common shade trees for the agro forestry are *Schima wallichii, Engelhardtia acerifolia, Ostodes paniculatus, Symplocus theifolia, Vibernum cordyfolium, Prunus nepalensis, Saurauia nepalensis, Eurya acuminata, Leucosceptrum canum, Maesa chisia, Quercus pachyphylla, Leucoseptrum cannum, Lyonia ovalifolia, Bauhinia purpurea, Osbeckia paniculata, Toona ciliata, Bassia butyracea, Celtis tentranda, Michelia excelsa, M. pustulata, M. indica, Quercus lamellosa, Q. lineata, Rhus semialata, Spondias auxillaris, Beilschmiedia sp, Cinnamomum sp, Ficus nemoralis, Ficus hookeri, Nyssa sessiliflora, Osbeckia paniculata, Viburnum cordifolium, Litsaea polyantha, Macaranga pustulata and Alnus nepalensis.* Hence, large cardamom agroforestry practice also supports conservation of tree biodiversity in the region (Mukherjee, 2008). Majority of cardamom plantations have Himalayan alder (*Alnus nepalensis*) as shade trees since the combination of Alnus and cardamom is sympatric and has proved to be ecologically and economically viable. During the last 5-6 decades, a large area of agricultural lands such as rice terraces were converted to *Alnus*-cardamom agro forestry using N_2-fixing *Alnus nepalensis* as shade tree (Sharma, *et al.*,2002 a). About 70 per cent of the cardamom-based agro forestry practices are under N_2-fixing *Alnus nepalensis* while 30 per cent are under the mixed-tree agro forestry species. The crop is predominantly cultivated between 600 and 2000 m that covers the subtropical to the cool temperate zones. The cardamom agro forestry stored 3.5 times more carbon than the rain fed agriculture showing potential mitigation possibilities of the agro forestry by sequestration of the atmospheric carbon. The agro forestry is an efficient management system where ratio of output to input is more than 13 compared to rain fed agriculture. These tree species are socio-culturally important for their multiple benefits such as wild edibles, timber trees, good quality fodder. Such traditional tree-based farming help ecological restoration, conservation and improvement of steep slopes into production zones, optimize the biomass and forest growth and commercially viable under storey cardamom crop. Large cardamom agro forestry thus supports conservation of tree biodiversity in the region though the use of *Alnus*-cardamom systems has recently proved more profitable. (Ghanashyam *et al.*, 2008).

(b) Jhum System of Cultivation

The alder-based *jhum* system, a unique and highly productive form of *jhuming* (shifting cultivation or slash and burn agriculture) has been developed. This system is usually practice at high altitude of Rimbhik and Lava region only by lepchas. Normally *ajhum* farmer cultivates *thejhum* fields for two years within a nine-year cycle (1:4 ratio of cropping to fallow). But the alder system allows two harvests in two out of every four to five years (1: 1 ratio of cropping to fallow). The farmers are able to improve the already declining jhuming system through the incorporation of a component, alder tree which is native and indigenous to the community (Ramakrishnan, 1992). This intervention results in minimized soil erosion, availability of more productive land, increased soil fertility and sustainable food production. The introduction of alder into the *jhuming* system under a five-year agricultural cycle could stabilize the system, with adequate nutrient recovery and make the system sustainable. Apart from nitrogen fixation, the production of nitrogen-rich litter and mineralization too contribute to biological build-up of soil fertility. This traditional '*Jhum*' cultivation is followed by most subsistence farmers.

(c) Zabo Farming System (under rain shadow zone)

The "Zabo" is an indigenous farming system in high altitude. *Zabo* means impounding of water. It has a combination of forest, agriculture and animal husbandry with well founded soil and water conservation base. It has protected forest land towards the top of hill, water harvesting tank in the middle and for irrigation during the crop period. Seepage water accumulates by internal drainage system which will enhance crop yield (Mukherjee, 2012).

(d) Management of Water

Darjeeling–Sikkim hills is a home of tribal groups (lepcha, bhutia, sikkimist and nepali) speaking a variety of languages with a strong tradition of social and cultural identity marked by diversity in customs, cultures and traditions. The people of the region following some indigenous farming system from time immemorial which are "Pani Kheti" system (Sharma *et al.*, 1992) In pani-kheti system of cultivation, water is diverted from hilltops and allowed to stand in the terrace, by making small bunds or grow water erosion control shrubs such as lemon grass and citronella. The weed and other plant biomass available were incorporated into the soil for nutrient management. Every stream rising from the hill is trapped soon after it emerges from forest, channelized at the rim of valley and diverted by network of primary, secondary and tertiary channels. This system is eco-friendly. (Mishra and Sharma, 1999).

Water application on hill slopes for irrigation of plantation crops such as ginger, turmeric, large cardamom and tea exposes to a serious problem of soil erosion. The hill farmers have developed the indigenous techniques of bamboo drip irrigation for irrigating crops in hill slopes (Mukherjee, 2010). This system of tapping natural streams, making water courses, application and harvesting of water behind the bunds on wet terraces, and safe disposal is a good example of the indigenous understanding of natural resource conservation and management.

(e) Livestock-based Farming

Livestock forms an integral part of village life of Darjeeling hills. The rearing of different species of animals (cattle, sheep and goat, yaks, pigs, poultry, etc.) for milk and meat purposes and these animals also provide manure to meet the crops requirement of nutrients. The government is also providing the necessary inputs through its various departmental schemes for the development of livestock. The production of dairy cattle on small land holdings in the rural area in conjunction with primary agriculture production creates employment and contributes substantially to domestic income and obtaining better utilization of farm resources. Cultivation of fodder crops on agricultural lands is impractical due to constraints of land availability and other inputs. Here number of natural feed resources (tree leaves, grasses, shrubs and vines) are available. The leaf of some fodder trees is almost as nutritious as that of leguminous fodder crops and offers an added advantage of producing fuel wood as a by-product. Leguminous fodder trees (*Albizia* sp; *Alnus nepalensis*) enrich the site through nitrogen fixation, which helps in effective soil and water conservation.

Production, Economic Potential and Market Chain

The finished product of Large cardamom is commercially graded as *Badadana* (big capsules) and *Chotadana* (small capsules), *Kainchi-cut* (capsules tails removed) or non-*Kainchi-cut* (capsule tails not removed). Pakistan is the single largest market importing large cardamom. It is also exported to UAE, Iran, USA, Afghanistan, UK, Malaysia, South Africa, Japan, and Argentina which are other potential markets. The major domestic markets in India are Kolkata, Delhi and Guwahati.

Impact of Climate Change

Climate change is causing many species to shift their geographical ranges, distributions and phenologies at faster rates than previously thought. Pollination and seed set in large cardamom depends on the activities of insects and weather. Temporal and spatial changes in weather would affect the yield of these crops. Upslope fog (hill fog) and valley fog are found in Sikkim especially when relative humidity rises to very high level. Fogs not only impair visibility but also reduce light interception by plants. Prolonged foggy condition could cause physiological injury to crop or affect photosynthetic activities as reported by Takemoto *et al.* (1988) and Mildenbergera *et al.* (2009).

Impact of Climate Change on Water Availability for Winter Crops

Over the years, there has been a reduction in the quantity of water available for irrigation in many parts of the country. This is happening in Sikkim as well. The season-wise rainfall data of past 30 years recorded at Tadong, Gangtok shows decrease in both number of rainy days and quantity of rainfall during monsoon season as well as in winter (post monsoon) season. However, the rate of decrease during winter was comparatively higher than during monsoon season. The number of rainy days has decreased at the rate of 4.50 days/30 years (0.15 day/year) during winter, whereas the decrease was higher during monsoon period *i.e.* 8.10 days/30 years (0.27 day/year). The average seasonal rainfall has decreased at the rate of 53.43 mm/30 years (1.78 mm/year) during winter, whereas the decrease was higher during monsoon period *i.e.* 139.01 mm/30 years (4.63 mm/year). Earlier report by Seetharam (2008) has revealed decreasing trend of winter rainfall at Gangtok at the rate of 0.7 mm per 10 years (one decade) from 1957 to 2005.

Major Threats and Challenges on Cardamom Farming and Biodiversity

Large cardamom based agro forestry has been an adaptive system in the mountain landscapes for ecosystem services and human well being. Recently, some irresistible problems on crop management, disease and pest control and lack of market intelligence have caused serious setbacks on livelihoods of people.

Large cardamom plantation area is declining over the years. Natural calamities such as drought, hailstorm, snowfall in plantations at higher agro ecological zones, widespread occurrence of fungal diseases and viral diseases are the major threats causing reduced production and reduction of agro forestry areas. Of these, the prime cause of cardamom plantation and agronomic yield decline is due to the infestation mainly by viral diseases *viz.*, *Chirkey* and *Foorkey*. Uprooting and burning of all the infected plants has been the only possible alternative to control these viral diseases. The *Dzongu Golsai* and *Seremna* varieties have been found to be tolerant to these diseases.

Current Issues in Large Cardamom Cultivation

☆ There is a change in rainfall pattern (*i.e.* number of rainy days reduced from ideal 200 rainy days to 171-160). Uneven distribution of rainfall and excess rainfall during very short period was also recorded.

☆ Lack of irrigation during dry period (October to March) and drought.

☆ Open cultivation practice followed in large cardamom which is not recommended.

☆ Soil erosion and nutrients depletion.

☆ Less organic nutrients inputs in large cardamom plantations.

☆ Reduction of bumble bee population due to destruction of cardamom ecosystem/developmental activities.

☆ Non adoption of scientific cultivation practices.

☆ Infestation mainly by viral diseases viz.,*Chirkey* and *Foorkey*.

☆ Non adoption of phytosanitary and IPM against pest and disease management in large cardamom.

☆ Hail storm and frost damage/chilling injury from December to January months.

☆ Lack of quality planting materials

Remedies Future Course of Action

☆ Creating awareness among large cardamom farming community about scientific cultivation practises in Sikkim.

☆ Adoption of scientific cultivation practises for large cardamom.

☆ Raising suitable local shade trees before planting of large cardamom. Provide shade to existing plantations.

☆ Developing water source/water harvesting ponds and irrigation facilities in large cardamom plantations.

☆ Irrigation and moisture conservation practices in large cardamom plantation during dry period (October to March).

☆ Production of organic nutrients inputs at local level and application in large cardamom.

☆ Promoting animal husbandry and on farm vermicomposting practices.

☆ Adoption of thorough phytosanitary measures and IPM against pest and diseases of large cardamom plantations.

☆ Conservation of bumble bees population by protecting the large cardamom ecosystem.

☆ Raising quality planting materials in certified nurseries at local level.

☆ Bringing back neglected large cardamom areas into scientific cultivation.

☆ Management of hailstorm and frost damage/chilling injury in large cardamom during December - January months.

☆ Practicing organic management of pests and diseases in large cardamom.

☆ Awareness programmes to encourage the large cardamom planters.

☆ Encourage youth to take up large cardamom farming.

Prospects

Large cardamom based agro forestry systems in the Eastern Himalayan region is a multifunctional system predominantly managed by the small holders as their adaptive traditional practice since time immemorial. Apart from Sikkim, large cardamom cultivation has been attempted in NER especially in Arunachal Pradesh, Manipur, Meghalaya, Mizoram and Nagaland states. Arunachal Pradesh and Nagaland states have large area with suitable agro-climatic conditions for the cultivation large cardamom. However, establishment, production and productivity are limited in Manipur, Meghalaya and Mizoram due to prolonged dry periods during winter and summer months. Hence introduction of hardy varieties along with provision for irrigation during dry months may be necessary for the successful cultivation of large cardamom in these states. Introduction of cultivars from the states of Sikkim and Darjeeling district of West Bengal, which are having wide adaptability, could help in area expansion programmes in North East Region.

Large cardamom showing location specificity, the production of quality planting material for the specific region is a prerequisite for area expansion. Large cardamom is affected by two serious viral disease and a devastating fungal blight disease, hence due care should be taken in plant quarantine, sanitation and establishment of certified nurseries in the respective states. Viability of use of Tissue culture technology in mass multiplication of disease free planting material of large cardamom and their successful cultivation in the field has been demonstrated by the Indian Cardamom Research Institute (ICRI), with the financial assistance of DBT, New Delhi, hence this technology could be used for the generation of disease free planting material production. There is very good scope for production of organic large cardamom in NER as most of the states use very little or no chemicals. The rich genetic diversity available in large cardamom could be effectively put to use by introducing appropriate cultivars/varieties in the North East Region.

References

Anonymous, 1977. Zingiberaceae. (In) *Encyclopedia Britannica Macropedia*, 15th Edn. 19.1150

Anonymous, 1999. The Ayurvedic pharmacopoeia of India. Government of India, 1 (2): 158-159.

Anonymous, 2007. *Annual Report* 2006-07, Spices Board India, Cochin –25.

Anonymous, 2016. *Annual Report* 2016-17, Spices Board India, Cochin –25.

Anonymous, 2011. *Annual Report* 2013-14 Department of Horticulture and Cash Crops Development Department, Govt.of Sikkim.

Anonymous, 2013. *Annual Report* 2013-14, Spices Board India, Cochin –25.

Baker, J.G., 1890-92. Scitaminae: (In) J.D. Hooker; The Flora of British India, London 6: 233-243

Biswas, A.K., Gupta R. K and Bhutia D.T. 1986 Characteristics of different plant parts of large cardamom. *Cardamom* 19 (3): 7-10.

Bisht, V. K; Negi, J. S, Bhandari, A. K. and Sundriyal, R. C. 2011. *Amomum subulatum* Roxb: Traditional, phytochemical and biological activities-An overview. *African Journal of Agricultural Research* 6 (24), pp. 5386-5390.

Buckingham, S. 2004. *Synthesis Report on Cardamom Cultivation.* Fauna and Flora International Community-based Conservation in the Hoang Lien Mountains, Darwin Initiative of the UK Government in Vietnam.pp. 76.

Chandrasekhar, R. 1987.Agro-Based Industrial Development in the Hills of Sikkim and Darjeeling District. *Cardamom 20* (3), 5-7.

Das, A. B., Rai, S. and Das, P. 1998. Karyotype Analysis and Cytophotometric Estimation of Nuclear DNA content in some members of the Zingiberaceae. *Cytobios* 384: 23 – 33.

Ghanashyam Sharma, Rita Sharma and Eklabya sharma 2008. *Indian Journal of Traditional Knowledge. 8 (1)* pp. 17-22.

Gupta, P.N.1983. Export Potential in Large Cardamom.25 (I): 3-9.

Gupta, U and John, T. D., 1987. Floral biology of large cardamom. *Cardamom* 20 (5): 8-15.

Gupta, U. 2000. Documentation of spike and capsule characterization in large cardamom. *J.Hill Res.* 13 (2): 122-124

Gyatso, K, Tshering P, and Basnet, B.S, 1980. Large Cardamom of Sikkim. *Krishi Samachar* 2 (4): 90-95

Jafri, M.A., Farah, K.J. and Singh, S. 2001. Evaluation of the gastric antiulcerogenic effect of large cardamom (fruits of *Amomum subulatum* Roxb.). *J. Ethnopharmocol.* 75: 89-94.

John, T. D. 1984. Large cardamom holdings in Sikkim. *Cardamom* 16 (2): 3 – 11.

Karibasappa, G. S., Dhiman, K.R, Biswas, A. K. and Rai, R. N. 1987. Variability association among quantitative characters and path analysis in large cardamom. *Indian J. Agric.Sciences.* 57 (12): 884-888.

Mildenbergera, K., Beiderwiedena, E., Hsiac, Y.J and Klemma, O. 2009. CO_2 and water vapor fluxes above a subtropical mountain cloud forest—The effect of light conditions and fog. *AgriForest Meteo.* 149 (10): 1730-1736

Mishra, A. K. and Sharma, U.C.1999. Traditional Water and Land Management System of the Apatani Tribe, *Asian Agri-History* 3: 185-94.

Mukherjee, D. and Chakraborty, S. 2010. Indigenous traditional knowledge in the context of conservation agriculture in eastern Himalaya range. *Int. Conf Traditional AgriPractices in Conservation Agri.* 18-20th September,

Mukherjee, D. 2008. Association of Medicinal Plants with important Tree species in hills of Darjeeling. *Env. Ecol.* 26: 1697-99.

Mukherjee, D. 2010 Indigenous traditional knowledge in the context of conservation agriculture in Eastern Himalaya Range. *Asian Agri-History Foundation* (AAHF) 45: 61-68.

Mukherjee, D. 2012. Resource conservation through indigenous farming system in hills of West Bengal. *Journal of Crop and Weed 8* (*1*): 160-164 (2012)

Nigam, S.S. and Purohit, R.M.1961 *Indian Perfum.*5: 3-7.

Pruthi, J. S. 1993. Major Spices of India – Crop Management and Post Harvest Technology, *ICAR Publications, New Delhi,* pp. 114-179.

Ramakrishnan, P.S. 1992. Shifting Agriculture and Sustainable Development: An Interdisciplinary Study from North Eastern India. *MAB Book Ser.,* UNESCO, Paris andParthenon Publishing Group, Carnforth, Lanes., U.K. 4, *Republished by Wiley Eastern*, New Delhi, India, pp. 42.

Rao, Y. S., Anand Kumar, Sujatha Charerjee, Naidu, R. and George, K., 1993.Large cardamom (*Amomum subulatum* Roxb.)- a review, *J.Spices and Aromatic Crops.* 2 (1 and 2): 1-15.

Roxburgh W 1820a. *Plants of the Coast of Coromandel. Vol. 3.* Mission Press, Serampore.

Roxburgh, W. 1820b. *Flora Indica. Vol. 1.* W. Bulmer and Co., London.

Seetharam, K. 2008. Climate change scenario over Gangtok. *Mausam* 59 (3): 361-366

Sharma, E., Sharma, R., Singh, K. K., and Sharma G, A. 2000. Boon to mountain populations, Large cardamom farming in the Sikkim Himalaya, *Mountain Res Dev.* 20: 108-111.

Sharma, E., Sundriyal, R. C., Rai, S. C., Bhatt, Y.K., Rai, L.K., Sharma, R. and Rai, Y.K. 1992. *Integrated Watershed Management: A Case Study in Sikkim Himalaya,* (Gyanodaya Prakashan, Nainital, India)

Sharma, G., Sharma, E., Sharma, R. and Singh, K. K.2002 a. Performance of an age series of *Alnus*-cardamom plantations in the Sikkim Himalaya: Productivity, Energetics and Efficiencies, *Annals Bot,* 89: 261-272.

Sharma, R., Jianchu Xu and Sharma, G. 2007. Traditional agro forestry in the eastern Himalayan region: Land management system supporting ecosystem services, *Tropical Ecol.* 48 (2) 129-136.

Sharma, R., Sharma, E. and Purohit, A.N. 1997a. Cardamom, mandarin and nitrogen-fixing trees in agro forestry systems in India's Himalayan region, Litter fall and decomposition, *Agro forestry systems,* 35 : 239-253.

Sharma, R., Sharma, E. and Purohit, A.N. 1994. Dry matter production and nutrient cycling in agroforestry systems of cardamom grown under *Alnus* and natural forest, *Agroforestry systems,* 27: 293-306.

Sharma, G., Sharma, R., Sharma, E., 2008. Traditional knowledge system in large cardamom farming: Biophysical and management diversity in Indian mountainous regions. (In) *Indian journal of traditional knowledge* Vol.8 (1), 17-22.

Shukla, S.H, Mistry H.A, Patel, V.G, Jogi, B.V. 2010. Pharmacognostical, preliminary phytochemical studies and analgesic activity of *Amomum subulatum* Roxb. *Pharm. Sci. Monit.*, 1 (1): 90-102.

Singh, V.B., and Singh, K. 1996. Large cardamom. *Spices*. Published by Indian Institute of Plantation Management, Bangalore and New Age International Publishers, New Delhi, pp 52-57.

Subba, J. R, 1984. Agriculture in the hills of Sikkim. Sikim Science Society., Gangtok, Sikkim. Pp.286.

Takemoto, B. K., Bytnerowicz, A. and Olszyk, D. M. 1988. Depression of photosynthesis, growth, and yield in field-grown green pepper (*Capsicum annuum* L.) exposed to acidic fog and ambient ozone. *Plant Physiol.* 88 (2): 477–482.

Tripathi, S. and Prakash,V 1999. A new species of *Amomum* Roxb. from Meghalaya, India. *Rheedea 9* (2): 177 – 180.

Upadhyaya, and Ghosh, S.P. 1983. Wild cardamom of Arunachal Pradesh. *Indian Horticulture.*, 27 (4): 25-27.

Thomas, V. P., Sabu, M. and Gupta, U. 2009. Taxonomic studies on cultivars of *Amomum subulatum* (Zingiberaceae) *Rheedea.* Vol. 19 (1 and 2) 25-36.

Verma, S.K., Rajeevan, V., Bordia, A. and Jain, V. 2010. Greater cardamom (*Amomum subulatum Roxb.*) – A cardio-adaptogen against physical stress. *J. Herb. Med. Toxicol.* 4(2): 55-58.

Chapter 12

Cocoa

S. Prasannakumari Amma, V.K. Mallika, E.K. Lalitha Bai and J.S. Minimol

Cadbury-KAU Co-operative Research Project,
College of Horticulture, Kerala Agricultural University,
Vellanikkara, P O KAU, Thrissur – 680 656 , Kerala
E-mail: sprasannakumari@gmail.com

The beans from the tree *Theobroma cacao*, the original source of chocolate, is one of the greatest treasures ever discovered by man. It is an ancient crop having been cultivated, harvested and used by the indigenous people of Central and South America for thousands of years. According to legend, the cocoa tree was discovered by the Mayan King Quetzalcoatle three thousand years ago. There are evidences to show that cocoa beans played a very vital role in the economy in Central and South America. The beans were used as currency (1 slave = 100 beans) during that period. Indians of Northern South America, Central America, and Mexico consumed a strong bitter, semisolid aromatic beverage made from the seed.

Cocoa was introduced to Europe during the 16th century. In 1828 technology to separate the cocoa butter from the dried beans was developed by Coenraad Van Houten, a Dutch chemist using hydraulic press, which led to the manufacture of cocoa powder. The inventor of chocolate as we know it today is unknown, but the first eating chocolate was sold in England in 1847 by Joseph Fry and Son. Two years later the Cadbury family began selling a similar product. In 1876 milk chocolate was invented by William Peter in Switzerland. This led to the worldwide multi-billion-dollar chocolate processing industry. Since then, cultivation of cocoa spread to many tropical countries of the world. It is now grown in 58 countries on more than 6.9 million ha worldwide (Figure 12.1) and is worth in excess of $4 billion to the world's economy annually. Cocoa is the commodity produced in the developing countries of the tropics and consumed by the affluent class in developed countries of the temperate zones. The major share (70 per cent) of global production of cocoa comes

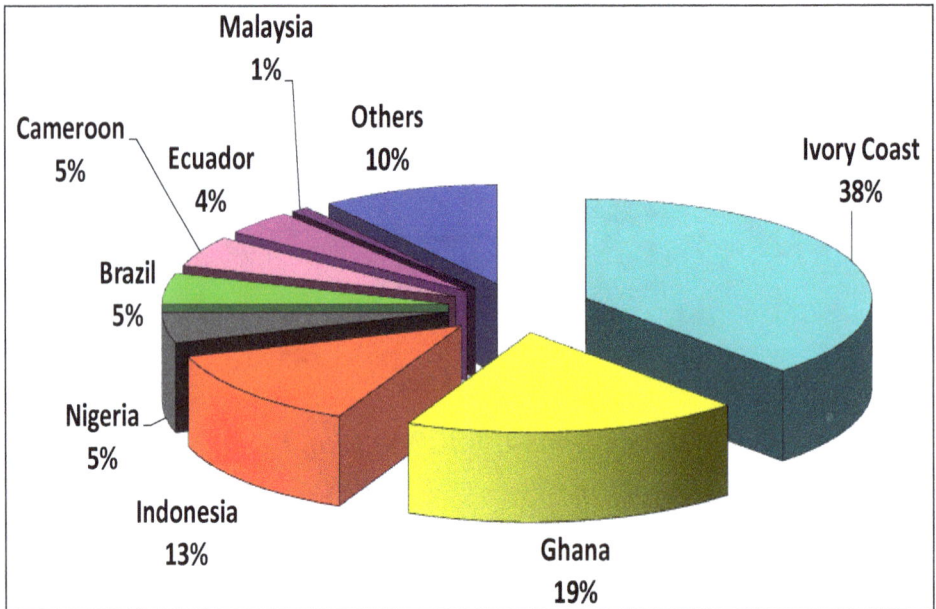

Figure 12.1: Leading Cocoa Producing Countries.

from West African countries like Côte d'Ivoire, Ghana, Nigeria and Cameroon. Of these, some countries rely heavily on cocoa exports for their economic development.

From the level of 1.5 million tones. in 1983-84, the world production of cocoa steadily increased and touched a peak of 3.5 million tones. in 2003-04. This significant increase in production arose mainly due to expansion in area. Over 90 per cent of the world cocoa is in the small holder sector and most of the farms are low yielding. The yield improvement, in spite of the genetic improvement programmes, is nominal. Apart from improvement in yield, varieties with resistance to pests and diseases are essential.

Most of the genetic diversity is found in the Americas, center of origin of the crop, whereas most of the production comes from West African countries. The genetic base in major producing countries is quite narrow and if the targeted goals are to be achieved, diverse genotypes are to be introduced from Amazon and neighbouring areas and exploited. A detailed assessment of biodiversity of cocoa in different areas of its cultivation, collection of valuable types, their conservation and utilization are relevant to bring about significant accomplishment through breeding.

Early History, Origin and Spread of Cultivated Cocoa

According to Cheesman (1944), cocoa is a native of tropical humid forests on the lower eastern equatorial slopes of the Andes in South America, occurring in primary and secondary regions of distribution in the Americas within a range between latitudes of 20°N and 20° S. The area drained by the river Amazon (70,00,000 km^2) includes the north- western tributaries of the river Amazon and western tributaries of the river Orinoco. Over 1000 watercourses make up the hydrologic systems of the river Amazon (Bartley, 2005). It is indicated that each component stream would represent a distinct cocoa population. The cocoa populations on the tributary rivers contain varieties that are unique to each river in terms of phenotypic characters. Prior to the arrival of Europeans, the evolution of cocoa diversity took place over the entire primary distribution area. Only part of the Amazon river basin and the adjacent areas on its periphery was occupied by cocoa at the start of European colonization. From the 17th century, expansion in area took place. The greater part of the primary cocoa diversity is located in the western and south- western quadrant of the region.

The first civilization to cultivate the tree was the Olmec, living in the heart of equatorial Central America, three thousand years ago. After the Olmec, the Maya who took up cultivation of this crop in fourth century AD believed that the tree belonged to Gods and that the pods growing from its trunk were an offering from the gods to man. The Toltecs and Aztecs, who later settled in the Mayan empire, also took interest to spread cultivation of cocoa. Cocoa travelled along the trade routes used by the Mayas, Aztecs and Pipil- Nicaraos (Young, 1994; Coe and Coe, 1996). Well-to-do Indians made a thick beverage by pounding roasted cocoa beans with maize and *Capsicum*. The first Europeans to drink cocoa were the Spanish who invaded and conquered the empire in Mexico in the 16th century. The Spanish learnt from the 'Aztecs' the technique of making `xocoatl', a drink made from cocoa beans after roasting and grinding. The word `chocolate' originated from `xocoatl'. The word `cacao' also was used by the Spanish and it has probably originated from `cacahuatl', a word that Aztecs used for cocoa beans. The present distribution of the diversity involves spontaneous and domesticated populations. In the course of time, as the inhabitants migrated to other areas, the species was extended to regions of the Americas beyond the range of spontaneous occurrence. Further expansion occurred some time after the conquest both within the Americas and the tropical regions of the Eastern Hemisphere.

Spanish explorers took cocoa to the Philippines, from where it spread throughout Southeast Asia, India and Sri Lanka. Amelonado cocoa was taken to West Africa. Criollo types spread to Central America and to large number of Caribbean Islands, including Trinidad in 1525 and thereafter to Jamaica. The dissemination to Venezuela and Costa Rica was made by the Spanish (Pittier, 1933). Introduction to Martinique and Haiti was by the French. Planting in Belem and Bahia in 1750, was attempted by Portuguese.

During 1822, cocoa seeds were taken from Portuguese colonies of South America to the island, Sao Tome of the West Coast of West Africa. It also spread to neighbouring island, Principe. Cocoa cultivation was started in Fernando Po in 1840. The most successful introduction into African mainland was made by the Ghanaian, Tetteh Quashie in 1879. He brought a pod from Fernando Po and the early population of cocoa in Ghana originated from this pod. From Ghana, it spread to other African countries, the most important of which are Ivory Coast, Nigeria and Cameroon.

Cocoa was introduced in the 16th century into Asia and the Pacific. Venezuelan Criollo was introduced in Celebes by the Dutch in 1560. They also introduced the crop into Java. The Spanish took Criollo types from Mexico to Philippines in 1614. It was introduced into Sri Lanka also from Trinidad at about 1798. From Sri Lanka, cocoa was taken to Singapore and Fiji in 1880, Samoa in 1883, Queensland in 1886 and Bombay and Zanzibar in 1887. Cocoa was introduced into Malaysia in 1778 and in Hawaii in 1831. Cocoa was introduced to India in1791; but its cultivation was started in a big way only in 1960's.

The Genus *Theobroma* and its Relatives

The scientific name of cacao tree is *Theobroma cacao* L. (2n=2x=20) as quoted by Linnaeus in 1753 in the first edition of *Species Plantarum*. The Mayas and Aztecs of Central America considered the tree and fruits to be of divine origin. The generic name *Theobroma* is derived from two Greek words *theos* (God) and *broma* (beverage) and thus the literal meaning is "drink of gods". Linnaeus felt that the tree deserved a name that reflected the Mayan belief that the tree belonged to the Gods.

The genus *Theobroma*, a member of family Sterculiaceae contains 22 species grouped into 6 sections (Cuatrecasas,1964). The family contains about 50 genera and 750 species of trees and shrubs, mostly tropical. Of these *T. cacao* is the most important species producing cocoa. *Cola* spp. are sometimes used for beverage purposes, but more usually as a masticatory in West Africa. Except *T. cacao* and *T. bicolour*, most of the spp. have a restricted geographical distribution. About 15 spp. are utilized for their edible pulp or seeds. *Theobroma grandiflorum* (cupuassu), *Theobroma gileri* (mountain cocoa), *T. bicolor* (macambo) and *T. subincanum* (wild cocoa) are other species utilized for their sweet, edible pulp and edible seeds. *T. bicolor* H and B is cultivated for edible pulp around beans and the beans are used like those of cocoa. Beans of *T. angustifolium* Moc. and Sesse. are mixed with cocoa in Mexico and Costa Rica and sweet pulp around the beans of *T. grandiflorum* (Willd. ex Spreng.) K.Schum. is used for making a drink in parts of Brazil and is also eaten. Some other species are *T.mammosum, T. obovatum, T. subincanum, T.simiarum* and *T.sylvestre*. Figures 12.2 and 12.3 provide a comparison of the fruits of *T.cacao* and *T. grandiflorum* respectively.

The different *Theobroma* species are crossable (Posnette, 1945, Martison, 1966, Jacob and Opeke,1971 and Williams,1977). The genus *Herrania* which has more than 10 species is closely related to *Theobroma*. Williams (1977) made about 10,000 reciprocal crosses among *Theobroma cacao, T.grandiflorum, T.microcarpum, T. speciosa, T. bicolor* and *Herrania*. Success of hybridization was unidirectional, being more marked when the cacao cultivar was used as the female parent. The butterfat content of *Herrania* x *T. cacao* was 80.4 per cent, whereas *Herrania* sp x *Herrania* sp gave only 67.7 per cent butter fat. Cuatrecasas (1964) discussed results of his studies on interspecific crosses involving *T. cacao, T. grandiflorum, T.angustifolium, T. microcarpum, T. speciosa, T. bicolor, T.mammosum, T. obovatum, T. subincanum, T.simiarum* and *T.sylvestre*, where there were indications of heterosis in several characters in some of the crosses.

Figure 12.2: *T. cacao.*

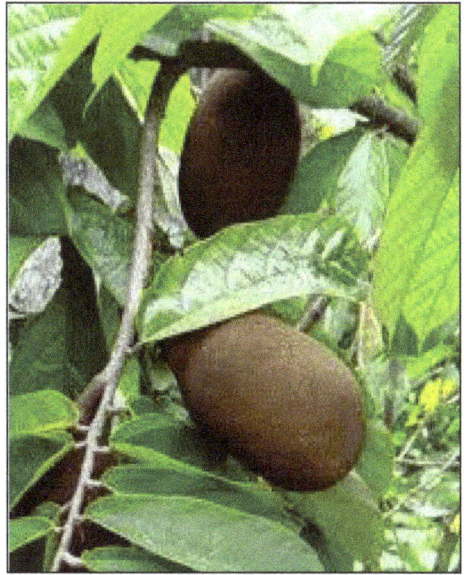

Figure 12.3: *T. grandiflorum.*

Ecological Adaptations

In wild state, cocoa grows to a height of about 12-15m. Seedling derived trees come to full bearing at the age of 5 or 6 years, though production starts at the age of 2-3 years. A well maintained plantation continues to be profitable for at least 25-30 years.

These are under-storey trees in the tropical rainforest and seasonally flooded sites from sea level to 1080 m. It usually grows in groups along river banks, where it may stand for 6 months in slow running water. As the cultivation started, cocoa was planted in environments similar to its natural habitat on the conviction that cocoa must be planted under such conditions. Experience over the last several years proved that heavily shaded cocoa produces low yield.

Purseglove (1968), Wood (1975) and Wood and Lass (1985) illustrated the ecological needs of the crop. Limits of cultivation are 20°N and S, with bulk of the crop lying between 10°N and S of equator. It is grown mainly at low elevations, usually below 300 m. But experience suggests realization of good yield in still higher elevations. Areas of tropical evergreen and semi- evergreen rain forests are the most suitable ecological zones for cocoa. The range in mean monthly temperature of majority of cocoa growing regions is 15 to 32° C. The absolute minimum for any reasonable period is about 10° C below which frost injury is likely. Thus, temperature sets the latitude limit for the best growth of cocoa to within 8° North and South of the equator.

The total annual rainfall ranges from 1500 to 3000 mm. Rainfall, below 1500 mm would necessitate regular irrigation to support unrestricted growth of crop.

Rainfall above 3000 mm with torrential and continuous rain for long spells favour incidence of diseases like black pod (*Phytophthora palmivora*) and vascular streak die back. For ensuring best growth of cocoa, even distribution of rainfall is more important than the total quantity. In most of the major South American, African and South-East Asian cocoa-producing countries, distribution is more or less even with minor peaks. It is so well distributed that around 10 cm of rain is received almost every month.

Cocoa survives in very heavy shade, which would kill many other species, but yield will be low. Photosynthesis is reduced by shading, but is partly compensated by large leaf area. The seedlings grow the best under 75 per cent shade and grown up plants must receive 50 per cent shade. Higher illumination rates produce higher yields, but during initial years of establishment, plants must be protected from scorching sun.

Cocoa is grown in a wide range of soils. The soil in high rainfall areas is relatively coarse-textured and acidic to neutral in reaction. Very coarse sandy soils with very low water holding capacity and low organic matter are not suitable. Usually virgin, freshly cleared forest soils are used for cultivation of this crop in most of the cocoa growing regions of the world, which are rich in organic matter and nitrogen, well-drained and acidic to neutral in reaction. Another basic soil requirement of this crop is a depth of upto 1.5 metres, free from any hard rock.

Morphology, Floral Biology and Fruit Set

Root System

The root system shows dimorphic growth characterized by the orthotropic tap root which goes deep into the soil and lateral plagiotropic ramification, which constitutes the feeding roots. The tap root of cocoa grows predominantly downwards with only a few branches. Under suitable growing conditions when the soil is deep, they grow to a depth of about 150–200 cm. The primary function of these roots is to provide anchorage. The main feeding roots arise from tap root at a depth of 15-20 cm and grow out in humic layer to a length of 5-6 m, giving rise to a dense mat of surface feeding roots. Middle of tap root is usually devoid of secondary roots, but these arise lower down and either grow upwards to surface mat or downwards into lower layers of soil. The rooting pattern is modified to an extent by environment.

Stem

Cocoa exhibits dimorphic branching and grows in tiers. Seedlings grow vertically with straight stem called "chupon" until it reaches an age of 12-14 months or attains a height of 150-200 cm. Growth of chupon ceases and terminal bud breaks up into 3-5 giving rise to three to five plagiotropic branches. These lateral branches are called 'fans' or 'fan branches'. The point at which fans arise is called 'jorquette' and the process of formation of fans from jorquette is called 'jorquetting'. The height at which jorquetting occurs varies with genotype and is also influenced to an extent by environment, with plants growing under dense shade recording high jorquettes. A layer of fans is called a 'tier'. If the plants are allowed to grow under natural conditions, axillary buds just below the jorquette grow and continue to grow vertically and jorquettes again. In this way, several successive chupons arise sympodially, each producing a tier of branches. Eventually, the lowermost tier of plagiotropic branches tends to dry and fall. Chupons can be distinguished from fans from their nature of growth and leaf arrangement. Fan growth will be predominantly to the sides and chupons grow vertically up.

Leaves

Cocoa leaves show dimorphic characters depending upon type of stem on which they are produced. Leaves of fans are arranged in one plane and are alternate. Leaf arrangement of chupons will be spiral with a phyllotaxy of 3/8. Leaves are large, simple, pubescent with pronounced pulvinus and dark green when mature. Petioles are 1-4 cm long and are longer on chupons. Stipules are lanceolate 5-20 x 1-2 mm, pubescent dropping off early. Leaf lamina is entire, simple, elliptic or obovate- oblong, usually glabrous, 12-60 x 4-20 cm, base rounded and obtuse, apex acuminate, main vein prominent, lateral veins pinnate, 9-12 pairs. The leaves exposed to full sun light are stronger and thicker than those under shade.

Leaves in the fan branches are produced in flushes, period between flushes depending upon temperature, tree health and soil moisture status. Under ideal conditions, new flushes appear only 2-4 times a year. During flushing, terminal bud grows out rapidly, producing 3 to 6 pairs of leaves. The young leaves are tender and delicate varying in shades from green to red. The leaves usually harden turning to green as they mature. Once the leaves are hardened, the terminal bud becomes dormant. Next flushing occurs when the environmental conditions become favourable. Flushing demands higher supply of nutrients, a part of which is met from older leaves. Hence flushing in cocoa is referred to as 'change of leaf'. The extent of leaf fall during flushing is an indication of health of tree. Leaves usually persists through two more flushes and drop on third. The period of active photosynthesis is the most marked during first 4-5 months of its existence.

Normally, buds arising on a chupon give rise to chupons, except that at the time of jorquetting. Similarly, fans generally produce fan branches. At times, fans produce chupons also.

Inflorescence

Cocoa produces flowers throughout the year, but flower production is not uniform with peaks during some months of year. These periods of peak flowering are often different for the different regions, indicating strong association with climatic factors. For example, in Ghanaian cocoa, normal flowering reaches its peak during May-June, starting from March and extending through April. In addition to this normal flowering, 'crazy' flowering also occurs during any period of year. Flowering starts after one or two years of planting.

Cocoa flowers are cauliflorous, borne on older leafless wood of main stem and fan branches and never on new flushes. Inflorescence is a much compressed cincinnial cyme with greatly reduced branch and is produced on thickened leaf axils on stem called "cushion" (Figure 12.4). The number of flowers/cushion/season

Figure 12.4: Cushion of Cocoa Tree.

is up to 50. A succession of flowering occurs, depending on the environmental conditions and health of the tree. Some cocoa trees may have pronounced two flowering peaks separated by small but continuous flowering periods as in Upper Amazonian selections. In West African Amelonados, the flowering peaks are separated by periods of no flowering.

Flowers

Flowers are regular, pentamerous and hermaphrodite. The size of the flower varies with genotype (Figure 12.5). Each flower has a pedicel of 1-3 cm length, greenish, whitish or reddish. Sepals are five in number, pink or whitish, 7-10x 1.5 x2.5 mm, triangular, valvate and shortly united at base. There are five petals, base of which is ovate, 3-4 mm long, expanding into a concave cup-shaped pouch, white with 2 prominent purple guidelines. Tip of the petal is spatulate, 2-3 mm long and attached to the pouch by a narrow connective. Androecium consists of 5 outer staminodes, which are erect, pointed with dark purple centres and white ciliate margins and form a ring around style. The five fertile stamens have 4 pollen sacks which dehisce longitudinally. The filaments bend outwards so that anthers are concealed in pouch of corresponding petal. Ovary is superior, five carpelled with numerous ovules. Ovule is anatropous with axile placentation at base and parietal above. The number of ovules/flower ranges from 20 to 60. Style is single, 2-3 mm long, hollow and shorter than the surrounding ring of staminodes. The style has five stigmatic lobes. The floral formula is $K_5C_5A_{5+5}G(5)$.

Fruits

The fruit of cocoa is botanically a berry, but commonly referred to as a pod. The pods are indehiscent and show wide variability in size, colour, shape and texture (Figure 12.6). Length of pods varies from 10-35 cm. The shape of the pod

Figure 12.5: Variability in Cocoa Flower.

is determined by the ratio between the pod length and width and also by the shape of the two ends of the pod. Pods may be spherical (Calabacillo), elongated and pointed (Cundeamor), oval (Angoletta) and regular melon shaped (Amelonado). Pod surface may be smooth or warty, with or without 5 or 10 furrows. Young pods are green or red and on ripening turn to yellow. Pericarp consists of three distinct

Figure 12.6: Variability in Cocoa Fruits.

layers; the hairy and thick epicarp, the mesocarp, which is thin and hard and more or less woody; and the hardy endocarp, which is of varying thickness. Pods attain full size in 4-5 months after pollination and take one more month for ripening. The fruit weight is variable and ranges from 200 to more than 1000 g, but the average pod weight is 300-500 g.

Self pollination of cocoa flowers is made practically impossible due to the peculiar flower structure (the fertile stamen is concealed in the pouched portion of the petal and stigma is surrounded by a ring of staminodes). The flowers are devoid of scent or nectar and the pollen grains are sticky. Natural pollination occurs only with the help of small crawling insects. The most important pollinating insect is the flying female midge (*Ceratopogonid* midges of the genus *Forcipomyia*). The insects are small and barely visible to the naked eye. The midges are attracted by the purple pigmented tissues of the staminodes and the guidelines of the petals. The midges moving on the guidelines near the anther pick up the pollen grains and when they crawl to the staminodes, some of the pollen get transferred from their body to the stigma. The midges fly to about 75 m only. Though the midges are the most important pollinating agents, other insects like ants, aphids, fruit flies and thrips are also implicated as probable pollinating agents. There is a probability, though slight, of wind pollination also. Flowers start to open late in the afternoon and are fully open by the forenoon the next day. As such, most of the pollination occurs in early hours of the day. Anthesis commences between 14.00 and 16.00 hrs and is complete between 14.00 and 16.00 hrs in the next day. Anther dehiscence starts between 04.00 and 05.00 hrs and is completed between 08.00 and 10.00hrs.

Taking into account the complicated flower structure and the problems of self and cross incompatibility, the extent of successful pollination varies considerably. Cocoa flowers are produced in large numbers but only a few of them (1-5 per cent) develop into fruits due to the above problems. Unfertilized flowers fall off within 24 hours. Instances of delayed self and cross incompatibility are also noticed.

Seeds

Seeds are called beans. Each pod contains 20-60 beans/pod. The beans/pod is higher in Forastero. Beans are arranged in 5 rows, variable in size (2-4 x 1.2-2 cm), ovoid or elliptic. A mature bean has no endosperm and from outside to the centre, consists of 1) a thin, resistant, pink- veined husk or a leathery testa, originating from tissues comprising seed coats of ovule 2) a fine translucent and shiny silver skin, which is the remains of endosperm and 3) two large convoluted cotyledons called 'nib' and a small embryo (germ). The cotyledons vary in colour based on the type of cultivar and it varies from white to dark purple (Figure 12.7). Fresh seeds are surrounded by mucilaginous, whitish, sugary, acid pulp which develops from outer integument of ovule. Fresh bean weight/pod varies from 60-250 g. The fresh weight varies from 1.3 to 4.0 g and the dry weight of bean, from 0.5-2.5 g. The international cocoa standards specify the bean count (Number of cured beans/100g.) to be 95. Beans constitute 25 per cent by weight of mature pod. The shell content of cured beans is around 10-12 per cent.

Figure 12.7: Variability in Cotyledon.

Self Incompatibility

A special feature of cocoa is self-incompatibility shown by a few cocoa types. This was first reported by Harland in 1925. Upper Amazon and Ecuador types introduced in Trinidad were self incompatible; but most of the self incompatible plants were cross-incompatible also. Many of the homozygous types like West African Amelonados are self-compatible. Even though the self compatible types may have the advantage of better fruit set under varied situations, self incompatibility is important in commercial hybrid seed production. Incompatibility in cocoa is unique in that, site of incompatibility is the embryo sac (Cope, 1962). After incompatible pollination, the pollen tube grows faster and delivers the gametes into embryo sac in a normal fashion. The embryo sac is in no way abnormal and the rejection is due to failure of male nuclei to unite with egg. This incompatibility is referred to as "prefertilization inhibition in the ovule" and it is genetically controlled. Fusion or non fusion is controlled by a series of alleles operating at a single locus (S), showing dominance or independence relationships (Purseglove, 1968). In incompatible matings, the flowers drop two to four days after pollination. If a population of cocoa is examined for self incompatiblity reactions, it could be observed that majority of the plants belong to self incompatible group. Cross incompatible types frequently occur between two individuals with different genotypes and it occurs only in diploid gametophytic systems when individuals share the same S genotype (Richards, 1986). Mallika *et al.* (2002) while working out compatibility relations among sixteen selected parents, observed 23 crosses to be cross incompatible among the 128 crosses attempted. Cross incompatibility is an indirect measure of the degree of closeness between the genotypes. When the parents used in crossing happen to be genetically similar, incompatibility mechanism operates.

Classification

Classification of cocoa is difficult as it has undergone a substantial genetic mixing over past several years. Like many species found in Amazon river basin, cocoa appears in a great variety of forms. Because of this, for a long time, there was lot of confusion on classification of species.

Pittier (1933) designated each group as a different species: *T. leiocarpum* for Forastero and *T. cacao* for Criollo. Since many genotypes are interfertile, inclusion of all wild and cultivated cocoa into a single species is well justified. The overall classification into two morpho-geographic groups, Criollo and Forastero, is in widespread use.

Cuatrecasas (1964) classified cocoa into two subspecies of *Theobroma cacao*

1. *T.cacao* subsp. *cacao*: The criollo like forms are included under this category and it has four forms.

   i) *T.cacao* subsp. *cacao* f. cacao: This is the original Central American Criollo, found in Mexico and British Honduras. The fruits are oblong, surface warty, 5 deep narrow furrows with five shallower furrows between, immature pods green or dark red, mesocarp thin, seeds round, cotyledons white and of the highest quality.

   ii) *T.cacao* subsp. *cacao* f. *pentagonum* (Bern.) Cuatr.: This is the original South Mexican and Central American Criollo, known as Alligator cocoa, characterized by highly rough and warty fruits. Fruits are oblong- oval, with prominent ribs, immature pods reddish yellow, mesocarp not woody, seeds large and round, cotyledons white and of the best quality.

   iii) *T.cacao* subsp. *cacao* f. *leiocarpum* (Bern.) Ducke.: This form is cultivated mainly on the Atlantic coast of Guatemala. Fruits are ovoid, shallowly furrowed, almost smooth, thin smooth shell, nibs plump, cotyledons white or pale violet and of high quality. Porcelain cocoa cultivated in Surinam belongs to this form.

   iv) *T.cacao* subsp. *cacao* f. *lacandonense* Cuatr. This is a wild half – wine known only in Chiapas in Mexico.

2. *T.cacao* subsp. *sphaerocarpum* (Chev.) Cuatr. (*T.sphaerocarpum* Chev.)

   This subspecies includes all Amazonian and Amelonado forasteros. These are native to South America found in the Hylaea from the Guianas and middle Amazon north and west ward to the Andes. Because of its adaptability, it spread throughout the tropics. Fruits are ovoid, smooth, shallowly 10 furrowed, rounded at both ends, unripe pods green, pericarp very thick, mesocarp woody, seeds ovoid, compressed, cotyledons dark purple. Quality is variable but lower than other subspecies.

The two subspecies and forms intercross freely to give fertile F_1 hybrids and hence contributed to evolution of a large number of distinct local populations.

Cheesman (1944) classified the cultivated and wild cocoas into three groups based on morphological characteristics of the pods, flowers or seeds.

1. The Criollo Group

Domesticated for a very longtime, probably by the Mayas, Criollo cocoas are cultivated in Mexico, Guatemala, Colombia, Venezuela, Madagascar, Nicaragua, Sri Lanka, Indonesia and Samoa islands. Criollo pods yellow or red when ripe, deeply 10 furrowed, markedly warty, conspicuously pointed, pod wall too thin, seeds large, plump and almost round and cotyledons white or pale violet (Figure 12.8). The beans ferment quickly, but yield is poor. It produces the highest quality cocoa and is very much sought for its strong aroma and low astringency. It is used in chocolate industry for luxury products. Criollo trees are found in isolated groups in very old plantations. The main types of Criollo include Pentagona or Lagarto cocoa, Real cocoa and Porcelana cocoa. It is susceptible to stress, major diseases and not adaptable to all situations. The trees are less vigourous and slow growing. This type constitutes only 5 per cent of the global production.

Figure 12.8: Criollo.

It can be subdivided into two

1. Central American Criollo: This is the original cocoa cultivated in Central America and Mexico. Selection over past several years resulted in fixing the recessive white cotyledons, with no astringency and requiring little fermentation. The unripe pod wall is green, turns yellow on ripening.

2. Venezuelan Criollo: This represents the population derived from Criollo introduced from Central America into Venezuela. This type exhibits wide

variability in colour, size and shape of pods. Unripe pod wall is usually red.

2. The Forastero Group

This is a very variable group, found in an indigenous or semi indigenous state in High Amazonia (Peru, Ecuador and Colombia) in Amazon river basin in the Guyanas and along the Orinoco river in Venezuela. These are now widely planted throughout the cocoa producing countries. This group constitutes bulk of world's cocoa and almost all of the production currently coming from Brazil, West Africa and South East Asia.

The Amazonian Forasteros are characterized by green unripe pods, which turn to yellow on ripening. The pods are inconspicuously ridged and furrowed, surface smooth, ends rounded or bluntly pointed, pod wall thick, seeds flattened, fresh cotyledons deeply pigmented and dark violet giving an astringent product (Figure 12.9). The trees give high yields and are hardy. Quality is not comparable with Criollo. The beans take 5-6 days for fermentation.

Figure 12.9: Forestero.

The West African type 'Amelonado', Ecuadorian type 'nacional' and Brazilian types 'Maranhao', 'Comum' and 'Para' types belong to this group. The Upper Amazonian cocoas include Forasteros collected during expeditions in twentieth century from upper part of Amazon basin, mainly to west of town of Iquitos. There is large variability among Upper Amazonians as compared to Lower Amazonian Forasteros.

3. The Trinitario Group

Both the above types were distributed throughout the Caribbean, where they hybridized in Trinidad, creating a distinct hybrid called Trinitario. These are very heterogeneous and exhibit a wide range of morphological and physiological characters. It is difficult to specify characters of Trinitarios as they may have pod and bean characters ranging from those of typical Criollos to those of Forasteros. Figure 12.10 depicts ICS 95, a trinitario tree with profuse yield. This type was introduced into Venezuela about 1825 and subsequently to most of the cocoa growing countries. There are some clones combining the vigour of Forastero with the quality of criollo. Trinitarios are of great importance in breeding and in trade, it is regarded as 'fine' cocoa. The Trinitarios are grown mainly in all the countries where the Criollos were formerly grown (Mexico and Central America, Trinidad, Colombia and Venezuela) and in many African and South East Asian countries.

Theories Associated with Origin of Cocoa Types

Numerous theories are put forward about origin of Criollo group. According to Cheesman (1944) Upper Amazon is the center of origin of Criollo and Forastero. It was suggested that the spread of Criollo throughout Central America began from a small population in the Upper reaches of Amazon, which

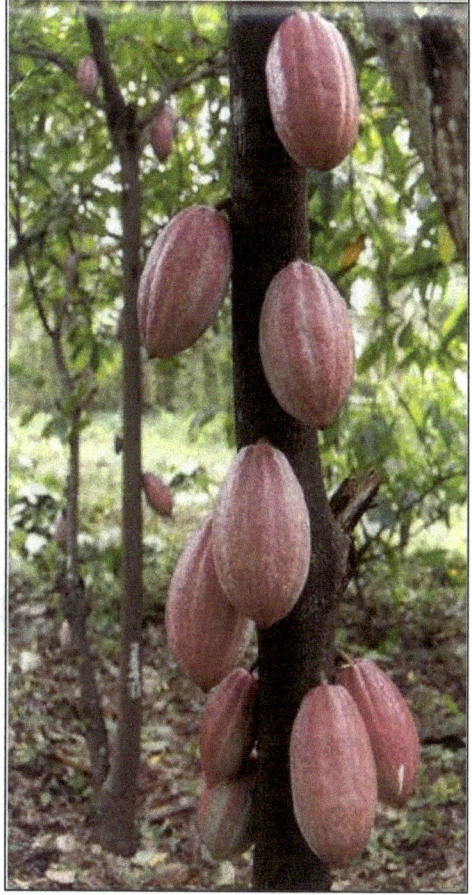

Figure 12.10: Trinitario.

may have crossed the Andes with help of man. Cuatrecasas (1964) on the other hand, suggested that the species was indigenous in the Amazon region to Mexico.

Alvim (1987) proposed that from the centre of origin (Amazon river basin), the species spread out naturally creating two main groups (1) the "Criollos" (*cacao dulce*) that developed north of Panama isthmus through the Andes towards lowlands of Venezuela, Colombia and Ecuador and towards North to Central America and Mexico and (2) the "Forasteros" (*'cacao amargo'*) which resulted from dissemination towards the Amazon valley in Northern Brazil and the Guayanas. Criollo types were cultivated by the indigenous people of Central and South America and Europeans were first exposed to these types. Commercial production commenced in Brazil using the Forastero types, mainly a uniform type called Amelonado. Both types were distributed throughout the Caribbean, where they hybridized in Trinidad, creating a distinct hybrid called Trinitario. It is recorded that Criollo population from Venezuela and the Amelonado type Forastero from Guayana could have been involved in hybridization leading to production of Trinitario. The general names used are based on Venezuelan terminology, namely Criollo = native, Forastero = foreign and Trinitario = native of Trinidad.

A map showing the origin and evolution of cultivated types of cocoa are provided in Figure 12.11.

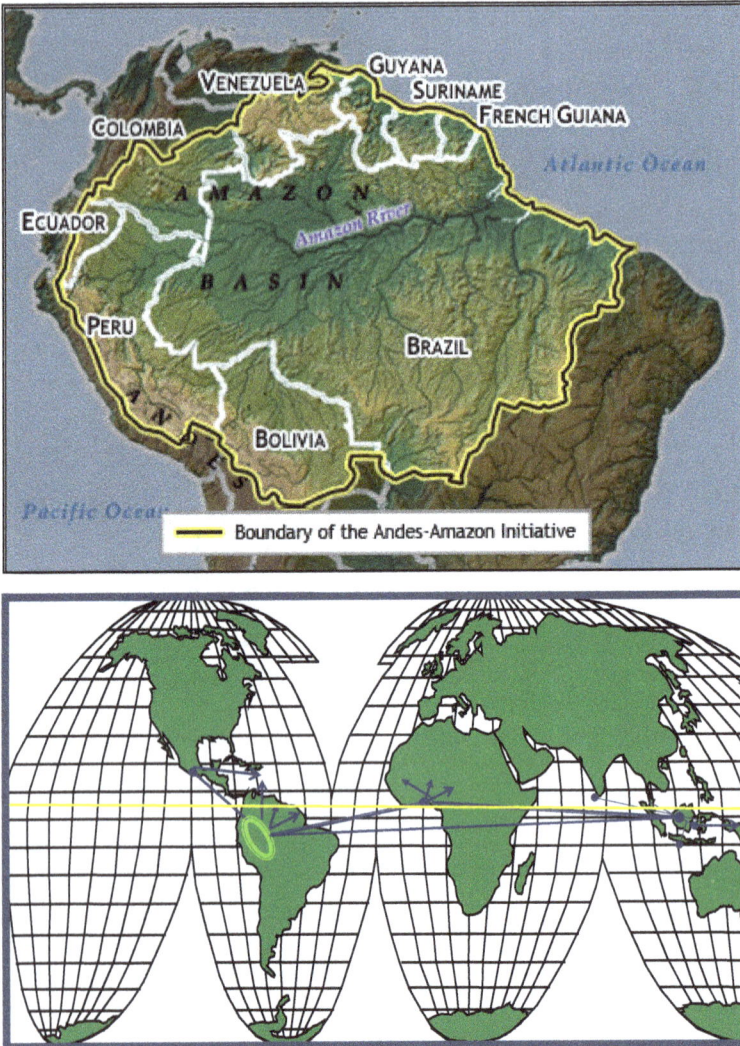

Figure 12.11: Map Showing Origin and Evolution.

Varieties/Clones

Each country has its own unique cultivar. The cultivars in different countries vary with the region from where they are introduced and the extent of hybridization. There are thousands of clones of cacao in field gene banks in different research institutions of world. Selection and hybridization work in the cocoa growing countries of the world have yielded a lot of clones under each category with distinct and unique features. Apart from this, a lot of expeditions are made mainly in centre of origin and neighbouring countries and a large number of useful clones are released. Wadsworth *et al.* (1997) provided the database of about 29,000 clones.

The clones bear name of place or of river in the region in which they have traditionally been harvested or the research station where it was evolved: Iquitos, Nanay, Parinari, Scavina, ICS *etc.* Guidelines for providing clone names to avoid duplication are provided by Wadsworth *et al.* (1997). Some of the important clones coming under each type are:

1. Criollo

1. RIM 189 (MEX)- Rosario Izapa Mexico
2. CAT 201 (ARA) – Cata Aragua State
3. ICS 39, 40, 45, 47, 60, 84, 89 and 91- Imperial College Selections

2. Forastero

1. AMAZ 2/1, 3/2 (Amelonado) – Amazonas
2. APA –4 (Amazonica Palmira)
3. BAL 209 (Borneo Albacca Plantation)
4. BE 3 (Belem)
5. CAS-1 (Campo Agricola do Santarem)
6. CATIE 11, 1000 (Centro de Agronomico de Investigacion, Turrialba, Costa Rica)
7. CC 18, 34 (Cacao Centre, Costa Rica)
8. UF 4, 29, 36, 700, 701, 703, 705, 710 (United Fruit Company)
9. SCA 6, 8, 9, 12 (Scavina)
10. CCRP 1, 2, 3, 4, 5, 6, 7, 8, 9,10 (Cadbury-KAU Co-operative Cocoa Research Project, Kerala Agricultural University). These clones are depicted in Figure 12.12.

The Scavina accessions, collected from the Upper Amazon in the 1930's, are the main source of resistance to witches' broom, cocoa swollen shoot virus, vascular streak die back and black pod diseases of cocoa (Soria, 1977; Kennedy *et al.*, 1987). These accessions, especially Scavina-6 and Scavina-12, transmit to their descendants not only resistance but also high hybrid vigor and good yielding performance. SCA 6 is the clone exploited very much in breeding programmes. It is indicated that this clone is currently used in 39 countries of the world for imparting resistance to major diseases of cocoa.

Trinitario

1. BL Z 9/R (MEX), BL Z 23/R (MEX), BL Z 34/R (MEX), BL Z 52/R (MEX), BL Z 56/R (MEX)- BeLize

Figure 12.12a: Clones.

Figure 12.12b: Hybrids.

2. C 26, 37, 38, 39, 40, 41, 42, 43, 44, 45, 46, 47, 48, 49, 52 (Cabruca, Brazil)

3. UF 1, 10, 168, 221, 296, 613, 650, 654, 666, 667, 668, 676, 677, 688, 707, 708, 709,711, 712 (United Fruit Company)

4. ICS 6, 8, 16, 43, 61, 95 (Imperial College Selections)

Biodiversity of Cocoa in Major Cocoa Growing Countries

Bartley (2005) elaborated the population structure of cocoa in major cocoa producing countries and a brief outline of the same is furnished below:

1. South America–Population Derived from Amazonian Region Germplasm Base

Ecuador

The cocoa populations comprise the oldest population cultivated for commercial purposes, outside Circum Caribbean region. The population up to the end of 19th century is from a single variety with a narrow genetic base. Germplasm was introduced near the end of the 19th century probably from foothills of the Andes and the name "Nacional/Arriba" was introduced to identify original variety that existed previously. This population was uniform except in terms of compatibility. The general characteristic is thick husk and large seeds, intermediate purple pigment of cotyledons, reflexed sepals and staminodes that bent outwards. Ecuador's arriba has long been prized for its earthy but floral aroma and flavour.

Several farmers introduced red fruited criollo from Venezuela, some time later and are referred to as Venzolano. Appearance of diseases in the Americas led to introduction of diverse types with disease resistance. A search was made for trees that escaped infection by *Crinipellis perniciosa* and some trees called refractarios could be identified. Disease free seedlings were distributed. Thus a complex population was established. Some original plantings of the "Nacional" types are still maintained as these produce quality product. The present population consists of hybrids between Nacional types and the introduced ones and their progenies. The population created by the private companies like United Fruit Company (UFCo) and Hacienda Clementia consist mainly of hybrid types.

One important feature of Ecuador derived genotypes is the short stature and small leaves. Estacion Experimental Tropicale, Pichilinge maintains a very diverse germplasm with Nacional and their hybrids.

Brazil

The area covered by the cacao population of Amazon extension zone is large and ill- defined. Knowledge about diversity is extremely scanty and only a small portion of genotypes could be collected for conservation and characterization. One important clone is BE2 with pale green flush leaves, pink staminodes, guidelines and cotyledons. Pound (1938) established that population of this region is composed of heterogenous nature of varieties. Le Cointe (1934) described four fruit types prevalent in the area before it got contaminated by introduced varieties.

1. Crioulo or Criollo: The variety produces large oblong fruits with well marked ridges with large seeds of 1.26g.

2. Jacare (Alligator): Fruits of this variety are elongate, sometimes with apex curved with a pronounced constriction below the base, thick husk, very rough with deep furrows and seeds weigh 0.95 g.

3. Amelonado: Fruits are ovoid with smooth surface and slightly prominent ridges. Seeds are well developed, more rounded with an average seed weight of 1.0g.

4. Calacinho (equivalent to Calabacillo): Fruits are almost rounded, smooth, ridges imperceptible, seeds small weighing 0.6 g. Trees are vigourous.

In the extra Amazon area, it is assumed that a single variety 'Comum' was involved during first century of establishment of crop in this region. The fruits are large and elliptic in shape, apex short, acute and straight and with a tendency for a constriction at the base. The fruit surface had fairly prominent ridges, and varied from being rough to smooth. Subsequent introductions were made using seeds from trees of unknown parentage on the Amazon river basin. This population is referred to as 'paras'. One of the varieties named ' Maranhao' introduced from Trinidad had different types *viz.*, Maranhao rugos, Maranhao Liso and Maranhao Melao. Maranhao and Paras types can be distinguished from Comum types by fruit characters and habit of trees. In contrast to erect habit of Comum, the other types had wide canopies with long branches that tended to curve towards the ground.

Several mutations are encountered in the population- mainly with undesirable traits. 'Folha de Louro' is characterized by slow growing plants of small stature and small leaves. Types with anthocyanin inhibitor alleles are called 'Catongo' and 'Almeida'. 'Maracuja' is another remarkable mutant with flowers resembling those of the genus *Passiflora* with sterile stamens replaced by normal ones. Significance of this mutant is that it would represent the wild ancestor of *Theobroma* with ten fertile stamens. 'Jaca' is another mutant genotype with only three whorls of plagiotropic branches as against five in *Theobroma cacao*. Another mutation which is entirely sterile but produces a few small red parthenocarpic fruits resembling *Capsicum* is called Pimenta.

Attempts to introduce Venezuelan and Nicaraguan cocoa in 1907 did not make any impact on the genetic composition of the Bahia population.

Surinam

The cocoa industry in Surinam is based on a single variety ' Porcelain'/'Suriname' resembling forasteros of Trinidad. This variety is uniform distinguished by a thin smooth shell and fullness of its beans. The unripe fruit colour is light green with short smooth pods tending to be spherical with a rounded apex. The variety is esteemed for its good qualities. Subsequently 'Caracas' cacao belonging to Venezuelan Criollo group and the Alligator Cacao were introduced and these entered into cultivation. At present, a number of variable types are available in the country.

2. Circum Caribbean Region and Neighbouring Territories–Population Evolving from a Criollo Germplasm Base

Trinidad and Tobago

The first plantations were raised from materials collected from Venezuela. The original Criollo populations had white cotyledons, smooth pod surface and thin husk. Some trees with rough fruit surfaces were also described. A phenomenon described as 'blast' occurred in the area which destroyed many of the cocoa plantations. Another variety was subsequently introduced from Amazon river valley. Hybridization between the two populations occurred. Later introductions were 'Ocumare'- a variety belonging to the Venezuelan Criollo group. Selection was practiced for better quality of Criollo and adaptability and high yield of Amazonian Forastero. Some varieties namely, 'Ochroe', Verdilico, Sangre Toro and Five grooved Cundeamor were reported from this area in early years.

Introductions were made from Nicaragua which comprised of Alligator, Pentagonum and *Cacao del Pais*. Other types introduced from Nicaragua included types with greater vigour and purple cotyledons segregating for Pentagonum fruit type and single plagiotropic fruit character.

In Tobago, the earlier plantations belonged to the Criollo group with distinct mamillate apex. Extensive cultivation started by beginning of 20th century and composition of population would have been similar to that of Trinidad.

Venezuela

Since 19th century, the criollo population of Venezuela has undergone a lot of changes with introduction of germplasm mainly of Calabacillo from Trinidad. The introduced variety would have hybridized with criollo and in time, supplanted the criollo and also spread to lands to which criollo was not adapted. Trees with Amelonado and Calabacillo fruits are also encountered. Trinitario forms with different degrees of intermixing and breeding with local populations are observed in the country and the gene base would be the same as in Trinidad.

Colombia

No evidence is found for existence of cocoa in the pre-Colombian times and it is considered that all the cocoa in Colombia have been introduced. The first plantations were derived from Criollo population originating in Venezuelan Andean foothills. Some genotypes with green fruits and single orthotropic branch character resembling Venezuelan types were observed. The genetic composition of the criollo underwent many changes with the introduction of varieties like 'Pajarito' from Trinidad and Calabacillo types from West Indian islands. The present population consists of mainly hybrids with prominent criollo characters.

Guyana

It is presumed that the varieties are introduced from Venezuela and therefore belonged to the Criollo group. Green fruited criollo types with white cotyledons are encountered in this region and are differing from normal concept of Venezuelan

Criollo. During the last two decades of 19th century, cocoa varieties were introduced from Trinidad, Surinam 'Porcelaine' and Grenada.

The Lesser Antilles

The major islands that make up the chain on the windward side of the Caribbean sea have cultivated cocoa during the past 200 or 300 years. Varieties such as 'Caracas" were introduced into the islands and the population developed from the introduced material have a large component of hybrid types. The principal varieties in Martinique and Guadeloupe were Criollo types with high variability. There are indications to show that Criollo, Calabacillo and Amelonado existed on this island.

The population structure in Grenada is based on presence of Amazonian variety or Calabacillo and Criollo from Venezuela. Trinidad was probably the source of seed for these plantings but some natural hybrids between Calabacillo and Criollo would have been produced in Grenada. Some genotypes with elongate green fruits classifiable as Criollo are found in a few locations and these may be the progenies of varieties introduced into Trinidad from Nicaragua.

In St. Lucia, Calabacillo types are found and there are indications to show that these have been introduced from Martinique. Criollo also occurs in other locations. Variable hybrids between Calabacillo and Criollo types are also recorded.

In the island of Dominica, the oldest trees were of Calabacillo, similar to those of 'Creole' of Martinique. Planting material was imported from Trinidad comprising of a sample of the Criollo and Forastero hybrids. Hence, the cocoa population on some farms comprise hybrid types similar to those of Trinidad.

The Greater Antilles

The largest of the islands is Hispaniola and cocoa cultivation was commenced using planting material introduced from the Atlantic coast cacao areas of Mesoamerica. Criollo types were introduced from Martinique and Venezuela during early days. Subsequently, hybrid types were introduced from Trinidad, Ecuador and Venezuela which resulted in the present population being highly variable.

In Jamaica, seeds of 'Caracas' variety were introduced from Venezuela, followed by Creole or Santo Domingo, Forastero from Amazon river basin and Calabacillo. Hybrids between these types are encountered at present.

Cuba has a long history of cocoa cultivation and the first plants were derived from trees on the main land, either on coast of Mexico or that of Gulf of Honduras.

In Puerto Rico, seeds for first plantations are from Venezuela. Plantations composed of introduced hybrids and Calabacillo types with red and green fruits are recorded. The Criollo plants are also encountered.

Mesoamerica

The original plantings along the entire length of the coast of Panama northwards and the interior of some countries were using Amazonian types same as the Martinique 'Creole'. The first plants along the Atlantic coast were Criollo types.

The self-compatible type identified as 'Matina' was also recorded in its original form and has some characteristics as Creole. This was followed by introductions from Trinidad. The United Fruit Company took up large scale cultivation of cocoa using 'Matina' variety followed by hybrids with Criollo types. Introductions were also made from Ecuador, especially, 'Nacional' variety.

The 'Creole' type got established along the coast of Honduras. This genotype possesses anthocyanin inhibitor gene.

Along the Atlantic coast of Mexico, Criollo, Forastero and Calabacillo existed. The traditional Criollo varieties were replaced to a large extent by an introduced type or hybrid between it and Criollo. Thus the population was described as being a mixture of native Criollo with an Amelonado of high productivity. The fruits were smaller in size and light green in colour.

3. Old World Countries

Various attempts were made to establish the crop in other regions. The first attempt to plant cocoa outside the American continent is attributed to the introduction of plants from Mexico into the Philippines in 1670. Growth of the plantings in the old world was gradual until end of 19th century. The planting materials for cultivation were derived from a few sources and with limited genetic content. Variability developed as a result of mutation and during the later periods from hybridization following the addition of other germplasm.

South Asian and Pacific Region

The cocoa cultivated in the first two centuries following the first introduction was the Mesoamerican Criollo group. Further introductions were made from Mexico and there is evidence to show presence of Amelonado genotypes.

Criollo from the Philippines and Indonesia probably is responsible for the first cocoa varieties in Sri Lanka in the beginning of 19th century. Cocoa arrived the island of Mauritius fairly early, from where it spread to Madagascar, Thailand and India.

The most important producer in the South East Asia in the beginning of the 20th century was Java and the major genotype was 'Java porcelaine'. A mutant with anthocyanin inhibitor gene is reported with green fruits, unpigmented flowers and flush leaves. A natural hybrid between Venezuelan Criollo and Trinidad varieties was introduced. Later Forastero cocoa was also introduced.

In Sri Lanka, cocoa trees existed in the Botanic Garden in 1819, which probably were derived from Philippines or Java or Trinidad. Hybrid populations imported from Trinidad and varieties like Cundeamor, Forastero, Criollo, Cayenne, Verdilico, Sangre Toro and Pentagonum were available in this country.

Introduction of cocoa into South India was made as seeds from Ambon island in Moluccas and was received in Madras in 1791. The progenies of these trees were distributed to other localities in South India. The best known of these plantations is one near Coimbatore, mainly of a pure Criollo type. Subsequent introduction was made in the early 20th century but its cultivation was limited to a few Government farms. Both Criollos and Forasteros were introduced into the country. In 1930s, it was decided to remove all Forastero plants in the country to maintain genetic purity of Criollos. The Criollos which were maintained in the farms, failed to come up well and many plants were damaged by pests and diseases. Some continued to survive, but yields were low. Cocoa cultivation resumed in a big way in 1960s with pods of Forastero type from Malaysia, followed by types from African countries.

Cocoa was planted in Peninsular Malaysia in 1777, but it did not gain importance until recent times. The first introductions belonged to Criollo group resembling Guatemalan and Colombian Criollo. Cocoa varieties were later brought from Sri Lanka which consisted varieties from Trinidad.

Cocoa cultivation in Papua New Guinea commenced in the beginning of 20th century. The first plantings would have comprised of Criollo, followed by Forastero and West African Amelonado.

In Fiji islands, the first introduction was made from varieties sent from Trinidad. After 1948, clones were introduced from Trinidad, Papua New Guinea, Western Samoa and Grenada. It was decided to grow Amelonado and Cundeamor types.

4. West African Countries

The start of cocoa cultivation in Africa dates from end of 18th century, although it was not a major success at first. The original Criollo types of cocoa – excellent in quality but very sensitive to pests – did not root very well on African soil. Some decades later, a stronger cocoa variety, though with a poorer quality – the Forastero – was introduced on the isles of São Tomé and Fernando Po. History of introduction of cocoa in West Africa is quite different from that of SE Asian countries. In this region, the first material introduced was of Amazonian origin, probably the State of Bahia in Brazil and planted in the island of Principe in 1822.

A single fruit was concerned in first introduction since it was stated that the original planting consisted of 30 plants. This provided the basic material for spread of cocoa cultivation in the African continent. In 1832, cultivation spread to Sao Tome. The variety existed in the region was self compatible and due to several generations of inbreeding, the variety became homozygous. This variety was related to 'Comum' of Brazil and was called 'Sao Tome Creoulo'. This variety was taken to the island of Fernando Po and it became the basis for the cocoa produced in this continent and it came to be known as ' West African Amelonado'. This introduction and the first encouraging crops from it coincided with the global rise in demand for cocoa in this era of first industrialization. At the same time, political instability and crop diseases threatened cocoa production in Latin America and so traders started looking for new territories where the cultivation of cocoa could be developed.

Varieties were later introduced into Sao Tome from Ecuador, Trinidad and Venezuela. Natural hybridization of all these genotypes with original variety developed a wide range of phenotypic variability. The Sao Tome population contains several unique genotypes that are probably unknown in other countries. Some mutants are also located in this country. One such is 'Laranja'. Another mutant is with thick tree trunks, short broad green fruits with short pointed apex and short peduncle with erect fruits. Another mutant carries a recessive gene in the homozygous state segregates for progenies with yellow leaves.

The islands of Sao Tome and Principe, on account of its central position played a significant role in spreading cocoa along coastal region of West Africa.

The first attempt to grow cocoa in Cameroon was in 1876 using 13 plants from the Royal Botanic Gardens. There are no records of these plants and subsequent introduction was from Sao Tome or Fernando Po by Baptist Missionaries and formed the source of planting material for beginning cocoa cultivation in Ghana in 1889. There are evidences to show that Forastero, Criollo, 'Puerto Cabello', 'Soconusco', 'Venezuela', 'La Guaira', 'Maracaibo', 'Guayaquil', 'Nueva Grenada' and 'Suriname' and other varieties from Trinidad existed in Cameroon. These varieties were planted mixed with the type known in Cameroon as 'Victoria-kakao'.

In Gabon, the variety cultivated in early years was 'Sao Tome Creoulo'. Preferred types like Soconusco were imported from Cameroon in 1898. Later studies indicated a large variability in the population.

In Angola, the first experiment with cocoa was initiated in 1860 utilizing seeds of 'Sao Tome Creoulo' and this variety continues to be the major one in areas where the crop was extended. In Belgium Congo, planting was taken up in 1884 with seeds obtained from Sao Tome, Venezuela and Colombia. In Ghana, Ivory Coast and Nigeria, origins of cultivation were based on import of seed of Creoulo from Sao Tome and the variety continued to occupy the prominent position for about six decades. Later introductions were made from Trinidad like Ocumare, Trinidad Criollo and Nicaraguan Criollo. Criollo types, Jamaican white variety and hybrid types were also introduced. Some mutants with albino leaf and anthocyanin inhibitor gene are also reported in Ghana. Hybrid varieties were later produced in these countries to overcome problem of low yield.

In Ivory Coast, cocoa became a cash crop only in 1912, when colonial authorities forced Africans to cultivate it. Cocoa cultivation initially occurred only in Southern forest zone. Then, cocoa was grown mainly on small family-owned farms with labor supplied principally by immigrants from other African countries. It produces 40 per cent of the world's cocoa and that previous success boosted the country's economic development.

Spread of Biodiversity: Germplasm Collection and Exchange

Distribution of cocoa germplasm is through internationally recognized agencies. The International Cacao Gene bank, Trinidad and the collection at CATIE Regional Germplasm Centre, Turrialba, Costa Rica are designated as "universal collection depositories". The core of Trinidad collection is Pound's Ecuadorian and Peruvian collection which forms 70 per cent of it, the 1952 Anglo Colombian collection, representatives of Chalmers' and Allen's material and selections from cultivated cocoa in Trinidad and other Caribbean Islands. The core of Turrialba collection is selections from cultivated cocoa, especially the United Fruit Company clones and their derivatives from Costa Rica, similar material from other American countries and Criollo. Large collections of primary material are also maintained in Colombia,

Ecuador, French Guiana, Venezuela and Brazil. Field collections are maintained in Puerto Rico, Cote d' Ivoire, Jamaica, Malaysia, Grenada, Nigeria, Papua New Guinea, Ghana and India.

There are thousands of clones of cacao in field gene banks in different areas of the world. Some of the largest collections are at the Cocoa Research Institute in Tafo, Ghana (6,000 accessions), the International Cocoa Gene bank in Trinidad (1,872 accessions), and CEPLAC in Brazil (1,749 accessions). Long distance distribution is done using intermediate quarantine facilities at the Royal Botanical Gardens, Kew (University of Reading, from 1983) and the United States Department of Agriculture (USDA) in Miami, Florida. These transfers are carried out by authorised organizations like IBPGR, now Bioversity International, Rome.

Some of the important cocoa germplasm repositories of the world as provided in ICGD (2003) by Wadsworth *et. al.* (2003) are furnished below:

1. CEPEC, Centrode Pesquisas do Cacau, Superindendencia Regional da Bahia do Espirito Santo, Rodovia Ileus- Itabuna Km 22, CEP 45 600-000 Ilheus, Bahia, Brazil
2. CEPLAC, Dep. Especial da Amazonia, Rodovia Augusto Montenegro, Km 7, Caixa Postal 1801, CEP 66, 823-110, Belem, Para, Brazil
3. IRAD, BP. 2067, Yaounde, Cameroon
4. N'koemvone Research Station, par Ebolowa, Cameroon
5. Instituto Colombiano Agropecuaria, Apartado Aero 233, Centro Nacional de Investigacion (CNI), Palmira, Valle, Colombia
6. CATIE Regional Germplasm Centre, 7170, Turrialba, Costa Rica
7. Centre National de Recherche Agricole, 01 BP 1827, Bingerville, Abidjan, 01, Cote d'Ivoire
8. Instituto Nacional de Reforma Agrarian, Cuba
9. Dominican Republic- Department of Coffee and Cacao, Center De Los Heroes, Santo Domingo, Dominican Republic
10. Instituto Nacional de Investigaciones Agropecuarias, Estation Experimental Tropical Pichilingue, Casilla Postale 24, Quevedo, Ecuador
11. Aburi Botanic Garden, P.O.Box 23, Aburi, Akwiapim, Ghana
12. Cocoa Research Institute of Ghana, Private Mail Bag, International Airport, Accra, Ghana
13. Ministry of Agriculture, Lands and Fisheries, Technical Department, Mt. Horne, St. Andrews, Greneda
14. Cadbury- KAU Co-operative Cocoa Research Project, Kerala Agricultural University, KAU P.O. Thrissur, 680 656, Kerala, India
15. ICAR-Central Plantation Crops Research Institute, Kasaragod, 671 124, Kerala, India
16. Estate Crops Research Institute, Jember Station, East Java, Indonesia

17. London Sumatra Plantations- PT PP Lon- Sum Plantations, Jin Jend. A. Yani No.2, P.O.Box 1154, Medan 20111, Sumatra, Indonesia

18. Agricultural Research Center Tuaran P.O, Box 3, Tuaran 89207, Sabah, Malaysia

19. Golden Hope Plantations, P.O, Box 11043, 50734 Kuala Lumpur, Malaysia

20. Malaysian Cocoa Board, Centre for Research and Production Hilir Perak, P.O. Box 25, 36307, Sungai Sumun, Perak, Malaysia

21. Campo Experimental Rosario Lzapa, Apdo Postal 96, Tapachula, Chiapas 30700, Mexico

22. Cocoa Research Institute of Nigeria, Idi- Ayunre, PMB 5244, Ibadan, Nigeria

23. Lowlands Agricultural Research Station, P.O Keravat, East New Britain, Papua New Guinea

24. Instituto Nacional de Investigacion y Promocion Agropecuaria, San Rogue, Iquitos, Peru

25. CIRAD- CP/E.E Agronomica de Poto, CP 375, Sao Tome and Principe

26. Agricultural Experiment Station- Ministry of Agriculture, Animal Husbandry and Fishery, Agricultural Experiment Station, P.O Box 1807, Paramaribo, Suriname

27. Cocoa Research Unit, University of West Indies, St. Augustine, Trinidad and Tobago

28. Imperial College of Tropical Agriculture, St. Augustine, Trinidad and Tobago

29. Trinidad Cocoa Board, Trinidad and Tobago

30. Royal Botanic Gardens, Kew, Richmond, Surrey, TW 9 3 AB, United Kingdom

31. The BCCCA/University of Reading Intermediate Quarantine Station, Department of Horticulture and Landscape, Plant Science Laboratories, Whiteknights, P.O Box 221, Reading RG6 6 AS, United Kingdom

32. Institut National pour l'Etude et la Recherche Agronomics, B.P.1513, Kisingani, Zaire

Conclusion

The future of world cocoa mainly depends on the ready availability of improved planting materials. A lot of improvement is warranted in resistance to major diseases like witches' broom, cocoa swollen shoot virus, vascular streak die back and *Phytophthora* pod rot, bean quality traits, bean size and adaptability to adverse weather conditions. To obtain any significant breakthrough in breeding, the cocoa breeders should have a very strong assembly of diverse germplasm. One of the serious limitations of cocoa breeding programmes in major producing countries is the narrow genetic base. Clones with unique characters are being collected through well organized explorations to the center of origin and neighbouring areas. Apart from this, many useful materials are being produced in different cocoa research institutes. All these improved and diverse genotypes should be accessible to the breeder to make effective genetic manipulations and evolve more desirable genotypes. Wonderful results through breeding can be achieved only through systematic collection, conservation and utilization of biodiversity.

References

Alvim, P. de. T. 1977. Cacao. (In)*Ecophysiology of Tropical Crops* (Eds.) Alvim, P.de.T. and Kozlowski, T.T. Academic Press, New York, pp.502

Bartley, B.G.D. 200. *The genetic diversity of cacao and its utilization.* CABI Publishing, CAB International, Wallingford, Oxfordshire OX 10 8DE, UK. pp 1-341

Cheesman, E.E., 1944. Notes on the nomenclature, classification and possible relationship of cacao populations. *Trop. Agric. Trin.* **21** (8): 144-159

Coe, S.D. and. Coe, M.D. 1996. *The true history of chocolate.* Thames and Hudson Ltd, London, United Kingdom

Cope, F.W. 1962. The mechanism of pollen incompatibility in *Theobroma cacao.* L. *J.Hered.* **17**: 157-182

Cuatrecasas, J. 1964. Cacao and its allies: a taxonomic revision of the genus *Theobroma. Contributions from the U.S National Herbarium* **35** (6): 379-614

Jacob, V.J. and Opeke, L.K. 1971. Interspecific hybridization in *Theobroma. Proc. 3rd Internat. Cacao Res. Conf.* Accra, Ghana, 1969: 552-555

Kennedy, A.J.; Lockwood, G.; Mossu, G.; Simmonds, N.W. and Tan, G.Y. 1987. Cocoa breeding: past, present and future. *Cocoa Growers' Bulletin*, Birmingham, **38**: 5-22, 1987.

Le Coinnte, P. 1934. *A Cultura do Cacau na Amazonia Ministerio de Agricultura*, Rio de Janeiro, Brazil pp.35

Mallika, V.K., Amma, S.P.K., Nair, RV and Namboothiri, R. 2002. Cross compatibility relationship within selected clones of cocoa. *National Seminar on Technologies for enhancing productivity in cocoa (Ext. Sum.)* Central Plantation Crops Research Institute, Regional Station, Vittal, Karnataka, India 29-30 November 2002. pp 44-46

Martison, V.A. 1966. Hybridization of cacao and *Theobroma grandiflora. J.Hered.* **57**: 134-136

Pittier H. 1933. Degeneration of cacao through natural hybridization. *J.Hered.,* **36**: 385-390.

Posnette, A.F. 1945. Interspecific pollination in *Theobroma. Tropic. Agric.* Trinidad, **22**: 188-190

Pound, F.J. 1938. Cacao and witches' broom disease of South America (*Marasmius perniciosus*). Report on a visit to Ecuador, the Amazon valley and Colombia. April 1937- April 1938. Yuilles' Printerie, Port-of- Spain, Trinidad and Tobago pp.58

Purseglove, J.W. 1968. *Theobroma cacao L.* (In) *Tropical Crops- Dicotyedons* (Ed.) Purseglove, J.W. John Wiley and Sons Inc. London, United Kingdom pp. 571-599

Richards, A.J. 1986. *Plant Breeding Systems.* London. George Allen and Unwin, Boston, Sydney pp.189-232

Soria, J. 1977. The genetics and breeding of cacao. *Proc. V Internat. Cocoa Res. Conf.,* Cocoa Research Institute of Nigeria, Ibadan, Nigeria 1975, pp.18-24.

Wadsworth, R.M., Ford, C.S., End, M.J and Hadley, P. 1997. *International Cocoa Germplasm Database Vol 1,2 and 3.* London International Financial Futures and Options Exchange (LIFFE), Cannon Bridge, London, EC 4R 3XX, U.K

Wadsworth, R.M., Ford, C.S., Turnbull, C.J. and Hadley, P. 2003. *International Cocoa Germplasm Database [ICGD]* v 5.2 Euronext. Liffe/University of Reading U.K

Williams, J.A. 1977. Hybridization between *Theobroma* and *Herrania* species. *Proc. V Internat. Cocoa Res. Conf.,* Ibadan, Nigeria, 1975: 171-175

Wood, G.A.R and Lass, R.A. 1985. *Cocoa.* 4th edn. Tropical Agriculture Series, Longman Publications, New York.

Wood, G.A.R. 1975. *Cocoa.* Longman Group Ltd., London. 3rd Edn.

Young A.M. 1994. *The chocolate tree: a natural history of cacao.* Smithsonian Institution Press, Washington, United States, pp. 200.

Chapter 13

African Oil Palm

P. Murugesan and G.M. Aswathy

ICAR-Indian Institute of Oilpalm Research,
Regional Station, Palode, Thiruvanandapuram – 695 562
E-mail: gesan70@gmail.com

Oil palm is a perennial tropical tree which is being propagated predominantly by seeds and seedlings are used as planting material for field planting (Murugesan *et al.*, 2005). The genus *Elaeis*, of the family Arecaceae, only contains two tropical species, *Elaeis guineensis* Jacq. (The oil palm) and *Elaeis* oleifera (HBK) Cortes found between 11° N and 15° S (Corley and Tinker, 2003). In all probability, *E.guineensis* originates from Africa (Zeven, 1964), whilst *E.oleifera* originates from Latin America (Hartley, 1988). Oil palm has originated from the West Coast of Africa. Genus *Elaeis* of the family Arecaceae contains two tropical species, *Elaeis guineensis* Jacq, and *Elaeis oleifera* (HBK) Cortes. Only *E.guineensis* is of economic interest due to the high oil content in the mesocarp and kernel. It is a pioneer species of African forests traditionally used by locals which exhibited several expansion phases. Oil palm is the most productive oil-bearing crop. Although it is planted in only 5 per cent of the total world vegetable oil acreage, palm oil accounts for 33 per cent of vegetable oil, 45 per cent of edible oil worldwide. The main oil palm growing countries in the world are Angola, Benin, Cameroon, Congo, Ghana, Cote d Ivoire, Ivory Coast, Nigeria, Sierra Leone, Brazil, Colombia, Costa Rica, Ecuador, Indonesia, Malaysia, Papua New Guinea, Sierra Leone and Thailand. In India, oil palm is cultivated in an area of 2.3 lakh ha across 11 states. Andhra Pradesh, Karnataka, Tamil Nadu, Mizoram and Kerala are the major oil palm growing states in the country, of which Andhra Pradesh alone has a share of more than 65 per cent in area under the crop. As per recent statistics, 12 million hectares have been planted with oil palm in the world with a production of over 45 million tons of palm oil. Oil palm is an exceptional example where selection progress is obtained through a single gene. The gene (Sh) controls shell thickness of the fruit are; 1. Dominant homozygote (Sh+ Sh+) formed the thick-shelled *dura* (D) 2. Recessive homozygote (Sh- Sh-) formed the shell-less *pisifera* (P) and 3. Heterozygote (Sh+ Sh-) formed the thin-shelled *tenera* (T). The *pisifera* is generally female sterile (no fertile bunch production), although there are some fertile *pisifera* in the population. The *tenera* is generated by crossing the *dura* (female) with the *pisifera* (male), *i.e.* Dura (D) × Pisifera (P) commonly referred as D × P. The D × P progenies produce 30 per cent more oil compared to the *dura*. The selection involves evaluation of the FFB yield and oil determination. The *dura* parents for use in hybrid seed production are selected on its own merits based family and individual palm selection. On the other hand, due to the sterile nature of the *pisifera*, its yield potential is determined indirectly through progeny test with elite *duras*. Pisiferas with a good general combining ability (GCA) are selected based on the *tenera* performance of the D × P or *Dura × Tenera* (D × T) progeny test experiment. In progeny testing, the *pisiferas* that give the best average performance of its half-sib progeny are selected. Though oil palm (*Elaeis guineensis*) is originated from West Africa, the commercial exploitation in the South East Asian countries was done from four seedlings introduced to Bogor, Indonesia in 1848. Intense breeding and selections on the Bogor descendents over several generations gave rise to the Deli *dura* population. Deli is one of the important Breeding Population of Restricted Origin (BPRO) used widely which was introduced from Africa and improved in South East Asia. Subsequently, Deli materials were extensively used to combine African sources like La Me and Congo. Deli material is unavoidable in breeding programmes as they complement good with African origins. Very few founders form the base of the origins that make up this commercial material: four for Deli (Rutgers, 1920), three for La Mé and around ten for Congo, one of which is over 50 per cent represented. The efficient genetic base is therefore very narrow. This is not only a drawback for improving yields, but especially for developing material tolerant of certain major endemic diseases. Moreover, the Deli origin is unavoidable in breeding schemes. It offers very good complement with the African origins and very good inherent agronomic value. Within Deli *dura* BPRO there is still apparently room to identify several breeding sub populations such as the Serdang, Ulu Remis, Banting, Johore Labis series About 13 deli sub populations are used worldwide for breeding programmes as reported by Rosenquist (1986). The other major groups reported by Billotte *et al.* (2001) are Congo, Cameroon/Cote d' Ivoire, Benin, Nigeria and Angola. AVROS is another historic population selected in Indonesia by Algemene Vereniging van Rubber-Planters Oostkust van Sumatra. They were produced through an open pollination process of a "Djongo" palm from the Botanic garden of Eala in Zaire (Democratic Republic of Congo) (Putri *et al*, 2009). Oil palm breeding has improved the planting materials. One of the factors leading to a significant increase in oil palm yields was the discovery of the better production potential displayed by crosses between the African "origins" and the Deli introduced and improved in Southeast Asia (Gascon and de Berchoux, 1964). The yield of *duras* has improved from 2.8 to 4.5 t/ha after four cycles of selections. The oil yield of *teneras* from subsequent *dura* selected and introgression with selected *pisiferas* has improved from 6.3 to 11.2 t/ha in the last four decades in Malaysia and other South East Asian countries. This was possible by use of improved and modern material developed from *deli dura* genetic base. But the genetic gain may not long last as many of the breeding programmes are continuously relay on restricted breeding base. In this context, knowledge on diversity of the germplasm and utilizing the specific traits of germplasm is required for incorporation of selected materials into the existing breeding programme to produce location specific hybrids for providing seed materials to diverse agro climatic conditions. Knowledge on genetic diversity among breeding materials could be an invaluable aid in oil palm (*Elaeis guineensis* Jacq.) improvement strategies.

Origin of Oil Palm

The oil palm, *Elaeis guineensis*, grows in the wild, semi wild and cultivated parts of the tropics, within 10° latitude of the equator, in Africa, South East Asia and South and Central America. However, it is endemic to the tropical lowlands of West and Central Africa, spreading from 16°N in Senegal to 15°S in Angola (Hartley, 1988). In the African forest, the oil palm is mainly pollinated by the weevil, *Elaeidobius kamerunicus* (Hartley, 1988). The oil palm is also found in East Africa, including the island of Madagascar where it is believed to have been introduced by Arab slave traders in the tenth century (Hartley, 1988).The African oil palm, (*Elaeis guineensis* Jacq.) is endemic to West and Central Africa and mostly found within the entire Guinean Zone extending inland for about 100 to 150 Km (Congo basin) from Latitude 10° N and 5° of the equator. Extensive natural or semi wild palm groves are distributed along the West Coast of Africa from Senegal to Angola (Zeven, 1967). Oil palm genetic resources from wild and semi wild populations are found in specific zones of the world such as the entire Guinean zone of West African for *Elaeis guineensis* and Central America for *Elaeis oleifera* (Ataga *et al.*, 2005). American oil palm (*Elaeis oleifera* HBK, Cortes) is found between 11° N and 15° S in Honduras, Nicaragua, Costa Rica, Panama, Ecuador and Brazil (Meunier, 1975; Ooi *et al.*, 1981; Escobar, 1982; Rajanaidu, 1983). On the level of genetic resources potentially usable in breeding programmes, an initial study was carried out as early as 1985 with isozyme markers on the collection existing in Côte d'Ivoire (Ghesquière, 1985). There is, however, another species in the genus, *E. oleifera*, found in the Amazon Forest and known in Brazil as 'caiaué'. This species has been incorporated into oil palm breeding programs (Meunier, 1975; Hardon *et al.*, 1985; Le Guen *et al.*, 1991; Santos, 1991). The interest in the 'caiaué' germplasm is due to some traits of the species that could be valuable to oil palm breeding, such as: slow growth, oil quality (mostly unsaturated oil) and disease resistance, including lethal yellowing, the major problem of oil palm cultivation in America (Bergamin Filho *et al.*, 1998), as well as *Fusarium* wilt, the major oil palm disease of Africa (Renard *et al.*, 1980; Hardon *et al.*, 1985). Caiaué' populations are spread throughout the humid areas of Central and South America. Extensive collections of 'caiaué' germplasm were done in the Amazon River Basin by EMBRAPA (Empresa Brasileira de Pesquisa Agropecuária, Brazil) and IRHO (Institut de Recherches pour les Huiles et Oléagineux, France), in the early 80s, constituting a valuable source of genetic material. In the Amazon Forest, 'caiaué' populations are usually found near rivers, on fertile and well-drained lands, known locally as 'Indian black lands' (Andrade, 1983; Barcelos, 1986).

Oil Palm Diversity

Collection and Evaluation of Germplasm

There are important germplasm collections of E. guineensis and improvement programmes exists in different countries and organizations. Researchers from various oil palm growing countries have collected germplasm and established field gene banks in Malaysia, Ivory Cost, Costa Rica, Brazil and Colombia etc. The major players involved in germplasm collections are Corporation Research Centre for Oil Palm (CENIPALMA), Indu Palma, Hacienda La Cabana in partnership with Palm Elite, La cabana/CIRAD France, Brazilian Agricultural Research Corporation, Agricultural Services Development (ASD), Malaysian Palm Oil Board (MPOB), NIFOR, Nigeria, Centre National De Recherche Agronomique (CNRA) (Ivory Coast), Institut National des Recherches Agricoles du Bénin (INRAB, Benin), The Institute of Agricultural Research for Development (IRAD), (Cameroon), Socfindo (Indonesia), Palmeras del Ecuador (Ecuador). The organizations viz., MPOB, ASD Costa Rica, Cenipalma, EMBRBA, UNIPALM, CIRAD are actively involved in crop improvement programme, introgressing new materials and supplying advance planting materials to meet the global expansion of the crop. India (DOPR) is having narrow genetic base and heavily relies on Thodupuzha duras and initiated introgression of new germplasm into existing breeding materials for evaluation and improvement. Some highlights of diversity studies of MPOB (Malaysia), IPORI (Indonesia), UNIVANCH (Thailand), ASD Costa Rica, NIFOR (Nigeria), CIRAD (France), CENIPALMA (Colombia), IIOPR (India) are presented below.

Diversity Study by MPOB, IPORI and UNIVANICH

MPOB has initiated an oil palm genetic resources programme which includes the collection, evaluation, utilisation and conservation of oil palm germplasm. To broaden the oil palm genetic base, extensive germplasm collections were made in its natural range in Africa (E. guineensis) and in tropical America (E. oleifera) (Rajanaidu, 1994). Malaysian Palm Oil Board (MPOB) carries out numerous explorations in Africa and South America since 1973 with cooperation of host countries and collects materials and shares equally between the host countries and MPOB. Most of the countries of Africa in which the oil palm can be found, including Nigeria, Cameroon, The Democratic Republic of the Congo, Tanzania, Madagascar, Angola, Senegal, Gambia, Sierra Leone, Guinea and Ghana were explored for oil palm populations (Rajanaidu and Jalani, 1994). Collections in Nigeria provided an important source of dwarf palms for breeding for dwarf palm population (Rajanaidu et al., 1998). These collections were planted in MPOB, Kluang Research Station and Kertong as field gene banks and they are used for evaluation and utilization. It is the largest germplasm collection in the world. Several trials laid out using materials from Tanzania, Senegal, Guinea Conakry and Ghana which were evaluated for Harvest Index (Malike et al., 2011). Tanzanian germplasm showed the highest mean of Harvest Index (HI) with significant values. Two expeditions were undertaken to Angola; the first 1991 collection has been planted and evaluated in Kluang Johar showed high potential for yield and long stalk. The second expedition was undertaken during 2010 and 125 accessions from 10 provinces of 25 sites (5 bunches per site) jointly by MPOB, Director General of Estate Crops, Indonesia and Nacional Do Café Angola (INCA) were collected which had several promising triats *viz.,* compactness, high mesocarp and long stalk (Marhalil and Rajanaidu, 2011). Population collected from Cameroon, Nigeria and Sierra Leone showed high genetic diversity. Cameroon and the DRC (Ex-Zaire) showed high tolerance to Ganoderma. Similarly, materials collected all over the world were characterized and their evaluation is still going on (Hayiti *et al.,* 2004). Indonesia and Malaysia are the largest consumers of oil palm planting materials in the world. About 500000 ha planted in Indonesia and 100,000 ha in Malaysia. Malaysia and Indonesia requires about 120 million seeds per year for new commercial planting and replanting demands which will be fulfilled through D×P seeds as the production of clonal seeds and tissue culture plantlets are limited. New generation of oil palm planting materials are mainly derived from Dami, Chemara, MARDI, Sofino, Socfin, Dabou, Banting Deli duras and sources of *pisiferas* are Nifor (Calabar), Ekona,Yangambi, La Me and AVROS (Rajanaidu *et al.,* 2007).Forty-two progenies of oil palm (*Elaeis guineensis* Jacq.) from the Malaysian Palm Oil Board (MPOB) germplasm collection from Angola were analyzed for their fatty acid composition and carotene content using gas chromatography and UV spectrophotometry, respectively. Their variation in the fatty acid traits and carotene content were considerably larger than in the current breeding materials in Malaysia. Their means for all the traits, except palmitic acid, were also higher. The phenotypic correlations were negative between palmitic/ stearic acid and palmitic/oleic acids, so increasing palmitic acid is likely to decrease stearic and oleic acids, and vice versa. A negative relationship was also found between oleic/linoleic acids. A number of progenies had carotene contents >1000 ppm and one progeny had iodine value >60. The heritability estimates for the individual fatty acids and carotene content were moderate to high, indicating good genetic control over the traits. The Angolan germplasm is therefore potentially useful breeding materials for improving Malaysian oil palm for commercial planting. However, further studies are needed before the breeding materials proper can be selected and used in actual breeding (Abdullah *et al.,* 2002). A total of 330 palms originating from 11 countries in Africa were screened using 10 EST-SSR primers. It was found that the germplasm exhibited a high level of genetic diversity. Most of the loci tested were 100 per cent polymorphic at 0.95 criterion. A total of 46 alleles were detected across all the germplasm populations. Of these, three were considered as rare alleles. The Nigerian germplasm showed the highest number of alleles per locus and the highest number of rare alleles, a high percentage of polymorphic loci and high heterozygosity, suggesting that Nigeria could be the centre of diversity of the wild oil palm. This study also revealed that the Madagascar germplasm is unique and different compared with the oil palm populations from the African mainland. Based on the dendrogram constructed, the germplasm populations could be divided into three major clusters; Cluster 1 consisting of (Angola, Tanzania, Cameroon, Nigeria, Democratic Republic of Congo formerly known as Zaire), Sierra Leone, Guinea and Ghana germplasm, Cluster 2 consisting of Gambia and Senegal germplasm, while the Madagascar germplasm was placed in Cluster 3. The mean genetic distance across the MPOB germplasm populations was 0.251. (Zulkifli *et al.,* 2012). The quality improvement of planting material needs a better understanding of the genetic relationships between genotypes from different populations used in the breeding programmes. In a study, 48 parents, representative of four populations used in Indonesian Oil Palm Research Institute (IOPRI) breeding programmes, were analysed with five selected. AFLP primer pairs and four isoenzymatic systems. One hundred and fifty eight scorable band levels were generated of which 96 (61 per cent) were polymorphic. AFLP allowed us to identify off-type descendants which were excluded from analysis. The use of unbiased Rogers distance clearly separated the four studied populations. The Neighbour-Joining method re-groups two African populations which are known as originating from different regions. Nevertheless, the variability revealed is in accordance with oil palm breeder's knowledge. The results obtained with AFLP showed that the crosses among the African sub-population, which is excluded in oil palm reciprocal recurrent selection (RRS) breeding programmes, may be more interesting than the crosses between the African and the Deli populations (Purba *et al.*, 2000). There are structuring into two groups, Centre and West African, and confirmed the merits of traditional "Deli × La Mé" crosses. In terms of the pool of parents used in second cycle RRS at the Indonesian Oil Palm Research Institute (IOPRI), Purba *et al.* (2000) conducted a study using isozymes and AFLP markers. Three groups were found: Deli, Congo and Cameroon/ Côte d'Ivoire. This analysis enabled them to propose a new selection strategy in accordance with the structure of the diversity discovered. The PCoA distinguished between three groups of populations (I, II and III) along discriminant axes 1 and 2 representing 19.8 per cent of the molecular variability studied. Group I contained the Côte d'Ivoire origins clearly separated from the other origins. Group II consisted of the origins from Brazil, Benin, Nigeria and origins from central Africa. It was not possible to distinguish between the different origins making up this group. Group III contained only the Deli origin. A total of 359 accessions of oil palm (*Elaeis guineensis* Jacq.) originating from 11 African countries (Nigeria, Cameroon, Congo DR, Tanzania, Angola, Senegal, Sierra Leone, Guinea, Ghana, Madagascar and Gambia) were characterized using the RFLP method using the standard Deli *dura* as the check. Genomic DNA from each sample was digested using five restriction enzymes and hybridized with four oil palm cDNA probes. Data were analyzed using Biosys-1 computer software to calculate the genetic variability parameters. In general, all the collections exhibited higher levels of diversity than the standard variety, Deli *dura*. The standard variety, Deli *dura*, lost 36 alleles as compared to the natural populations indicating a reduction in genetic variability. Material from Nigeria showed the highest mean number of alleles per locus (1.9) and percentage of polymorphic loci (67.2 per cent). These findings, combined with others, suggest that Nigeria may be the center of diversity of wild oil palm. It further suggests that oil palm natural populations maybe possessing adequate genetic variability that are potentially useful for improvement programs (Maizura *et al.*, 2006). The oil palm germplasm with different characteristics were identified and the genetic distance within and between the groups was estimated. (Li-Hammed *et al.*, 2015). This study reports on the detection of additional expressed sequence tags (EST) derived simple sequence repeat (SSR) markers for the oil palm. A large collection of 19243 *Elaeis guineensis* ESTs were assembled to give 10258 unique sequences, of which 629 ESTs were found to contain 722 SSRs with a variety of motifs. Dinucleotide repeats formed the largest group (45.6 per cent) consisting of 66.9 per cent AG/CT, 21.9 per cent AT/AT, 10.9 per cent AC/GT and 0.3 per cent CG/CG motifs. This was followed by trinucleotide repeats, which is the second most abundant repeat types (34.5 per cent) consisting of AAG/CTT (23.3 per cent), AGG/CCT (13.7 per cent), CCG/CGG (11.2 per cent), AAT/ATT (10.8 per cent), AGC/GCT (10.0 per cent), ACT/AGT (8.8 per cent), ACG/CGT (7.6 per cent), ACC/GGT (7.2 per cent), AAC/GTT (3.6 per cent) and AGT/ACT (3.6 per cent) motifs. Primer pairs were designed for 405 unique EST-SSRs and 15 of these were used to genotype 105 *E. guineensis* and 30 *E. oleifera* accessions. Fourteen SSRs were polymorphic in at least one germplasm revealing a total of 101 alleles. The high percentage (78.0 per cent) of alleles found to be specific for either *E. guineensis* or *E. oleifera* has increased the power for discriminating the two species. The estimates of genetic differentiation detected by EST-SSRs were compared to those reported previously. The transferability across palm taxa to two *Cocos nucifera* and six exotic palms is also presented. The polymerase chain reaction (PCR) products of three primer-pairs detected in *E. guineensis, E. oleifera, C. nucifera* and *Jessinia bataua* were cloned and sequenced. Sequence alignments showed mutations within the SSR site and the flanking regions. Phenetic analysis based on the sequence data revealed that *C. nucifera* is closer to oil palm compared to *J. bataua*; consistent with the taxanomic classification. (Ngoot-Chin Ting *et al.*, 2010). Species-specific simple sequence repeat (SSR) markers are favored for genetic studies and marker-assisted selection (MAS) breeding for oil palm genetic improvement. This report characterizes 20 SSR markers from an *Elaeis oleifera* genomic library (gSSR). Characterization of the repeat type in 2000 sequences revealed a high percentage of di-nucleotides (63.6 per cent), followed by tri-nucleotides (24.2 per cent). Primer pairs were successfully designed for 394 of the *E. oleifera* gSSRs. Subsequent analysis showed the ability of the 20 selected *E. oleifera* gSSR markers to reveal genetic diversity in the genus *Elaeis*. The average Polymorphism Information Content (PIC) value for the SSRs was 0.402, with the tri-repeats showing the highest average PIC (0.626). Low values of observed heterozygosity (H_o) (0.164) and highly positive fixation indices (F_{is}) in the *E. oleifera* germplasm collection, compared to the *E. guineensis*, indicated an excess of homozygosity in *E. oleifera*. The transferability of the markers to closely related palms, *Elaeis guineensis, Cocos nucifera* and ornamental palms is also reported. Sequencing the amplicons of three selected *E. oleifera* gSSRs across both species and palm taxa revealed variations in the repeat-units. The study showed the potential of *E. oleifera* gSSR markers to reveal genetic diversity in the genus *Elaeis*. The markers are also a valuable genetic resource for studying *E. oleifera* and other genus in the Arecaceae family. (Noorhariza Mohd Zaki *et al.*, 2012). A total of 5,521 expressed sequence tags (ESTs) from oil palm were used to search for type and frequency of simple sequence repeat (SSR) markers. Dimeric repeat motifs appeared to be the most abundant, followed by tri-nucleotide repeats. Redundancy was eliminated in the original EST set, resulting in 145 SSRs in 136 unique ESTs (114 singletons and 22 clusters). Primers were designed for 94 (69.1 per cent) of the unique ESTs (consisting of 14 consensus and 80 singletons). Primers for 10 EST-SSRs were developed and used to evaluate the genetic diversity of 76 accessions of oil palm originating from seven countries in Africa, and the standard Deli *dura* population. The average number of observed and effective alleles was 2.56 and 1.84, respectively. The EST-SSR markers were found to be polymorphic with a mean polymorphic information content value of 0.53. Genetic differentiation (F_{ST}) among the populations studied was 0.2492 indicating high level of genetic divergence. Moreover, the UPGMA (unweighted pair-group method with arithmetic mean) analysis revealed a strong association between genetic distance and geographic location of the populations studied. The germplasm materials exhibited higher diversity than Deli dura, indicating their potential usefulness in oil palm improvement programmes. The study also revealed that the populations from Nigeria, Congo and Cameroon showed the highest diversity among the germplasm evaluated in this study. The EST-SSRs further demonstrated their worth as a new source of polymorphic markers for phylogenetic analysis, since a high percentage of the markers showed transferability across species and palm taxa (Rajinder Singh *et al.*, 2008). Sex ratio and shell-thickness type are among the main components determining yield in oil palm. An integrated linkage map of oil palm was constructed based on 208 offspring derived from a cross between two *tenera* palms differing in inherited sex ratio. The map consisted of 210 genomic simple sequence repeats (SSRs), 28 expressed sequence tag SSRs, 185 amplified fragment length polymorphism markers, and the *Sh* locus, which controls shell-thickness phenotype, distributed across 16 linkage groups covering 1,931 cM, with an average marker distance of 4.6 cM. Quantitative trait locus (QTL) analysis identified eight QTLs across six linkage groups associated with sex ratio and related traits. These QTLs explained 8.1–13.1 per cent of the total phenotypic variance. The QTL for sex ratio on linkage group 8 overlapped with a QTL for number of male inflorescences. In most cases a specific QTL allele combination was responsible for genotype class mean differences, suggesting that most QTLs in heterozygous oil palm are likely to be segregating for multiple alleles with different degrees of dominance. In addition, two new SSRs were shown to flank the major *Sh* locus controlling the fruit variety type in oil palm. Kittipat Ukoskit *et al.*, 2014). The development of an oil palm RFLP marker map has enabled marker-based QTL mapping studies to be undertaken. Information from 153 RFLP markers was used in combination with phenotypic data from an F_2 population to estimate the position and effects of quantitative trait loci (QTLs) for traits including yield of fruit and its components and measures of vegetative growth. The mapping population consisted of 84 palms segregating for the major gene influencing shell thickness. Marker data were analysed to produce a linkage map consisting of 22 linkage groups. The QTL mapping analysis was carried out by interval mapping and single-marker analysis for the unlinked markers; significance thresholds were generated by permutation. Using both single-marker and interval-mapping analysis significant marker associated QTL effects were found for 11 of the 13 traits analysed. The results of interval-mapping analysis of fruit weight, petiole cross section and rachis length, and ratios of shell:fruit, mesocarp:fruit and kernel:fruit indicated significant ($P<0.05$) QTLs at the genome-wide threshold. The putative QTLs were associated with between 8.2 per cent and 44.0 per cent of the phenotypic variation, with an average of 27 per cent for the single-marker analysis and 19 per cent for the interval-mapping analysis. The higher percentage of phenotypic variation explained in the single-marker analysis, when compared to the interval-mapping analysis, is likely to be due to the lower stringency associated with the single-marker analysis. Large dominance deviations were associated with a sizeable proportion of the putative QTLs. The ultimate objective of mapping QTLs in commercial populations is to utilise novel breeding strategies such as marker-assisted selection (MAS). The potential impact of MAS in oil palm breeding programmes is discussed (Rance *et al.*, 2001). Very few reports for cataloguing and descriptor development are available in the case of oil palm. The two leading palm oil producing countries *viz.*, Malaysia and Indonesia are not members of International Union for the Protection of New Varieties of plants (UPOV) system. India is also not a UPOV member but enacted Plant Variety Protection and Farmers Right Act, 2001 (PPV and FR Act, 2001) for variety protection. Costa Rica has variety protection for all the species including oil palm. Colombia has practical experience in the examination of Distinctness, Uniformity and Stability for oil palm varieties. Recently, thirteen countries have formed East Asia Plant Variety Protection Forum (EAPVPF) for developing test guidelines for designated crop plants including oil palm (http://www.eapvp-forum.org/). Malaysia has developed criteria draft guidelines for DUS and PVP regulation has been published in gazette on 20th Oct, 2008 (http://pvpbkkt.doa.gov.my/). Feyt (2001) has grouped oil palm varieties for the purpose of variety protection. In the past, oil palm descriptor was published by Rajanaidu (1989) and yield, vegetative measurements, certain physiological traits, bunch characters and oil quality were included as indicators of variety or genotype. Enzyme system is considered as phenotypic traits and isozymes provide good opportunities for finding quality descriptors (Chan *et al.*, 1988). Existing evaluation and experimental results may not be sufficient for description of variety for the purpose of registration under PPV and FR Act. Therefore, additional parameters are required for complete characterization of variety in view of recent developments in variety registration and protection mainly to encourage crop improvement.

c. Diversity Studies by NIFOR and CIRAD

Initial research on oil palm in Nigeria started in the period between 1930 and 1950, when this country was one of the most important supplier of palm oil in the world. Several research stations were created where important steps were taken to develop outstanding genotypes now used in several breeding stations in Ghana and in Costa Rica. First attempts to develop an oil palm research program in Nigeria come from 1930, given the importance this country had played in the supply of palm oil for world trade up to WWI. However, due to WWII, The West African Institute for Oil Palm Research (WAIFOR) did not start its functions until 1952. The main station (1735 ha) was located 32 km from Benin, and a substation was established at Abak in the Calabar province (182 ha). Some Agriculture Department stations utilized for the program were the Moor Plantation (Ibadan), Ogba (Benin), Nkwele (Onitsha), Umudike (Umahia) and Obio-Akpa (Abak). The institute became eventually The Nigerian Institute for Oil Palm Research (NIFOR). The existing collections of Nigerian Research Institute for Oil Palm includes 1954 Ufuma and Aba collections, 1973 NIFOR/MARDI collections from 45 Nigerian locations from the marginal zone of the old Nsukka province (Okwuagwu, 1985). Apart from this, various African, deli origins have been made and exploited for breeding and selection. Very detailed evaluation of these collections has been carried out both in Nigeria and elsewhere (Rajanaidu *et al.*, 2001). NIFOR had extensive collections from the main oil palm belt, Eastern part of Nigeria (High lands 200-400 m above sea level) was covered during recent expedition where greatest diversity exists. The area included was Afikpo, ABAKALIKI, Okigwe and Umuahia which were unexploited (Okwuagwu *et al.*, 2011). The Nigerian breeding program was initiated with germplasm of poor bunch composition, since Aba and Calabar origins formed the major proportion of the material utilized. They recommended the use of molecular markers to get information on germplasm diversity. Ataga, (1994) reported high kernel variation in Nigerian germplasm collections at Benin City, Nigeria. A study was conducted to assess the extent of genetic variability, broad-sense heritability, and correlation between yield and yield components in three Deli/dura x tenera (DxT) breeding populations. Populations 1, 2, and 3 were made up of 14 Deli x tenera progenies, 16 dura x tenera progenies, and 21 Deli/dura x tenera progenies respectively. The combined analysis of variance for number of bunches (BN), average bunch weight (ABW), and fresh fruit bunch yield (FFB) revealed significant genotypic differences. The phenotypic coefficient of variation, however, was generally greater than the genotypic coefficient of variation, implying the influence of genotype x environment interaction in the expression of these traits. Broad-sense heritability estimates for the three traits varied considerably from population to population. Estimates of heritability were high in population 1 (78, 88.6, and 70.7 respectively for BN, ABW, and FFB yield). The corresponding figures were 27.5, 41.5, and 24.3 for population 2, and 5.3, 32.7, and 20.5 for population 3. High genotypic coefficient of variation (31.4, 27.4, and 26.5), heritability, and genetic advance as percent of mean (57.2, 52.7, and 45.9) for the three bunch yield traits in population 1 imply the potential for improvement of these traits through selection. On the whole, population 1 is an appropriate starting point for the next cycle of breeding and selection. Highly significant positive ($p < 0.01$) relationships were noted between FFB and BN suggesting that BN is a major yield contributing component. Strong negative correlations between BN and ABW (–0.220**, –0.260**, and –0.368**), however, denote that selection for high BN may result in lower ABW and vice-versa, which would hinder the exploitation of high heritabilities. Accordingly, any form of selection that takes into account the additive genetic variation may neglect other pathways, such as heterosis which can be identified by progeny testing only (Okwuagwu *et al.*, 2008). Test cross population of Deli *dura* parents namely; NIFOR ex UP Malaya, NIFOR ex Serdang Avenue, Ulu Remis Deli ex Sabah, NIFOR ex Serdang Avenue x La Me and Equador Deli were evaluated for bunch yield and fruit quality characters in rainforest ecology of Nigeria. The objective of the experiment is to select new Deli *dura* parents for inclusion in the next cycle of NIFOR oil palm breeding programme. The results showed that the progenies from the Equador Deli parent had the highest bunch yield due to the complementation of number of harvested bunch and average bunch weight. The reverse was however, the case with the progenies from NIFOR ex UP Malaya parent despite the excellent fruit quality characters. The existence of transgressive segregants in some of the progenies is an indication that selection would be effective in the improvement of these quantitative traits. Therefore, new parents must be limited to recombination of individuals of similar mean bunch yield and fruit quality to build up favourable genes for each component (Okoye *et al.*, 2009). The results of the development of oil palm (*Elaeis guineensis* Jacq.) microsatellite markers are given step by step, from the screening of libraries enriched in (GA)*n*, (GT)*n*, and (CCG)*n* simple-sequence repeats (SSRs) to the final characterisation of 21 SSR loci. Also published are primer sequences, estimates of allele size range, and expected heterozygosity in *E. guineensis* and in the closely related species *E. oleifera*, in which an optimal utility of the SSR markers was observed. Multivariate data analyses showed the ability of SSR markers to efficiently reveal the genetic diversity structure of the genus *Elaeis* in accordance with known geographical origins and with measured genetic relationships based on previous molecular studies. High levels of allelic variability indicated that *E. guineensis* SSRs will be a powerful tool for genetic studies of the genus *Elaeis*, including variety identification and intra- or inter-specific genetic mapping. PCR amplification tests on a subset of 16 other palm species and allele-sequence data showed that *E. guineensis* SSRs are putative transferable markers across palm taxa. In addition, phenetic information based on SSR flanking region sequences makes *E. guineensis* SSR markers a potentially useful molecular resource for any researcher studying the phylogeny of palm taxa. (Billotte *et al.*, 2001).

d. Diversity Study by South and Central American Organizations

Colombia is a main oil palm producer in Latin America. In order to increase the narrow genetic base of the existing oil palm cultivars collection activities are carried out for *Elaeis guineensis* in African counties including Angola and Cameroon. Additionally old *duras* present in commercial stations have been studied. Prospecting exercises and collection of genotypes were done in the Colombian Amazon region for *Elaeis oleifera* (Rocha and Rey, 2007). American oil palm populations are spread throughout the humid areas of central and South America. Extensive collections of *Oleifera* germplasm were done in Amazon River basin by EMBRAPA (Empresa Brasileira de Pesquisa Agropecuaria, Brazil). In Amazon forest *Oleifera* populations are found near rivers on fertile well drained lands (Moretzsohn, *et al*, 2002). *Elaeis oleifera* or 'caiaué', a close relative of oil palm (*E. guineensis*), has some agronomic traits of great interest for the oil palm genetic breeding such as slow growth, oil quality (mostly unsaturated) and disease resistance. An analysis of a Brazilian oil palm germplasm collection was carried out using RAPD (Random Amplified Polymorphic DNA) markers with the objective of understanding the genetic variation of 'caiaué' accessions collected in the Amazon Forest in the last two decades. A sample of 175 accessions obtained along the Amazon River Basin was analyzed and compared to 17 accessions of oil palm from Africa. Ninety-six RAPD markers were used in the analysis, of which fourteen were shown to be specific to oil palm, while twelve were specific to 'caiaué'. Results showed that the Brazilian 'caiaué' accessions studied have moderate levels of genetic diversity as compared to oil palm accessions. The data allowed the establishment of similarity groups for 'caiaué' accessions, which is useful for selecting parental plants for population breeding. Cluster analysis showed that, in general, genetic similarities are not correlated to geographical distances, but consistent with geographical dispersal along the Amazon River network. AMOVA showed that most of the genetic variation is found within populations, as expected for anallogamous and long-lived perennial species. The study provides important information to define strategies for future collection expeditions, for germplasm conservation and for the use of *E. oleifera* in breeding programs (Moretzsohn *et al.*, 2006). The species *E. oleifera* is a promising genetic resource for oil palm breeding programs. The main purpose of this study was to quantify the genetic diversity within and between accessions collected in the Ecuadorian Amazon. Nine microsatellite markers were genotyped in 40 plants from the *E. oleifera* germplasm bank of INIAP in Ecuador. The number of alleles varied from two to five, with a total of 26 alleles and a mean number of 2.89. The polymorphism information content was 0.35, indicating that all markers were informative and enough to access the variability within and between *E. oleifera* plants. The average inbreeding coefficient was -0.03, the mean expected heterozygosity 0.41, the average observed heterozygosity 0.42 and seven of the nine markers were in Hardy–Weinberg equilibrium (HWE). This result shows that the analyzed population was close to the assumed HWE, showing high variability between plants and no inbreeding or sampling effect. The 40 plants were clustered in seven groups differentiated by the Tocher method. Seedlings derived from a same accession were grouped separately, indicating variability within the sampled accessions. This variability was illustrated by grouping the 40 plants by the UPGMA method and confirmed by the molecular variance analysis (AMOVA). Of the total variation, 72 per cent was detected among plants within the sampled accessions. These results have implications for breeding purposes and the collection of new *E.oleifera* germplasm. (Digner Ortega Cedillo *et al.*, 2016). The objective of this work was to evaluate the genetic diversity, its organization and the genetic relationships within oil palm (*Elaeis oleifera* (Kunth) Cortés, from America, and *E. guineensis* (Jacq.), from Africa) germplasm using Restriction Fragment Length Polymorphism (RFLP) and Amplified Fragment Length Polymorphism (AFLP). In complement to a previous RFLP study on 241 *E. oleifera* accessions, 38 *E. guineensis* accessions were analyzed using the same 37 cDNA probes. These accessions covered a large part of the geographical distribution areas of these species in America and Africa. In addition, AFLP analysis was performed on a sub-set of 40 accessions of *E. oleifera* and 22 of *E. guineensis* using three pairs of enzyme/primer combinations. Data were subjected to Factorial Analysis of Correspondence (FAC) and cluster analysis, with parameters of genetic diversity being also studied. Results appeared congruent between RFLP and AFLP. In the *E. oleifera*, AFLP confirmed the strong structure of genetic diversity revealed by RFLP, according to geographical origin of the studied material, with the identification of the same four distinct genetic groups: Brazil, French Guyana/Surinam, Peru, and north of Colombia/Central America. Both markers revealed that genetic divergence between the two species is of the same magnitude as that among provenances of *E. oleifera*. This finding is in discrepancy with the supposed early tertiary separation of the two species. (Edson Barcelos *et al.*, 2002). ASD has 450 hectares for field experiments in Coto and owns one of the most diverse oil palm germplasm collections in the world. They had introduced several BPRO from Malaysian Research Stations and intensively use MAR 559 for seed production due to high additive genetic effect in relation to several economic traits. Angola *dura* is commonly used as female parents which were selected from

Ivory Coast and Kade Research Station in Ghana. Apart from above, ASD use Bamenda (Cameroon) and Kigoma (Tanzania) (Barbosa and Chinchilla, 2003). It is reported that Deli dura types had better oil and kernel yields than African types (Richardson, 1995). Calabar materials have a shorter leaf length and higher leaf area than other traditional materials. Ekona materials have greater oil to bunch ratio. ASD Costa Rica breeding programme has been producing commercial oil palm planting materials since 1974. The seed production programme is oriented towards the exploitation of the genetic potential present within Deli and AVROS BPROs. They recently included new sources of germplasm *viz.*, Djongo, La Me, Nigeria and Ekona. ASD has been supplying oil palm planting materials to meet the global expansion of the crop as commercial basis.

e. Diversity Study by ICAR-IIOPR (India)

Systematic cultivation of oil palm was started in India in the 1960s under the aegis of the Department of Agriculture, Kerala by establishing a hundred acre plantation at Thodupuzha, Idukki District. Planting materials were introduced from Nigeria and Malaysia. It was established that Indian genetic base of oil palm "Thodupuha duras" had high variations for shell weight (52 per cent) and kernel (47 per cent) characteristics (Murugesan *et al.*, 2011). Research work on oil palm improvement was initiated during the year 1974 (Rethinam, 1996). The *Dura* population consisting of over 700 palms available at Thodupuzha formed the base to start initial selection programme. The palms were screened for number and weight of Fresh Fruit Bunch (FFB) and 40 superior palms were selected. Prior to this exercise, *tenera* hybrids were produced by crossing 11 superior *dura* palms with *pisifera* pollen grains imported from Nigeria (*Pisifera* was not identified during this period). These hybrids were planted at the then Central Plantation Crops Research Institute (ICAR-CPCRI), Research Centre, Palode during 1976 (Pillai and Nampoothiri, 1981) and monitored for their performance over the years and two hybrids were selected as most promising ones with a potential of yielding 4.6 tons of palm oil per hectare under rainfed condition (Nampoothiri *et al.*, 1992). Systematic collection of oil palm germplasm was started at Palode in the 1980s and presently the field collection consists of 62 accessions. Performance of all these accessions is being monitored for introgression of superior palms in the breeding programme. During the year 1992-96 *Elaeis oleifera*, another species was introduced from unknown origin and they are being maintained at Palode. Their performance was exceedingly good and interspecific hybridization was initiated during 1998 and the hybrids are under evaluation for identifying the promising ones and backcrossing with the cultivated species, *E. guineensis*. Dwarf stature in oil palm is a desirable character since harvesting is a problem when the palms grow old. A search for dwarf gene was made in the available population and one dwarf palm was identified at Palode. This palm was selfed and the progenies have segregated into dwarfs and talls. Efforts are being made to introgress this gene through hybridization. During 1995-96, a joint programme by Central Plantation Crops Research Institute, Food and Agriculture Organization and Technology Mission for Oil Seeds and Pulses (TMOP) was launched to collect cold and drought tolerant germplasm from the African countries. Primary collections were made from Cameroon, Tanzania, Zambia and

Guinea Bissau and they are being tested for their variability and stress tolerance in seven locations (Palode, PCKL, Pedavegi, Adilabad, Nellore, Mulde, Mohitnagar) of India Pillai *et al.* (2000). Ten African *duras* planted under two environments namely irrigated and stress were evaluated for their performances at Pedavegi and inferred all the physiological parameters except inter cellular CO_2 concentration and leaf canopy temperature were higher under irrigated conditions when compared to stress (Mathur *et al.,* 2001). Higher photosynthetic rates were recorded in the Guinea Bissau germplasm which coincide with their high stomatal conductance and had tolerance to water stress (Suresh *et al.,* 2004). Thirty one accessions of such resources were planted during 1998 at DOPR Research centre Palode were evaluated which resulted in identification of one *tenera* and 83 *dura* and three *pisiferas* (Murugesan and Mandal, 2010). All the germplasm accessions were conserved as field gene bank. Germplasm from primary/secondary centres of origin were collected during 1994 under FAO programme and planted in the gene bank at ICAR-IIOPR, Research Centre, Palode during 1998 were studied for genetic diversity. Twenty six accessions representing three African countries (Guinea Bissau, Tanzania and Zambia) were studied using 50 individual palms. Thirty numbers of vegetative and bunch component traits recorded during 2008 to 2013 were analysed in this study. An attempt was made using Shannon-Weaver Diversity Index (SWDI) with an objective to understand the level of diversity in these traits. In general, mean of all the accessions exhibited high levels (0.694) of diversity. Mean diversity estimate (0.778) was highest in Tanzanian source closely followed by Zambia (0.727) and least value (0.576) was observed in Guinea Bissau. Low diversity values (<0.32) for bunch weight, shell thickness, single fruit, and nut weight noticed in Guinea Bissau when compared to other sources. Highest level of homozygosity (SWD=0) for spine length was observed in Guinea Bissau population and similar trend of homzygosity noticed in other traits also in GB accessions. These findings combined with other evaluation results suggest that 'Tanzanian population' possess adequate genetic variability that is potentially useful for oil palm improvement program in India. More palms should be preserved for populations that have higher diversity and those with rare traits. (Murugesan *et al.,* 2015). The research work of the Oil Palm improvement in India was started with selections from *dura* palms planted at Oil Research Station at Thodupuzha, Kerala during 1961. Initially, the performance of the population was assessed on the basis of Fresh Fruit Bunch (FFB) yield and number of bunches for nine years (1974-82). Limited number of palms was exploited for hybridization and population improvement. Selected palms were utilized for hybridization and production of hybrids. Although this population is widely utilized for seed production and genetic improvement, there was no information about its phenotypic variations especially for fruit quality components. A total of 341 *dura* palms of Thodupuzha materials were assessed for different fruit and seed characteristics. Fruit form analysis revealed that seven palms (2 per cent) are *teneras* out of 341 palms and rest of the palms are *duras*. The percentage of co-efficient of variation was high for shell weight followed by kernel weight and lowest variation was recorded for percentage of mesocarp and kernel oil per fruit. This study also unearths potentiality of unexploited *dura* palms (US356 US225, US147, US239, US380, US297, S285 and US375) mainly on the basis of mesocarp content and oil per fruit

(>84 per cent). Promising palms could be effectively utilised for introgression into the current breeding programme (Murugesan *et al.*, 2011). Oil palm breeding programme in India has been based on Thodupuzha *Dura ´* NIFOR *Pisifera* (Murugesan *et al.*, 2011). Several exotic germplasm collections were made from different oil palm growing countries to India (Namboothiri, 1994). Minimum descriptor was studied for different genotypes and some descriptor related investigations have been reported (Anonymous, 2009), Murugesan *et al.* (2009), Murugesan and Gopakumar (2010a), Murugesan and Gopakumar (2010b) and Murugesan (2010). As a part of improvement work, evaluation was undertaken in the field gene bank consisting several exotic germplasm collections planted during 1981- 1994 at ICAR-Directorate of Oil Palm Research, Research Centre, Palode. A tenera palm from exotic collection planted during 1981 and *Elaeis oleifera* of Surinam origin planted during 1988 showed low stem elongation when compared to their counterparts. Nigerian materials were introduced vide code No. E130756 through National Bureau of Plant Genetic Resources (ICAR-NBPGR), New Delhi and Surinam *oleifera* was a chance introduction. These palms were characterized based on IBPGR oil palm descriptor and evaluated for their yield potential. Surinam oleifera was also compared with other exotic accessions for their descriptive characteristics. Dwarf tenera had bunch yield of 118kg/palm/year with 9.1 bunch numbers and 24 cm height increment, whereas Surinam had 75kg/year/palm with 6 bunches and 15 cm height increment. It was proved by characterization that both of them had dumpiness *viz.*, short trunk, short leaves and other vegetative characteristics. As per bunch analysis, dwarf tenera recorded 20.4 per cent and 1.08 per cent of Mesocarp oil to Bunch and Kernel oil to Bunch, respectively, where as Surinam *oleifera* had 9.25 per cent and 3.1 per cent. In order to improve the yield and bunch quality components of these compact materials, selfing and inter se crossing and back crossing (with *E. guineensis*) was attempted for *tenera* and *oleifera*, respectively. (Murugesan *et al.*, 2009).Thirty one accessions of four different sources planted during 1998 in the field gene bank at ICAR-Indian Institute of Oilpalm Research (ICAR-IIOPR), Research Centre, Palode Kerala were evaluated. Three *pisiferas*, one *tenera* and 83 *dura* palms were identified out of 87 palms as per fruit form analysis. The section of results pertaining to *pisiferas* is presented in this paper. Characterization of *pisifera* palms from genetic resources was carried out for possible utilization of individual palms. The characterization revealed that one *pisifera* (DOPRG-53-E66) showed fertile character which recorded normal bunch and fruit development with 25 per cent fruit to bunch whereas, other *pisiferas* (DOPRG-53-E-75 and DOPRG-54-E65) showed aborted bunches throughout the evaluation period. However, DOPRG-54-E65 found to sets few fruits twice with *virescence* fruit pigmentation and shell-less kernel. DOPRG-53-E66, DOPRG-54-E65, and DOPRG-53-E-75 palms had average height increment of 65.1, 69.2, and 55.1 cm per year, respectively. Suitable sterile *pisifera* palms could be selected as promising palm after progeny testing for commercial hybrid seed production (Murugesan and Goutam Mandal, 2010). Multivariate statistical tools like cluster analysis have proved useful in characterizing and studying genetic diversity of germplasm resources. Thus, this study was aimed at classifying the diversity pattern in oil palm germplasm using two hierarchical clustering methods (single linkage clustering method and Ward's method). 595 oil palm genotypes grouped into 44 accessions were morphologically characterized for yield traits, bunch quality traits, morphological and physiological traits and fatty acid traits. The two clustering methods classified the accessions into eight groups and differ slightly in the assigning of the accessions into groups.

Summary

Oil palm breeding has improved the planting materials. But the genetic gain may not long last as many of the breeding programmes are continuously relay on restricted breeding base. Deli is one of the important Breeding Population of Restricted Origin (BPRO) used widely which was introduced from Africa and improved in South East Asia. About 13 deli sub populations are used worldwide for breeding programmes. Knowledge on diversity of the germplasm and utilizing the specific traits of germplasm is required for incorporation of selected materials into the existing breeding programme to produce location specific hybrids for providing seed materials to diverse agro climatic conditions. There are important germplasm collections of *E. guineensis* and improvement programmes exists in different countries and organizations. Researchers from various oil palm growing countries have collected germplasm and established field gene banks in Malaysia, Ivory Cost, Costa Rica, Brazil and Colombia *etc.* The major players involved in germplasm collections are Corporation Research Centre for Oil Palm (CENIPALMA), Indu Palma, Hacienda La Cabana in partnership with Palm Elite, La cabana/CIRAD France, Brazilian Agricultural Research Corporation, Agricultural Services Development (ASD), Malaysian Palm Oil Board (MPOB), NIFOR, Nigeria, Centre National De Recherche Agronomique (CNRA) (Ivory Coast), Institut National des Recherches Agricoles du Bénin (INRAB, Benin), The Institute of Agricultural Research for Development (IRAD), (Cameroon), Socfindo (Indonesia), Palmeras del Ecuador (Ecuador). The organizations *viz.,* MPOB, ASD Costa Rica, Cenipalma, EMBRBA, UNIPALM, CIRAD are actively involved in crop improvement programme, introgressing new materials and supplying advance planting materials to meet the global expansion of the crop. While extensive collections of germplasm have been made only a few were made to use this material to increase diversity in breeding populations and commercial seed is still consists predominantly Deli in case of female parent (*duras*). The different studies clearly reported that there are three major groups available in oil palm in wild/semi wild conditions. They are West African countries (Senegal, Guinea, and Sierra Leone), the West Central East Africa (Ghana, Nigeria, Cameroon, Zaire, Angola, Tanzania, Bahia including deli duras) and Madagascar. For yield improvement and breeding programme intra and intergroup combinations are suggested with wide genetic distance among accessions. Some highlights of diversity studies of MPOB (Malaysia), IPORI (Indonesia), UNIVANCH (Thailand), ASD Costa Rica, NIFOR (Nigeria), CIRAD (France), CENIPALMA (Colombia), and IIOPR (India) are reported here.

References

Abdullah, Noh and Nookiah, Rajanaidu and Din, Ahmad Kushairi and Amiruddin, Mohd Din and Yusop, Mohd Rafiiand Saleh, Ghizan and Zainal Abidin, Mohd Isa. 2002. Variability in Fatty Acid Composition, Iodine Value and Carotene

Content in the MPOB Oil Palm Germplasm Collection from Angola.Journal of Oil Palm Research. 14 (2): 18-23.

Andrade, E.B. 1983. Relatório de expedição para coleta de germoplasma de caiaué (*Elaeis oleifera* (H.B.K.) Cortés), na Amazônia brasileira. Manaus, EMBRAPA/ CNPSD (atual CPAA).

Anonymous. 2009. Annual report (2008-09). Directorate of Oil Palm Research, Andhra Pradesh, Pedavegi, pp.16-17.

Ataga, C.D. 1994. Variation in kernel size in Nigerian oil, palm (*Elaeis guineensis* Jacq.) germplasm collection. Plant Genetic Resources Newsletter, (100): 24

Ataga, C.D., Okolo, E.C. and Okwuagwu, C.O. 2005. Evaluation of an Oil Palm (*Elaeis guineensis*, Jacq) genetic material collected from a marginal zone of Nigeria. (In) Proceedings of Agriculture, Biotechnology and Sustainability Conference, PIPOC, 2005, 25-29, September, Selangor, Malaysia, pp 733-740.

Barbosay, R. and Chinchilla, C. 2003. ASD Oil palm germplasm from Nigeria, ASD Oil Palm papers N? 26, pp 33-44.

Barcelos, E. 1986. Características genético-ecológicas de populações naturais de caiaué (*Elaeis oleifera* (H.B.K.) Cortés) na Amazônia brasileira. Instituto Nacional de Pesquisas da Amazônia, Fundação Universidade do Amazonas. Dissertação de mestrado.

Bergamin Filho, A., Amorim, L., Laranjeira F. F., Berger, R.D., and Hau, B. 1998. Análise temporal do amarelecimento fatal do dendezeiro como ferramenta para elucidar sua etiologia. Fitopatologia Brasileira. 23: 391–396.

Billotte, N., Risterucci, A.M., Barcelos, E., Noyer, J.L., Amblard, P. and Baurens, F.C. 2001. Development, characterisation, and across-taxa utility of oil palm (*Elaeis guineensis* Jacq.) microsatellite markers. Genome. 44: 413–425.

Chan K.W., Yong Y.Y., Ahmad Alwi and Goh, K.H. 1988. Comparison of the yield, bunch and oil characteristics and their heritability before and after introduction of pollinating weevils (*E.kamerunicus*) in the oil palm (*E.guineensis*) in Malaysia. (*In*) Proceedings.1987 Int. Oil Palm conference Progress and prospects (Ed by A. Halim Hassan *et al.*), Palm Oil Research Institute of Malaysia, Kula Lumpur, pp 557-567.

Corley, R. H. V. and Tinker, P. B. 2003. The Oil Palm. Blackwell Science Ltd, Oxford, pp 562.

Digner Ortega Cedillo, Carlos Felipe Barrera, Eduardo Morillo Velastegui, Leonardo Quintero Roman, Jorge Daniel Ortega, Jorge Orellana Carrera, Victor Cevallos, Caio Cesio Salgado, Pedro Crescêncio Souza Carneiro, Cosme Damião cruz 2016. Genetic diversity within and between accessions of *Elaeis oleifera* from the ecuadorian amazon *International Journal of Agriculture and Environmental Research*, 2: 1480- 1493.

Edson Barcelos, Philippe Amblard, Julien Berthaud, and Marc Seguin. 2002. Genetic diversity and relationship in American and African oil palm as revealed by

RFLP and AFLP molecular markers. Pesq. agropec. bras., Brasília. 37 (8): 1105-1114.

Escobar, C. 1982. Preliminary results of the collection and evaluation of the American oil palm *E. oleifera* (HBK, Cortes) in Costa Rica. (In) *The oil palm in agriculture in the eighties* (Eds.) E. Pushparajah and Chew P.S. Vol.1. Incorp. Soc. Planters, Kuala Lumpur, pp. 79-93.

Gascon, J. P. and De Berchoux, C. 1964. Characteristics of the production of some origins of Elaeis guineensis (Jacq.) and their cross breedings. Application to selection of the Oil Palm tree. Oleagineux, 19 (2): 75-84.

Ghesquiere, M. 1985. Enzyme polymorphism in oil palm (*Elaeis guineensis* Jacq.). Oleagineux,pp 529-540.

Hardon, J.J., Rao, V. and Rajanaidu, N. 1985 A review of oil palm breeding. (In) Russel GE (Ed) Progress in plant breeding. Butterworths, London, pp 139–163.

Hartley, C.W.C. 1988. The oil palm (*E. guineensis* Jacq.) 3rd edn. Longman Group Limited New York

Hayiti, A., Wickneswari, R., Maizura, I. and Rajanaidu, N. 2004. Genetic diversity of oil palm (*Elaeis guinensis,* Jacq.) germplasm collections from Africa: implications for improvement and conservation of genetic resources. *Theo and Applied Genetics* 108: 1274-1284.

http: //pvpbkkt.doa.gov.my/

http: //www.eapvp-forum.org/

Kittipat Ukoskit, Vipavee Chanroj, Ganlayarat Bhusudsawang, Kwanjai Pipatchartlearnwon, Sithichoke Tangphatsornruang and Somvong Tragoonrung. 2014. Oil palm (*Elaeis guineensis* Jacq.) linkage map, and quantitative trait locus analysis for sex ratio and related traits. Molecular Breeding. 33 (2): 415-424.

Le Guen, V., Amblard, P., Omore, A., Koutou A. and Meunier, J. 1991. Le programme hybride interspécifique *Elaeis oleifera* ?*Elaeis guineensis* de l'IRHO. Oléagineux, 46: 479–487.

Li-Hammed, M. A., Ahmad Kushairi, Rajanaidu, N., Mohd Sukri Hassan, Che Wan Zanariah, Ngah, C.W., Jalani, B. S., and Elegbede, Isa Olalekan. 2015. Genetic diversity in oil palm germplasm as shown by hierarchical clustering methods. International Journal of Recent Scientific Research. 6 (6): 4866-4872.

Maizura I., Rajanaidu, N., Zakri, A.H. and Cheah, S. C. 2006. Assessment of Genetic Diversity in Oil Palm (*Elaeis guineensis* Jacq.) using Restriction Fragment Length Polymorphism (RFLP). Genetic Resources and Crop Evolution, 53 (1): 187-195.

Malike, F.A, Abdullah, N. and Ahmad, N. 2011. Evaluation of Harvest Index in the MPOB germplasm collections. (*In*) Proceedings of Agriculture, Biotechnology and Sustainability Conference, PIPOC, 2011, 15-17, November, KLCC, Malaysia, pp 172-175.

Marhalil, M and Rajanaidu, N. 2011. Oil Palm germplasm (*Elaeis guineensis*, Jacq): Second prospection. International seminar on breeding for sustainability in oil palm 18th Nov 2011 KLCC, Kula Lumpur, pp 133-140.

Mathur, R.K., Suresh, K., Nair, S., Parimala, K. and Sivaramakrishna, V.N.P. 2001. Evaluation of exotic dura germplasm for water use efficiency in oil palm (*Elaeis guineensis*, Jacq.). Indian J of Plant genetic Resources, 14: 257-259.

Meunier, J. 1975. The Americain (oil palm) *Elaeis melanococca*. *Oleagineux*, 30. No. 2 pp. 51-61.

Moretzsohn, M. C., Ferreira, M.A., Amaral, Z.P.S., Coelho, P.J.A., Grattapaglia, D. and Ferreira, M.E. 2006. Genetic diversity of Brazilian oil palm (*Elaeis oleifera* H.B.K.) germplasm collected in the Amazon Forest. Genetic Resources and Crop Evolution. 53 (1): 187-195.

Moretzsohn, M. C., Ferreira, M.A., Amaral, Z.P.S., Coelho, P.J.A., Grattapaglia, D. and Ferreira, M.E. 2002. Genetic diversity of Brazilian oil palm (*Elaeis oleifera* H.B.K) germplasm collected in the Amazon Forest. *Euphytica*, 124: 35-45.

Murugesan, P. and Goutam Mandal. 2010. Identification and Characterization of Pisifera Palms from Different Oil Palm Genetic Resources. International Journal of Oil Palm, 7 (1 and 2): 33-37.

Murugesan, P., Maryrani, K.L., Ramajayam, D., Sunilkumar, K., Mathur, R. K., Ravichandran, G., Naveen kumar, P. and Arunachalam, V. 2015. Genetic diversity of vegetative and bunch traits of African oil palm (*Elaeis guineensis*) germplasm in India. *Indian Journal of Agricultural Sciences* 85 (7): 892–895.

Murugesan, P., Gopakumar, S., Shareef, M.V.M. and Pillai, R.S.N. 2009. Evaluation and Characterization of Dwarf Tenera and oleifera Genetic Resources of Oil Palm. International Journal of Oil Palm, 6 (2): 13-19.

Murugesan, P. and Mandal, G., 2010. Identification and characterization of *pisifera* palms from different oil palm genetic resources *Int. J. OF Oil Palm*, 7 (1 and 2): 33-37.

Murugesan, P. and Gopakumar, S. 2010a. Variation in phenotypic characteristics of ASD Costa Rica hybrids of oil palm in India, *Indian J. Hort.* 67 (2): 152-155.

Murugesan, P. and Gopakumar, S. 2010b. Preponderance of female bunches in African oil palm germplasm at Athirapilli, Kerala. (In) Proceedings of 22nd Kerala Science Congress 28-31, January, KFRI, Peechi, pp 75-77.

Murugesan, P., Bijimol, G. and Gopakumar, S. 2009. Nut Component Analysis of Exotic and Indigenous Sources of Oil Palm (*Elaeis guineensis*, Jacq.) planting materials. *International Journal of Oil Palm*, 6 (1): 19-21, June 2009.

Murugesan, P., Meenu Merlin, J., Dipu Joseph, Bindu, S.J., Pillai, R.S.N.and Nampoothiri, K.U.K. 2011. Yield potential and phenotypic variation of fruit size and seed characteristics of oil palm *duras* at Thodupuzha. *Journal of Plantation Crops*, 39 (1)

Murugesan, P., Gopakumar, S. and Haseela, H. 2011. Performance of *tenera* x *tenera* progenies derived from Thodupuzha (Kerala) oil palm germplasm II. Bunch quality components. *Indian Journal of Horticulture.* 68 (3): 303-306.

Murugesan, P, Gopakumar, S., Shareef, M.V.M. and Pillai, R.S.N. 2009. Evaluation and characterization of dwarf tenera and oleifera genetic resources of oil palm. *International Journal of Oil Palm,* 6 (2): 13-19.

Murugesan, P., Mathur, R.K., Pillai, R.S.N. and Babu, M.K., 2005. Effect of accelerated aging on seed germination of oil palm (*Elaeis guineensis* Jacq. var. dura Becc.). *Seed Technology,* pp.108-112.

Murugesan, P. 2010. Enriching oil palm industry. Indian Horticulture, *Jan-Feb 2010: 16-18*

Murugesan, P., Haseela, H., Gopakumar S., Shareef, M.V.M. and Gopakumar, S. 2011. Fruit and seed development in *Elaeis oleifera* (HBK) Cortes under tropical climate of Kerala. *Journal of Plantation Crops,* 39 (1): 1-5.

Murugesan, P., Haseela, H., Gopakumar, S. and.Shareef, M.V.M. 2011. Fruit and seed development in *Elaeis oleifera* (HBK) Cortes of Surinam origin. *Indian Journal of Horticulture.* 68 (1): 28-30.

Namboothiri, K.U.K., Pillai, R.S.N. and Ravindran, P.S. 1992. Evaluation of certain oil palm hybrids under rainfed conditions, *Journal of Plantation Crops* 20 (supplement), pp 166-169.

Ngoot-Chin Ting, Noorhariza Mohd Zaki, Rozana Rosli, Eng–Ti Leslie Low, Maizura Ithnin, Suan-Choo Cheah, Soon-Guan Tan and Rajinder Singh. 2010. SSR mining in oil palm EST database: application in oil palm germplasm diversity studies. Journal of Genetics. 89 (2): 135-145.

Noorhariza Mohd Zaki, Rajinder Singh, Rozana Rosli and Ismanizan Ismail. 2012. *Elaeis oleifera* Genomic-SSR Markers: Exploitation in Oil Palm Germplasm Diversity and Cross-Amplification in Arecaceae. International Journal of Molecular Science. 13 (4): 4069-4088.

Okoye, M. N., Okwuagwu, C.O. and Uguru, M.I. 2009. Performance of 5 Deli Dura Parents in the NIFOR Oil Palm Breeding Programme. *Academic Journal of Plant Sciences* 2 (3): 139-149.

Okwuagwu, C. O., Okoye, M.N., Okolo, E.C., Ataga C.D. and Uguru, M.I. 2008. Genetic variability of fresh fruit bunch yield in Deli/dura x tenera breedingpopulations of oil palm (Elaeis guineensis Jacq.) in Nigeria. Journal of Tropical agriculture 46 (1-2): 52-57.

Okwuagwu, C.O, Ataga, C.D, Okoye, M.N. and Okolo, E.C. 2011. Germplasm collection of highland palms of Afikpo in Eastern Nigeria. Bayero *Journal of pure and applied Sciences,* 4 (1): 112-114.

Okwuagwu, C.O. 1985. The genetic base of the NIFOR Oil palm breeding programme. (In) Proceedings of International Workshop oil palm germplasm and utilization. Kula lumpur, PORIM 10: 228-237.

Ooi, S.C., Da Silva, E.B., Muller, A.A and Nascimento, J.C. 1981. Oil Palm Genetic Resources. Native *E. oleifera* populations in Brazil offer promising sources. *Pesquis. Agropecuaria Brasisileira 16 (3): 385-395.*

Pillai, R.S.N, Blaak, G. and Paul Closen, H. 2000. Collection of Oil Palm (*Elaeis guineensis* Jacq.) germplasm from Africa, *International Journal of Oil Palm*, 1 (1 and 2): 23-37.

Pillai, R.S.N and Nampoothiri, K.U.K. 1981. Preliminary investigations on *pisifera* with special reference to genetic improvement of oil palm in India. Placrosym IV, pp 308-313.

Purba, A. R., Noyer, J.L., Baudouin, L., Perrier, X., Hamon, S. and Lagoda, P.J.L. 2000. A new aspect of genetic diversity of Indonesian oil palm [*Eleis guineensis* Jacq.) revealed by isoenzyme and AFLP markers and its consequences for breeding. Theoretical Applied Genetics. 101: 956-961.

Putri, L.A.P, Rivallan, R., Puspitaningrum, Y., Sudarsono, Purba, R., Perrier, X., Asmono, D. and Billotte, N. 2009. Assessing genetic diversity of oil palm (*Elaeis guineensis*, Jacq.) germplasm collections using microsatellite markers. (In) Proceedings of Agriculture, Biotechnology and Sustainability Conference, PIPOC, 2009, KLCC, Malaysia, pp 823-835.

Rajanaidu, N. 1994. PORIM oil palm gene bank, collection, evaluation, utilization and conservation of oil palm genetic resources. In conjunction with the release of Elite oil palm planting materials and launching of oil palm Genebank. PORIM, Bangi, Selangor.

Rajanaidu, N. and Jalani, B. S. 1994c. Oil palm genetic resources -collection, evaluation, utilization and conservation. *A paper presented at PORIM Colloquium on Oil Palm Genetic Resources. PORIM, Malaysia 13 September 1994*

Rajanaidu, N., Kushairi, A., Chan, K.W. and Mohd Din, A. 2007. Current status of oil palm planting material in the world and future challenges. (In) Proceedings of the PIPOC 2007 International Palm Oil Congress (Agriculture, Biotechnology and Sustainability. KLCC, Kula Lumpur, pp. 503-520.

Rajanaidu, N., Kushairi, A., Rafii, M.Y., Moh'd Din, A., Maizura, I. and Jalani, B.S. 2001. Oil palm breeding and genetic resources. (In) Basiron, Y; Jalani B.S, and Chan, K.W) Mlaysian Palm Oil Board, Kula Lumbur, pp 171-237.

Rajanaidu, N. 1983. *Elaeis oleifera* Collection in South and Central America. pp 42-51. Plant Genetic Resources Newsletter. No. 56, pp. 42-51.

Rajanaidu, N., Jalani, B.S. and Kushairi A. 1998. Oil palm genetic resources, the development of novel planting materials. (In) International Oil Palm conference, Nusa Dua Bali, September 23-25, pp.208-220.

Rajinder Singh Noorhariza Mohd Zaki, Ngoot-Chin Ting, Rozana Rosli, Soon Guan Tan, EngTi Leslie Low, Maizura Ithnin and Suan-Choo Cheah. 2008. Exploiting an oil palm EST database for the development of gene-derived SSR markers and their exploitation for assessment of genetic diversity.63 (2): 227-235.

Rance, K. A., Mayes, S., Price, Z., Jack, P. L. and Corley, R. H. V. 2001. Quantitative trait loci for yield components in oil palm (*Elaeis guineensis* Jacq.). Theoretical and Applied Genetics. 103 (8): 1302-1310.

Renard, J.L., Noiret, J.M. and Meunier, J. 1980. Sources and ranges of resistance to *Fusarium* wilt in the oil palms *Elaeis guineensis* and *Elaeis melanococca*. Oléagineux 35: 387–392.

Rethinam, P. 1996. Oil palm cultivation in India, present status and future potentialities. Paper presented at National Seminar on oil palm production and processing held at Mysore, Karnataka from 13-14 Feb, pp. 1-11.

Richardson, D.L. 1995. The history of Oil palm breeding in the United Fruit Company, ASD Oil Palm Papers 11: 1-23.

Rocha, P.J and Rey, L. 2007. Oil Palm breeding programme in Cenipalma, Colmbia: A way to sustainability. (In) Proceedings of the PIPOC 2007 International Palm Oil Congress (Agriculture, Biotechnology and Sustainability). KLCC, Kula Lumpur, PP. 927- 952.

Rosenquist, E.A. 1986. The genetic base of oil palm populations. In: Proceedings of international Workshop of oil palm germplasm and utilization, Kulala Lumpur, Malaysia, pp 27-56.

Rutgers, A.A.L. 1920. De Opkomst der oliepalmcultuur. Batavia. Drukkerijen Ruygrok and Co

Santos, M.M. 1991. Polimorfismo isoenzimático de população subespontânea de dendê (*Elaeis guineensis* Jacq.) do estado da Bahia e sua relação genética com seis procedências africanas. PhD thesis Universidade de São Paulo, Ribeirão Preto, São Paulo, Brazil

Suresh, K., Ch. Nagamani, Sivasankar Kumar, K.M. and Vinod Kumar, P. 2004. Variations in the photosynthetic rate and associated parameters in the different Oil Palm germplasm. *Journal of Plantation Crops* 32 (Suple): 67-69.

Zeven, A. C. 1964 On the Origin of the Oil Palm (*Elaeis guineensis* Jacq.), Grana Palynologica, 5: 1, 121-123.

Zeven, A.C. 1967. The semi-wild oil palm and its industry in Africa. Agrica. PUDOC. Wageningen.

Zulkifli, Y., Maizura, I., Rajinder, S. 2012. Evaluation of MPOB oil palm germplasm (*Elaeis guineensis*) populations using EST-SSR. 24. 1368-1377.

Chapter 14

Biodiversity and Genomics of Oil Palm

B.K. Babu and R.K. Mathur

ICAR-Indian Institute of Oilpalm Research,
P O Pedavegi – 534 450, West Godavari District, A.P.
E-mail: rkmathur1967@gmail.com

The oil palm (*Elaeis guineensis* Jacq.) belongs to the family Arecaceae which is commonly known as African oil palm. Over 5000 years ago, it was initially traded for culinary purposes (Zeven, 1972). The oil palm is known to be originated from West Africa, however its exact centre of origin within Africa is not known (Corley and Tinker, 2003). Based on the recent findings, as per the highest allelic diversity Nigeria could be the centre of origin of oil palm in Africa (Bakoumé *et al.*, 2015). The genus of oil palm *Elaeis* consists of two taxonomically well defined species, *i.e.,* one is African oil palm (*E. guineensis*) and second is American oil palm (*E. oleifera*). Oil palm cultivation has been expanded rapidly in recent years and has become a major source of the world supply of oils and fats. It is the highest edible oil yielding crop (Crude Palm Oil – 4-6 and Palm Kernel Oil - 0.4 to 0.6 tones per ha per year) which is much higher than that of any major oil producing crop. Oil Palm, although an introduced crop in India, is grown in an area of 3.06 lakh ha in 18 states by the year 2015-16. Indonesia followed by Malaysia, Thailand and Nigeria are the major leading countries in area and production of oil palm. Major exporters of palm oil in the world are Indonesia and Malaysia. The major palm oil importing countries are India, China, Netherlands and Pakistan.

Genetic Resources in Oil palm

The success of any crop improvement programme depends on the availability of wider spectrum of genetic variability (germplasm) in the species. It's more important especially in respect to crops like oil palm where narrow genetic base is the major constraint in achieving genetic progress through breeding. The present day oil palm breeding mainly depends on the Deli duras derived from four seedlings planted in Bogor Botanical garden in Java, Indonesia during 1848 as a source of females (Rajanaidu and Jalani, 1999). Further, the source of Pisifera (males), in oil palm breeding is also limited to a few palms. Realizing the importance of genetic resources, many countries started prospection programme and the earliest and important was that carried out in Congo in 1920s. After World War-II, prospection in Congo was done in estates planted with Yangambi material, among palms of local origin on estates and in grove areas. In Nigeria, collection and exploitation of oil palm was started in 1912 and 1939. Nigerian Institute for Oil Palm Research (NIFOR) played important role in aggressive exchange programme with other countries to collect Deli Dura from as many sources as possible.

In India oil palm was first introduced as an ornamental palm at the National Botanical Garden, Kolkata in the later part of 19th century (Rethinam, 1998). During 1947-48, Maharashtra Association for Cultivation of Science (MACS) introduced African oil palm in Pune. Systematic research work on oil palm started in 1960 when Kerala State Department of Agriculture had undertaken a 40 ha plantation at Thodupuzha (now under OPIL) using Deli Dura materials introduced from Malaysia and Tenera x Tenera population from Nigeria. Later some more introductions received from Nigeria, Malaysia, Cote de Ivoire, Papua New Guinea and Zaire were planted in Oil Palm India Ltd., Kerala between 1971 and 1983. In Andaman and Nicobar Islands, the Andaman forest and plantation development corporation planted with materials from Nigeria, Malaysia, Ivory Coast, Papua New Guinea and Zaire during 1976 to 1985. ICAR has taken up systematic collection of oil palm germplasm during 1979 at Central Plantation Crops Research Institute, Research station, Palode, Kerala. Accordingly, materials from different sources were collected consisting mostly of random sample of Tenera.

Table 14.1: Availability of Oil Palm Germplasm and their Sources

Desirable Traits	Name of Species	Available Sources	
		Region	Name of Country
High yield, Drought tolerance, short stature, High oil extraction ratio	*Elaeis guineensis*	West Africa	Ghana, Nigeria, Ivory Coast, Benin, Togo, Sierra Leone
		Central Africa	Congo region, Gabon, Zaire, Uganda
		South Africa	Angola, Mozambique
Dwarfness, compact canopy, superior oil quality, long bunch stalk	*Elaeis oleifera*	South America	Brazil, Ecuador, Peru, Columbia, Surinam, Panama, Amazonian belt
		Central America	Costa Rica, Nicaragua and Honduras
Drought tolerance, Dwarf with Compact canopy, High oil extraction ratio, superior Oil quality, big kernel, long bunch stalk	*Elaeis guineensis, Elaeis oleifera*	SE Asia	Malaysia, Indonesia, Papua New Guinea, Thailand

Sources of Genetic Resources of African Oil Palm

Deli: This is thick shelled Dura derived from the original four Bogor palms in Java. Distribution of subsequent progenies to other countries followed by local selection led to the development of the Elmina, Serdang, Avenue and Ulu Remis. Deli Dura sub-population/selection in Malaysia and the Dabou and Le Me Dura sub-population/selections in Ivory coast. The rather high yielding uniform populations lead to speculation of common progenitors for the four Bogor palms. Deli dura provides the mother palms for all major palm commercial hybrid seed production programmes. The Dumpy and Gunung Melayu palms are short variants of the Deli.

AVROS (*Algemene Vereniging Rubberplanters ter Oostkust van Sumatra*)

Seeds from the Dejongo from Eala Botanical Garden in Zaire were obtained and planted in 1923 by AVROS at Sangai Panchur to give rise to the well known the SP 540 Tenera palms. AVROS pisifera are noted for vigorous growth, precocious bearing, thin shell, thick mesocarp and high yielding. Major seed production programme in Colombia, Costa Rica, Indonesia, Malaysia, Papua New Guinea are based on Deli x AVROS pisifera lineage.

Yangambi

Breeding programme started at INEAC, Congo at Yangambi, Zaire with open pollinated seeds of Dejongo palm and tenera of Yawenda and developed Yangambi population. This population characterised by excessive vigour, bigger fruit and high oil content.

La Me

IRHO developed the La Me populations from 21 tenera palms from seeds collected from wild grooves of Ivory Coast. It is used in seed production in West Africa and Indonesia. La Me progenies and teneras are characteristically smaller palms with smaller bunch and fruits, but they appear to be more tolerant to suboptimal growing conditions.

Binga

This sub-population derived from F2 and F3 of Yangambi progenies planted in Binga plantation, Yangambi, Zaire. Palms Ybi 69MAB and Bg 312/3 are the parent palms of breeding interest.

Ekona

Ekona population derived from wild palms of the Ekona area of Cameroon and bred further in the Unilever plantation of Crown Estate, Ndian Estate, and Lobe estate. It is noted for its high bunch yield, good oil content and wilt resistant.

Calabar

Breeding population of NIFOR are much broader based collections from Aba, Calabar, Ufuma,Umuabi.

Genomics of Oil Palm

Although the traditional plant breeding based on phenotypic selection is very effective, it has suffered from several limitations for complex traits. Unlike morphological and biochemical markers, DNA markers are basically unlimited in number and are not affected by environmental factors and/or the developmental stage of the plant (Winter and Kahl, 1995). The molecular markers were transformed from the earlier RFLP markers to the highly variable and effective SNP markers. The most widely used markers in the recent times are SSR and SNP markers for several purposes like genetic diversity, linkage maps, and for GWAS studies. The advances in omics technologies that are, comprehensive and integrated genomic, transcriptomic, and proteomic analyses can elucidate genetic architecture of plant genomes and the relationships between genotype and phenotype. The rapid advances in DNA sequencing technology have made whole-genome sequencing (WGS) both technically and economically feasible. More than 25 economically important plants' genomes have been sequenced (Hamilton and Buell, 2012). The next-generation sequencing (NGS) technologies are used not only for WGS but also to allow applications related to target region deep sequencing, epigenetics, transcriptome sequencing (RNA-seq), megagenomics, and genotyping. Oil palm is a diploid (2n=32) with an estimated genome size of 1.8 giga bases. The full draft genome sequence of 1.535 Gb of E. guineensis was recently published (Singh *et al.*, 2013a) and freely available. In the present review we focussed on the status of molecular breeding and its applications in oil palm like genetic diversity, construction of linkage maps, mapping of QTLs for different traits and marker assisted selection approaches and future perspectives for the oil yield enhancement and for quality improvement.

Genetic Diversity Analysis of Oil Palm Germplasm

Since only four bogour palms laid basis for the development of the present industrialized development of oil palm, there exists a narrow genetic diversity among the oil palm germplasm. However, against to that a considerable amount of variation also present among the different sources of oil palm germplasm. The development of modern plant breeding techniques has greatly facilitated the wider use of a wealth of diversity from many sources including landraces. A comprehensive exploration of potential genetic resources and exploitation of natural genetic variations are proven source of useful genomic information. A rich diversity of germplasm can be explored for their desirable traits like yield and can be further utilised to develop new varieties through molecular plant breeding approaches. Molecular marker techniques have revolutionised the tree genomics and understanding of structure and behaviour of palm genomes. This will pave the way towards the detection of novel and superior genotypes.

Initially several genetic diversity works were based on using RAPD, RFLP and AFLP molecular markers. However, due to certain drawbacks these markers were replaced by SSR and SNP markers. Use of RAPD for genetic diversity study of oil palm was reported for the first time by Shah (1994). Oil palm germplasm accessions collected from Africa (Cameroon, Tanzania, Nigeria and Zaire) were studied using

20 primes and recorded high levels of genetic variation among the accessions. Rival *et al.* (1998) studied the suitability of RAPD markers for detection of soma-clonal variants in oil palm. The results from the 387 arbitrary primers showed no intra clonal variability and no difference between mother and regenerated palms. The authors opined that RAPD approach is not suitable for the detection of the mantled variant phenotype. Later Mayes *et al.* (2000) used RFLP markers (40 probes covering 60 per cent oil palm genome) to assess genetic diversity within 54 palms of a specific oil palm breeding program. A new aspect of genetic diversity was studied by Purba *et al.* (2000) among the Indonesian oil palm germplasm using Isozyme and AFLP markers and reported its consequences for oil palm breeding. The findings from these results showed that crosses between the Africa sub population may be more interesting than the African and Deli cross population. Barcelos *et al.* (2002) studied genetic diversity and relationship of American and African germplasm using AFLP and RFLP markers. Both markers revealed that genetic divergence between the two species is of the same magnitude as that among provenances of *E. oleifera*. Later the genetic diversity was focussed on trait specific diversity of oil palm accessions using different molecular markers. Arias *et al.* (2015) studied genetic and phenotypic diversity of natural American oil palm germplasm. The results from SSR markers and agro-morphological traits showed that analyses of variance for yield and bunch components demonstrated statistically significant differences among countries and geographical regions for several of the traits evaluated. SSR marker analyses revealed high genetic diversity (HT=0.797) and the presence of specific alleles by each country of origin from *E. oleifera*. Recently, Okoye *et al.* (2016) studied genetic relationship between elite oil palms from Nigeria and Malaysia using SSR markers. Likewise different workers used different molecular markers on different oil palm germplasm for various purposes of genetic diversity studies. A comprehensive list of few genetic diversity studies were given in Table 14.2.

Table 14.2: The List of Genetic Diversity Studies Conducted in Oil Palm

Sl.No.	Marker Used	Population	Reference
1	Isozymes	Pollen of seven accesions of *E. oleifera* and hybrid between *E. oleifera* and *E. guineensis*.	Ataga and Fatokun, (1989)
2	SSRs	194 oil palm from 49 populations	Bakoume *et al.* (2014)
3	RFLP	11 oil palm germplasm collections *viz.*, Nigeria, Cameroon, Congo DR, Tanzania Angola, Senegal, Sierra leone, Guinea, Ghana, Madagascar and Gambia	Maizura *et al.* (2001)
4	AFLP	687 accessions belonging to 11 African countries and Deli *dura*	Kularatne *et al.* (2001)
5	AFLP and RFLP	Within oil palm germplasm (both *E. oleifera* and *E. guineensis*)	Barcelos *et al.* (2002)
6	RAPD	*E. oleifera* accession collected from the Amazon forest	Moretzsohn *et al.* (2002)
7	RAPD	Five *dura* germplasm accession	Mandal *et al.* (2004)
8	SSR	Elite Oil Palms from Nigeria and selected breeding and Germplasm Materials from Malaysia	Okoye *et al.*, 2016

Sl.No.	Marker Used	Population	Reference
9	SSR	E. oleifera and E. guineensis	Zaki 2012
10	SSR	NIFOR Oil	Okoye et al. (2016)
		Palm Main Breeding Parent Genotypes	
11	96 SSR markers	121 breeding plants from 3 different populations in Thailand.	Taeprayoon et al. (2015)

Functional Genomics in Oil Palm

Simple sequence repeat (SSR) markers offers several advantages like high polymorphic ability, co-dominant inheritance, poly allelic nature, integrating the genetic, physical and sequence-based physical maps in plant species, and simultaneously have provided molecular breeders with an efficient tool to link phenotypic and genotypic variation. However, the construction of SSRs are often tedious and costly cloning and enrichment procedures required for their generation (Zane *et al.*, 2002; Squirrell *et al.*, 2003; Weising *et al.*, 2005). The EST databases have become particularly attractive resources for such In-silico mining, as was demonstrated in, *e.g.*, citrus (Chen *et al.*, 2006), coffee (Poncet *et al.*, 2006), and particularly in the cereals [Yu *et al.*, 2004]. The SSRs developed from expressed sequence tags (ESTs), popularly known as EST–SSRs or genic microsatellites, represent functional molecular markers as a putative function for a majority of such markers can be deduced by database searches and other *In-silico* approaches. Furthermore, they represent genic regions of the genome. In oil palm a considerable amount of EST sequences (nearly 40,979) were available till June, 2016 in the NCBI website. Few reports were available on the insilico identification of EST-SSRs and their use in characterization of oil palm germplasm. The first report of a systematic study of genes expressed by means of expressed sequence tag (EST) analysis in oil palm was done by Jouannic *et al.* (2005). A total of 2411 valid EST sequences were thus obtained from five different cDNA libraries were generated from male and female inflorescences, shoot apices and zygotic embryos. Timothy *et al.* (2012) 289 EST-SSRs were tested to detect polymorphisms in elite breeding parents and their crosses. 230 of these amplified PCR products, 88 of which were polymorphic within the breeding material tested. Detailed analysis and annotation of the EST-SSRs revealed that mostly they related to transcription and post-transcriptional regulation. Ting *et al.* (2010) did SSR mining in the EST sequences of 19243 *Elaeis guineensis* which were available to that date. They found that di-nucleotide repeats formed the largest group (45.6 per cent) consisting of 66.9 per cent AG/CT motifs. This was followed by trinucleotide repeats, which is the second most abundant repeat types (34.5 per cent) consisting of AAG/CTT (23.3 per cent). Singh *et al.* (2008) exploited the EST database of oil palm for assessment of genetic diversity. A total of 5521 EST sequences were mined and developed 145 SSRs.

QTL Mapping in Oil Palm

The basic requirement for MAS is identification of markers associated with the trait being targeted. For this development of linkage map is required. By screening a large population of sibs of many different markers, pair or groups of markers that are linked, and tend to be inherited together, can be identified. Such groups are expected to be on the same chromosome, and the closeness of the linkage, calculated statistically, shows the relative position of the markers along the chromosome (Corely and Tinker, 2003). In many important agronomic plant species, a large number of DNA markers and linkage maps have been developed. Many QTL for important traits have been mapped on the whole genomes, setting up the basis for rapid genetic improvement through MAS.

Linkage Maps in Oil Palm

Mayes *et al.* (1996) developed first AFLP marker based genetic linkage map from a mapping population generated by the selfing of an important breeding material segregated for the shell thickness character. Billotte *et al.* (2005) developed a first high density linkage map using microsatellite markers. They used a tenera palm from the La Me´ population (LM2T) and a dura palm from the Deli population (DA10D) and a set of 390 SSR markers. They constructed a linkage map consisting of 255 microsatellites, 688 AFLPs and identified locus of the Sh gene, near an AFLP marker E-Agg/M-CAA132 was mapped at 4.7 cM from the Sh locus. The use of relatively new Diversity Array Technology "Genotyping-by-Sequencing (DArTSeq) platform through genotyping of two closely related *tenera* self-pollinated F2 populations, generated a total of 11675 DArTSeq polymorphic markers of good quality. These markers were used in the construction of the first reported DArTSeq based high density maps for oil palm (Gan, 2014). Ting *et al.* (2014) developed High density SNP and SSR-based genetic maps of two independent oil palm hybrids. A 4.5K customized oil palm SNP array was developed using the Illumina Infinium platform. The SNPs and 252 SSRs were genotyped on two mapping populations, an intraspecific cross with 87 palms and an interspecific cross with 108 palms. Parental maps with 16 linkage groups (LGs), were constructed for the three fruit forms of *E. guineensis* (dura, pisifera and tenera). Map resolution was further increased by integrating the dura and pisifera maps into an intra-specific integrated map with 1,331 markers spanning 1,867 cM.

QTL Mapping for Yield Traits

Pootakham *et al.* (2015) reported genome wide SNP discovery and identification of QTLs associated with agronomic traits in oil palm using genotyping – by-sequencing (GBS) out of 3417 fully informative SNP markers, they were able to place 1085 on a linkage map, which spanned 1429.6 cM and had an average of one marker every 1.26cM. Recently Teh *et al.* (2016) performed GWAS for oil-to-dry-mesocarp content (O/DM) on 2,045 genotyped *tenera* palms using 200K SNPs. They found that eighty loci were significantly associated with oil-to-dry mesocarp yield ($p = 10–4$) and three key signals were found. They reported the most comprehensive use of high density SNP genotyping in oil palm to date, the use of a GWAS approach to identify SNP variants associated with differences in the key oil-to-dry mesocarp yield trait, and confirmation of their action in an independent cross. Jeennor and Volkaert (2014) estimated the position and effects of QTLs linked to oil yield traits in African oil palm. Co-dominant microsatellites (SSR) and candidate gene-based sequence polymorphisms were applied to construct a linkage map for a progeny showing large differences in oil yield components. The progeny was genotyped for 97 SSR markers, 93 gene linked markers, and 12 non-gene-linked SNP markers. From these, 190 segregating loci could be arranged into 31 linkage groups while 12 markers remained unmapped. Using the single marker linkage, interval mapping and multiple QTL methods, 16 putative QTLs on seven linkage groups affecting important oil yield related traits such as fresh fruit bunch yield (FFB), ratio of oil per fruit (OF), oil per bunch (OB), fruit per bunch (FB) and wet mesocarp per fruit (WMF) could be identified in the segregating population with estimated values for explained variance ranging from 12.4 per cent to 54.5 per cent. Ukoskit *et al.* (2014) generated an integrated linkage map of oil palm based on 208 offspring derived from a cross between two tenera palms differing in inherited sex ratio. The map consisted of 210 genomic simple sequence repeats (SSRs), 28 expressed sequence tag SSRs, 185 amplified fragment length polymorphism markers, and the Sh locus, which controls shell thickness phenotype, distributed across 16 linkage groups covering 1,931 cM, with an average marker distance of 4.6 cM. Quantitative trait locus (QTL) analysis identified eight QTLs across six linkage groups associated with sex ratio and related traits. These QTLs explained 8.1–13.1 per cent of the total phenotypic variance.

QTL Mapping for Oil Quality Parameters

A map was constructed using AFLP, RFLP and SSR markers for an interspecific cross involving a Columbian *E. oleifera* (UP 1026) and a Nigerian *E. guineesis* (T 128) by Singh *et al.* (2009). At a 5 per cent genome wide significance threshold level, QTLs associated with Iodine value (IV), myristic acid (C14:0), palmitic acid (C16:0), palmitoleic acid (18:0), oleic acid (18:1), and linoleic acid (18:2) content were detected. Significant QTL for C14:0, 16:1, 18:0and 18:1 content was detected around the same locus on Group 15, thus revealing another major locus influencing fatty acid composition in oil palm. The oil palm fruit mesocarp contains high lipase activity that increases free fatty acids (FFA) and necessitates its post-harvest inactivation by heat treatment of fruit bunches. The mesocarp lipase activity causes consequential oil losses and requires costly measures to limit FFA quantities. Morcillo *et al.* (2015) demonstrated that elite low-lipase lines yield oil with substantially less FFA than standard genotypes, allowing more flexibility for post-harvest fruit processing and extended ripening for increased yields. They identified the lipase and its gene co-segregation with the low-/high-lipase trait, providing breeders a marker to rapidly identify low lipase lines and introgress the trait into major cultivars.

QTL Mapping for Biotic Stress Resistance

The oil palm is badly affected by basal stem rot (BSR) disease in Southeast Asia. BSR disease is caused by the fungus Ganoderma boninense, which is a major threat to oil palm compared with other Ganoderma spp. Ali *et al.* (2015) used 58 simple sequence repeat markers were utilized with three progeny types, namely, KA4G1, KA4G8, and KA14G8, to perform a comparative molecular mapping for association with BSR. A total of 319 alleles were identified with an average of 5.51 alleles per locus. Five markers, mEgCIR0793:180, mEgCIR0894:200, mEgCIR03295:210, mEg-

CIR3737:146 and mEgCIR3785:299 were found to be associated with Ganoderma disease with P values of 0.018, 0.033, 0.037, 0.034 and 0.037, respectively, in single progeny analysis. However, in pooled data (KA4G1, KA4G8 and KA14G8), only two alleles, mEgCIR0804:213 (P value = 0.001) and mEg- CIR3292:183 (P value = 0.001), were found to be associated with Ganoderma disease. Mandal *et al.* (2014) developed PCR based early detection of *Ganoderma* sp. causing basal stem rot of oil palm in India.

Association Mapping in Oil Palm

Till now very association mapping studies have been conducted in oil palm. Babu *et al.* (2017) initiated work on association mapping for different oil yield related traits using SSR markers. Association of trait-marker data resulted in identification of seven significant QTLs by GLM approach, where as four significant QTLs were detected by MLM approach at a very significant threshold (P) level of 0.001 and 0.01. However, no QTLs identified for bunch weight and oil to wet mesocarp. This high cut-off level of significance was given to reduce the false QTLs which will give wrong interpretations. By GLM approach, two QTLs were identified for oil to bunch (per cent) trait linked by SPSC00033 and SPSC00067 SSR markers. One QTL each linked to fruit to bunch (per cent), kernel to fruit (per cent), mesocarp to fruit (per cent), oil to dry mesocarp (per cent) and shell to fruit (per cent) traits. The oil to bunch trait was linked by two markers SPSC00033 and SPSC00067. Out of these two, SPSC00033 was significantly linked to the oil to bunch at P of 0.0002, explained $R2$ of 11.5 per cent. Kernel to fruit and shell to fruit traits were found to be linked by SOTIG14040 and SMG00156 markers respectively.

Identification of Genes/QTLs for Important Traits

A major breakthrough which revolutionised the oil palm industry was started with the invention of single gene inheritance for shell thickness (Sh gene) by the plant breeders at Yangambi Research Station, The Democartic Republic of Congo (Congo DR), Africa during 1920s. The *SHELL* gene is responsible for identification of oil palm fruit forms *viz.*, Dura, Pisifera and Tenera (Biernaert and Vanderwayen, 1941). The dura genotype has thick shell, consisted of dominant Sh allele (Sh/Sh), and contributes 15 percentage of oil, whereas pisifera genotypes was shell less, consisted of recessive shell alleles (sh/sh), usually female sterile (Corley and Tinks, 2003), which contributes 25 percentage of oil. Singh *et al.* (2014) revealed the identification of the VIRESCENS (*VIR*) gene, which controls fruit exocarp colour and is an indicator of ripeness. VIR is a R2R3-MYB transcription factor with homology to *Lilium LhMYB12* and similarity to *Arabidopsis PRODUCTION OF ANTHOCYANIN PIGMENT1 (PAP1)*. They identified five independent mutant alleles of *VIR* in over 400 accessions from sub-Saharan Africa that account for the dominant negative *virescen* phenotype. Each mutation resulted in premature termination of the carboxy-terminal domain of VIR, resembling McClintock's C1-I allele in maize. The identification of *VIR* will allow selection of the trait at the seed or early nursery stage, 3-6 years before fruits are produced and thus greatly advancing introgression into elite breeding material.

Lee *et al.* (2015) developed a consensus linkage map of oil palm using co-dominant markers (*i.e.* microsatellite and SNPs) and two F1 breeding populations generated by crossing dura and pisifera individuals and identified a major QTL for stem hight. Four hundreds and forty-four microsatellites and 36 SNPs were mapped onto 16 linkage groups with a total coverage of 1565.6 cM, with an average marker space of 3.72 cM. They mapped a major QTL for stem height on the linkage group 5. For short stature of oil pal, at our institute (ICAR – IIOPR), Bulk Segregant Analysis (BSA) analysis with 400 SSR markers (both genomic and genic) among dwarf and tall bulks DNA was attempted and a total of 50 SSR markers were able to find polymorphism between the bulks of dwarf, and tall genotypes. However, only two SSR markers were able to clearly differentiate the dwarf from tall genotypes, which can be used in selection of dwarf palms at early stages.

Gan (2014) carried out marker development studies in oil palm for genetic linkage mapping and QTL analysis for use in MAS. The use of AFLP method identified 29 primer pairs that yielded 49 putative shell-thickness related polymorphic bands. The use of relatively new Diversity Array Technology "Genotyping-by-Sequencing (DArTSeq) platform through genotyping of two closely related *tenera* self-pollinated F2 populations, generated a total of 11675 DArTSeq polymorphic markers of good quality. These markers were used in the construction of the first reported DArTSeq based high density maps for oil palm. Saturation of the shell thickness (*Sh*) region with all available DArTSeq markers as well as map integration around the *Sh* regions for both the populations resulted in identification of 32 SNPs and DArT markers mapped within a 5cM flanking region of the *Sh* gene. Homology search of the DArTSeq marker sequence tag (64bp) against the recently published oil palm genome assembly confirmed that 23 out of the 32 (72 per cent) DArTSeq markers were located on the p5_sc00060 scaffold in which the SHELL gene was identified. Besides above reports, several linkage map and QTL mapping studies were performed which were given in Table 14.3.

Genomic Selection

Oil palm (*Elaeis guineensis* Jacq.) requires 19 years per cycle of phenotypic selection. One cycle of selection, which includes phenotypic evaluation of testcrosses and inter-crossing of the best palms to form the next cycle, requires approximately 19 years. Oil-palm breeding is expensive not only because of the length of time required per cycle, but also because large planting areas are required. Wong and Bernardo (2008) evaluated the response to phenotypic selection, marker-assisted recurrent selection (MARS), and genome wide selection with small population sizes in oil palm, and assess the efficiency of each method in terms of years and cost per unit gain. With population sizes of N = 50 or 70, responses to genome wide selection were 4–25 per cent larger than the corresponding responses to phenotypic selection, depending on the heritability and number of quantitative trait loci. Cost per unit gain was 26–57 per cent lower with genome wide selection than with phenotypic selection when markers cost US $1.50 per data point, and 35–65 per cent lower when markers cost $0.15 per data point. With population sizes of N = 50 or 70, time per unit gain was 11–23 years with genome wide selection and 14–25 years with phenotypic selection.

Table 14.2: The List of QTL Mapping and Linkage Analysis Studies Conducted in Oil Palm

Sl.No.	Marker Used	Population	Number of Palms	Trait (s) for QTL Mapping	Salient Findings	Reference
1	SSRs and gene-based markers	A controlled cross progeny population established from a cross between a female Topi Deli dura and a male Yangambi pisifera parents	52	oil yield related traits like fresh fruit bunch yield (FFB), ratio of oil per fruit (OF), oil per bunch (OB), etc	16 putative QTLs on seven linkage groups affecting important oil yield related traits such as fresh fruit bunch yield (FFB), ratio of oil per fruit (OF), oil per bunch (OB), etc	Jeennor and Volkaert (2014)
2	SSRs	The *E. guineensis* cross LM2T × DA10D	116 full-sibs	Palm Oil Fatty Acid Composition	Sixteen QTLs affecting palm oil fatty acid proportions and iodine value were identified	Montoya *et al.* (2014)
3	SSR and SNPs	Elaeis interspecific pseudo-backcross of first generation (*E. oleifera × E. guineensis*) × *E. guineensis*	134 full-sibs	palm oil fatty acid composition	19 QTL associated to the palm oil fatty acid composition were evidenced	Montoya *et al.* (2013)
4	SSRs, EST-SSRs, AFLP	From a cross between two tenera palms	208	Sex ratio and related traits	Eight QTLs across six linkage groups associated with sex ratio and related traits	Ukoskit *et al.* (2014)
5	SNP, RFLP and SSR markers	From the self-pollination of the tenera palm, T128	240 palms	VIRESCENS gene	Five independent mutant alleles of VIR	Singh *et al.* (2014)
6	SNPs and 252 SSRs	On two mapping populations, an intraspecific cross with 87 palms and an interspecific cross with 108 palms.	-	Linkage map	Integrated map with 1,331 markers spanning 1,867 cM.	Ting *et al.* (2014)
7	SSR, AFLP, and RFLP	The mapping population is a high-yielding dura6pisifera cross	-	Linkage map	Integrated map was 2,247.5 cM long and included 479 markers	Seng *et al.* (2011)
8	SSR	From a cross between a tree (D1) from the Deli population and a tree (L1) from the La Me population	-	lipase gene	Identified the lipase and its gene co-segregates with the low/high lipase trait	Morcillo *et al.* (2015)
9	AFLP, RFLP and SSR	An interspecific cross involving a Colombian *Elaeis oleifera* (UP1026) and a Nigerian *E. guinneensis* (T128).	118	Fatty acid composition	Significant QTL for C14:0, C16:1, C18:0 and C18:1 content was detected around the locus on Group 15,	Singh *et al.* (2009)

Next Generation Sequencing in Oil Palm

The genome sequence of oil palm will be a rich resource for oil palm breeders, geneticists and evolutionary biologists alike. Singh *et al.* (2013) reported the 1.8 Gb genome sequence of the African oil palm, *Elaeis guineensis*, the predominant source of worldwide oil production. A total of 1.535Gb of assembled sequence and transcriptome data from 30 tissue types were used to predict at least 34,802 genes, including oil biosynthesis genes and homologues of WRINKLED1 (WRI1), and other transcriptional regulators. They also reported the draft sequence of the South American oil palm *Elaeis oleifera*, which has the same number of chromosomes (2n=32) and produces fertile interspecific hybrids with *E. guineensis* but seems to have diverged in the New World. It has revealed that palms are ancient tetraploids, and that the African and South American species probably diverged in the Old and New Worlds. Over represented genes in lipid and carbohydrate metabolism are expressed differentially in mesocarp and kernel, accounting for the different properties of palm fruit and palm kernel oils.

Status of Oil Palm Genomics in India

Simple sequence repeats and RAPD primers for assessment of genetic uniformity among field planted clones of oil palm were studied by Jayanthi *et al.,* 2008. Forty plants were selected at random from six different clones. Eight plants of fran, six plants of zeus, eight plants of Emerald from two locations, five plants from Ruby and eight plants of Tornado and five plants of Conte were collected. RAPD was performed using 10 nucleotide random primers, PCR with microsatellite primers also carried out using the optimized protocol (Jayanthi *et al.,* 2009). The absence of variation using microsatellite primers reveals that there is no variation in these repeat sequence sites amplified with these primers.

The preliminary investigation on genetic diversity analysis of the *E. oleifera* palms by RAPD using random 10 nucleotide primers are also been reported (Mandal *et al.,* 2003c). The maximum similarity was recorded between Eg-10 and Eo-11 (0.952) and minimum (similarity: 0.710) was between two pairs of palms (Eo-05 and Eo-08 and Eo-12 and Eo-21). The oleifera palms formed three major groups by UPGMA method cluster analysis. In India, five of the dura germplasm accessions available, namely ASD1 (98C-254 D) and ASD2 (98C-208D) from ASD Costa Rica, and PLD1 (GDD3), PLD2 (240D x 281D) and PLD3 (80D x 281 D) from Palode, Kerala, India were analyzed for their genetic diversity using RAPD (Mandal *et al.,* 2004). This study RAPD analysis showed six different groups, each consisting of palms from the same accessions, although no two palms from any accessions were completely similar. Palms from guinea Bissau accession were highly homogenous in comparison to other groups and the same accession was genetically more distant from others. Cameroon and ASD Costa Rica accessions were found closer each other (Mandal *et al.,* 2003a; Mandal *et al.,* 2003b; Mandal and Susmitha, 2006).

In India, negligible amount of work was done on the molecular characterization in oil palm germplasm in comparison to the international status. There were few reports available from India. Riju *et al.* (2007) reported mining of oil palm (Expressed Sequence Tag) EST sequences from dbEST of NCBI. CAP3 program was used to assemble EST sequences into contigs. Candidate SNPs and Indel polymorphisms were detected using the perl script auto_snip version 1.0 which has used 576 ESTs for detecting SNPs and Indel sites. They found 1180 SNP sites and 137 indel polymorphisms with frequency 1.36 SNPs/100 bp. Among the six tissues from which the EST libraries had been generated, mesocarp had high frequency of 2.91 SNPs and indels per 100 bp whereas the zygotic embryos had lowest frequency of 0.15 per 100 bp. Satish and Mohan Kumar, (2007) used RAPD markers for determining the DNA polymorphism among the oil palm (*Elaeis guineensis*) varieties *dura*, *pisifera* and *tenera*, and monitoring the specificity of the primers for identifying each genotype. The three varieties were evaluated using thirty, 10-mer primers. Of the 30 primers, 26 yielded significant polymorphic DNA bands. Recently from our institute we published a detailed report on development, identification and validation of CAPS marker for SHELL trait which governs dura, pisifera and tenera fruit forms in oil palm (*Elaeis guineensis* Jacq.) and association mapping studies. We identified one cleaved amplified polymorphic site (CAPS) marker for differentiation of oil palm fruit type which produced two alleles (280 and 250bp) in dura genotypes, three alleles in tenera genotypes (550, 280, and 250bp) and one allele in pisifera genotypes (550bp). Association mapping of marker data with phenotypic data of eight oil yield related traits resulted in identification of seven significant QTLs by GLM approach, four by MLM approach at a significant threshold (P) level of 0.001. Significant QTLs were identified for fruit to bunch and oil to bunch traits, which explained R^2 of 12.9 per cent and 11.5 per cent respectively.

References

Ali, H.E.O.,; Panandam, J.M.; Tan, S.G.; Sharifaha, S.S. A.; Tan, J.S.; Ling,H.C.; Namasivayam, P. and Peng, H.B. 2015. Association between basal stem rot disease and simple sequence repeat markers in oil palm, *Elaeis guineensis* Jacq. *Euphytica.*, **202**: 199-206.

Arias, D.; Gonzalez, M.; Prada, F.; Ayala- Diaz, L.; Montoya,C.; Daza, E. and Romero, H.M. 2015. Genetic and phenotypic diversity of natural American oil palm (*Elaeis oleifera* (H.B.K.) Cortes) accessions. *Tree genet Gen.*, **11**: 122

Ataga, C.D. and Fatokun, C.A. 1989 Disc polyacrylamide gel electrophoresis of pollen proteins in the oil palm (*Elaesis*). *Euphytica.*, 40: 83-88.

Babu BK, Mathur RK, Kumar PN, Ramajayam D, Ravichandran G, Venu MVB, *et al.* (2017) Development, identification and validation of CAPS marker for SHELL trait which governs dura, pisifera and tenera fruit forms in oil palm (*Elaeis guineensis* Jacq.). PLoS ONE 12 (2): e0171933. doi: 10.1371/journal.pone.0171933

Bakoume, C.R.; Wickneswari, Siju. S.; Rajanaidu, N.; Kushairi, A. and Billotte, N. 2014. Genetic diversity of the word largest oil palm (*E. guineensis* Jacq.) field gene bank accession using microsatellite markers. *Gnet. Reso. Crop Evol.*, **62**: 156-168.

Barcelos, E.; Amblard, P.; Berthaud, J. and Seguin, M. 2002. Genetic diversity and relationship in American and African oil palm as revealed by RFLP and AFLP molecular markers. *Pesquisa Agropecuaria Brasileira.*, **37**: 1105-1114.

Beirnaert, A. and Vanderweyen R. 1941. Contribution à l'étude genetique et biometrique des variétés d' *Elaeis guineensis* Jacquin. Institut National Pour l'étude Agronomique du Congo Belge (INEAC), Brussels.

Billottte, N.; Marseillac, N.; Risterucci, A. M.; Adon, B.; Brottier, P.; Baurens, F.C.; Sing, R.; Herran, S.; Asmady, H.; Billot, C.; Amblard, P.; Durand-Gasselin, T.; Courtois, B.; Asmono, D.; Cheah, S.C.; Rohde, W.; Ritter, E. and Charrier, A. 2005. Microsatellite- based high density linkage map in oil palm (*Elaeis guineensis* Jacq.). *Theor. and Appl. Genet.*, **110**: 754-765

Chen, C.; Zhou, P.; Choi, Y.A.; Huang, S. and Gmitter, F.G. 2006. Mining and characterizing microsatellites from citrus ESTs. *Theor. Appl. Genet.*, **112**: 1248–1257

Corley, R.H.V. and Tinker, P.B. 2003. The oil palm (Ed.4), World Agriculture Series, Oxford, UK: Blackwell Publishing, xxviii+562p

Gan, S.T. 2014. The development and application of molecular markers for linkage mapping and quantitative trait loci analysis of important agronomic traits in oil palm (*Elaeis guineensis* Jacq.) PhD thesis submitted to the University of Nottingham, UK.

Hamilton, J. P. and Buell, C. R. 2012. Advances in plant genome sequencing. *Plant J* **70** (1): 177–90.

Jayanthi, M., Sujatha, G. and Mandal, P.K., 2009, Optimization of PCR reagents for amplification of microsatellites in oil palm. *Indian Journal of Horticulture*, **66**: 147-148.

Jayanthi, M.; Mandal, P. K.; Sujatha, G.; Jayasri, K. S.; Srinivas Rao, G.; Sunitha, B. and Kochu Babu, 2008. M. Simple Sequence Repeats (SSR) an RAPD primers for assessment of genetic uniformity among the field planted clones of oil palm. Paper submitted for presentation in the PLACROSYM-2008, to be held at NRCC, Puttur, Karnataka from 4-8 December.

Jeennor, S. and Volkaert, H. 2014. Mapping of quantitative trait loci (QTLs) for oil yield using SSRs and gene-based markers in African oil palm (*Elaeis guineensis* Jacq.). *Tree Genet. and Genom.*, **10**: 1–14

Jouannic, S.; Argout, X.; Lechauve, F.; Fizames, C.; Borgel, A.; Morcillo, F.; Aberlenc-Bertossi, F.; Duval, Y. and Tregear, J. 2005. Analysis of expressed sequence tags from oil palm (*Elaeis guineensis*). *FEBS-Letters.*, **579**: 2709-2714

Kularatne, R. S.; Shah, F.H. and Rajanaidu, N. 2001. The evaluation of genetic diversity of Deli dura and African oil palm germplasm collection by AFLP technique. *Trop. Agri. Res.*, **13**: 1-12

Lee, M.; Xia, J. H.; Zou, Z.; Jian, Y.; Rahmadsyah. and Yuzer, A. *et al.*, 2015. A consensus linkage map of oil palm and a major QTL for stem height. *Sci. Rep.*, **5**: 8232 | DOI: 10.1038/srep08232

Maizura, I. and Rajanaidu, N. 2001. Genetic diversity of oil palm germplasm collections using RFLPs. In : Proceedings of the 2001 PIPOC International Palm Oil Congress, Agriculture Conference on 'Cutting edge technologies for sustained competitiveness, Kuala Lumpur, Malaysia, 20-22 August, 2001. pp526-535.

Mandal, P.K and Susmitha,D., 2006, Biochemical and molecular characterization of oil palm (*Elaeis guineensis* Jacq.) germplasm using biochemical and molecular parameters. *Journal of Plantation Crops,* **34**: 534-539.

Mandal, P.K., Aruna, C., Mallaiah, M., Shamila, S., Sivasankar Kumar, K.M and Sireesha, K. 2003a, Selection of index leaf of oil palm for Biochemical analysis. *International Journal of Oil palm,* **3 and 4**: 17-21.

Mandal, P.K., Malliah, P., Sireesha, K., Shamila, S and Aruna, C. 2004, the use of RAPD markers for molecular characterization of Oil palm (*Elaeis guineensis* jacq.) germplasm. *Journal of Plantation crops,* **32**: 131-133.

Mandal, P.K., Srinivas Rao, B. and Sailakshmi, A., 2003b, Genetic diversity study of oil palm (*Elaeis guineensis* Jacq) germplasm collected from Cameroon and ASD Costa rica along with indigenously developed material by Randomly Amplified Polymorphic DNA (RAPD) assay. (In) *Proc.of National seminar on modern biology.* August 28-30, 2003.

Mandal, P.K.; Kochu Babu, M.; Jayanthi, M. and Satyavani, V. 2014. PCR based early detection of *Ganoderma* sp. causing basal stem rot of oil palm in India. *J. Plant. Crops.,* **42 (3):** *392-394*

Mayes, S.; Jack, P. L. and Corley, R.H.V. 2000. The application of molecular markers in a specific breeding programme for oil palm. *Heredity.,* **85**: 288-293

Mayes, S.; James, X.M.; Horner, S.F.; Jack, P.L. and Corely, R.H.V. 1996. The application of restriction fragment length polymorphism for the genetic fingerprinting of oil palm (*Elaeis guineensis* Jacq.). *Mol. Breed.,* **2**: 175-180

Montoya, C.; Cochard, B.; Flori, A.; Cros, D. and Lopes, R. *et al.,* 2014. Genetic Architecture of Palm Oil Fatty Acid Composition in Cultivated Oil Palm (*Elaeis guineensis* Jacq.) Compared to Its Wild Relative *E. oleifera* (H.B.K) Corte´s. *PLoS ONE* **9 (5):** e95412. doi: 10.1371/journal.pone.0095412

Montoya, C.; Lopes, R. and Albert, F. *et al.,* 2013. Quantitative trait loci (QTLs) analysis of palm oil fatty acid composition in an interspecific pseudo-backcross from *Elaeis oleifera* (H.B.K.) Cortés and oil palm (*Elaeis guineensis* Jacq.) *Tree Genet. Genom.,* **9**: 1207–1225

Morcillo, F.; Cros, D.; Billotte, N.; NgandoEbongue, G.F.; Domonhédo, H.; Pizot, M.; Cuéllar, T.; Espéout, S.; Dhouib, R.; Bourgis, F.; Claverol, S.; Tranbarger, T.J.; Nouy, B. and Arondel, V. 2015. Improving palm oil quality through identification and mapping of the lipase gene causing oil deterioration. *Nat. Communications.,* **2160** doi: 10.1038/ncomms3160

Moretzsohn, M.C.; Ferreira, M.A.; Amaral, Z.J.A.; Grattapaglia, D. and Ferreira, M.E. 2002. Genetic diversity of Brazilian oil palm (*Elaeis oleifera* H.B.K.) germplasm collected in the Amazon Forest. Euphytica., 124: 35-45

Okoye M. N., C. Uguru BM.I, Singh R. and Okwuagwu C. O. 2016 Genetic Relationships between Elite Oil Palms from Nigeria and Selected Breeding and Germplasm Materials from Malaysia via Simple Sequence Repeat (SSR) Markers Journal of Agricultural Science; Vol. 8, No. 2; 2016 159-178

Poncet V, Rondeau M, Tranchant C, Cayrel A, Hamon S, de Kochko A and Hamon P 2006 SSR mining in coffee tree EST databases: potential use of EST-SSRs as markers for the Coffea genus. Mol Gen Genomics 276: 436–449

Pootakham, W., Jomchai, N., Areerate, P., Jeremy R. Shearman., Sonthirod, C., Sangsrakru, D., Tragoonrung, S. and Tangphatsornruang, S.2015 Genome-wide SNP discovery and identification of QTL associated with agronomic traits in oil pal, using genotyping –by- sequencing (GBS). Genomics, 105: 288-295

Purba, A. R., Noyer, J. L., Baudouin, L., Perrier, X., Hamon, S., and Lagoda, P. J. L. 2000 A new aspect of genetic diversity of Indonesian oil palm (Elaeis guineensis Jacq.) revealed by isoenzyme and AFLP markers and its consequences to breeding. Theor. And Appl. Gent, 101: 956-961

Rajanaidu, N. and Jalani, B. S. 1999. Proc. Seminar Sourcing of oil palm planting materials for local and overseas joint ventures. Palm Oil Research Institute of Malaysia, Kuala Lumpur.

Rethinam, P. 1998. Oil palm research and development in India. (In) Oil Palm Research and Development (Edited by P. Rethinam and K. Suresh). Proceedings of the National Seminar on "Opportunities and challenges for the oil palm development in the twenty first century" from 19-21 January 1998 at Vijayawada (A.P.).

Riju, A., Arumugam, C and Arunachalam, V., 2007, Mining for single nucleotide polymorphisms and insertions/deletions in expressed sequence tag libraries of oil palm. Bioinformation 2 (4): 128-131.

Rivall, A. Bertrandt,. Beul~M,. C. Combeps. and Trousloantd P. Lashermes. 1998. Suitability of RAPD analysis for the detection of somaclonal variants in oil palm (Elaeis guineensis Jacs) Plant Breeding 117 (1), 73-76

Seng, T. Y. *et al.* 2011. Genetic linkage map of a high yielding FELDAdeli x yangambi oil palm cross. Plos One 6, e26593.

Shah, F. H., Rasid, O., Simon, A. J. and Dunsdon, A. 1994. The utility of RAPD markers for the determination of genetic variation in oil pal, (Elaeis guineensis). Theor. and Appl. Genet., 89: 713-718

Singh, R., Low, E. T., Ooi, L. C., Abdullah, M. O., Rajanaidu, N., Ting, N. C., Marhalil, M., Chan, P.L., Maizura, I.; Manaf, M.A., *et al.*, 2014 Nature Communications. DOI: 10.1038/ncomms5106.

Singh, R., Low, E.T., Ooi, L.C., Ong-Abdullah, M., Nookiah, R., Ting, N.C., *et al.*, 2014. The oil palm VIRESCENS gene controls fruit colour and encodes a R2R3-MYB. Nat. Commun. 5,4106.doi: 10.1038/ncomms5106

Singh, R., Ong-Abdullah, M., Low, E.T., Manaf, M.A., Rosli, R., Nookiah, R., *et al.*, 2013a. Oil palm genome sequence reveals divergence of inter fertile species in Old and New worlds. Nature 500, 335–339.doi: 10.1038/nature 12309

Singh, R., Tan, S.G., Panandam, J.M., Rahman, R.A., Ooi, L.C., Low, E.T., *et al.*, 2009. Mapping quantitative trait loci (QTLs) for fatty acid composition in an inter specific cross of oil palm. BMC Plant Biol. 9: 114. doi: 10.1186/1471-2229- 9-114

Squirrell J, Hollingsworth PM, Woodhead M, Russell J, Lowe AJ, Gibby M, PowellW 2003. How much effort is required to isolate nuclear microsatellites from plants? Mol Ecol 12: 1339–1348

Taeprayoon, P., Tanya, P., Lee, S.H., and Srinivas, P.2015 Genetic background of three commercial oil palm breeding populations in Thailand revealed by SSR markers. Australian journal of Crop Science, 9 (4): 281-288.

Teh CK, Ong AL, Kwong, Apparow S, Chew FT, Mayes S, Mohamed M, David A and Harikrishna K 2016 Genome-wide association study identifies three key loci for high mesocarp oil content in perennial crop oil palm. Scientific Reports. 6: 19075, DOI: 10.1038/srep19075

Timothy J T, Wanwisa K, Duangjai S, Fabienne M, James W T, Somvong T and Norbert B 2012 SSR markers in transcripts of genes linked to post-transcriptional and transcriptional regulatory functions during vegetative and reproductive development of Elaeis guineensis. BMC Plant Biology 2012, 12: 1

Ting, N.C., Jansen, J., Mayes, S., Massawe, F., Sambanthamurthi, R., Ooi, L.C., *et al.*, 2014. High density SNP and SSR-based genetic maps of two independent oil palm hybrids. BMC Genomics 15: 309. doi: 10.1186/1471-2164-15-309

Ting, N.C., Zaki, N. M., Rosle, R., Low, E. T., Maizura, I., Cheah, A. C., Tan, S. G., and Singh, R. 2010. SSR mining in oil palm EST databse: application in oil palm germplasm diversity studies. J. of Genetics.

Ukoskit K, Vipavee C, Ganlayarat B, Kwanjai P, Sithichoke T and Somvong T, 2014 Oil palm (*Elaeis guineensis* Jacq.) linkage map, and quantitative trait locus analysis for sex ratio and related traits Mol Breeding (2014) 33: 415–424

Weising K, Nybom H, Wolff K and Kahl G 2005 Application of DNA fingerprinting in plant sciences. DNA Finger printing in plants principles, methods, and applications. CRC Press, Boca Raton, pp 235–276

Winter P. and Kahl G. 1995. Molecular marker technologies for plant improvement. World Journal of Microbiology and Biotechnology, 11: 438-448.

Yu J-K, Dake TM, Singh S, Benscher D, Li W, Gill B and Sorrells ME 2004 Development and mapping of EST-derived simple sequence repeat markers for hexaploid wheat. Genome 47: 805–818

Zaki, N.M., Singh, R., Rosli, R., and Ismail, I. 2012 *Elaeis oleifera* Genomic- SSR markers: Exploitation in oil pal, germplasm diversity and cross- amplification in Arecaceae. Int. J. Mol. Sci. 13: 4069-4088

Zane L, Bargelloni L and Patarnello T 2002 Strategies for microsatellite isolation: a review. Mol Ecol 11: 1–16

Zeven, A. C. 1968. Oil palm groves in Southern Nigeria. Part II. Development, deterioration and rehabilitation of groves. J. Niger. Inst. Oil Palm Res., 5, 21-39.

Previous Volumes–Contents

Biodiversity in Horticultural Crops Vol 2/*Peter, K V ed*

Part I: General

Part II: Biodiversity of Vegetables

Part III: Biodiversity of Spices

Part IV: Biodiversity of Ornamentals

Biodiversity in Horticultural Crops Vol 4/*Peter, K V ed*

Biodiversity in Horticultural Crops Vol 5/*Peter, K V ed*

Index

D

Figure 2.2: Landscapes of different Eastern Himalayan Region.

A: Upland forest at Chele la, Bhutan; B: Alpine region at Dochula, Bhutan; C: Alpine zone at Gnathang valley, Sikkim and D: Lowland forest at Pedong, Sikkim. (p. 63)

Acid Lime–Leaves with Small Wing (p. 177)

Acid Lime–Thorny Fruiting Branch

Acid Lime–Branch with Fruit

Acid Lime–Fruit Developmental Stages

Acid Lime–Leaves with Small Wing (p. 96)

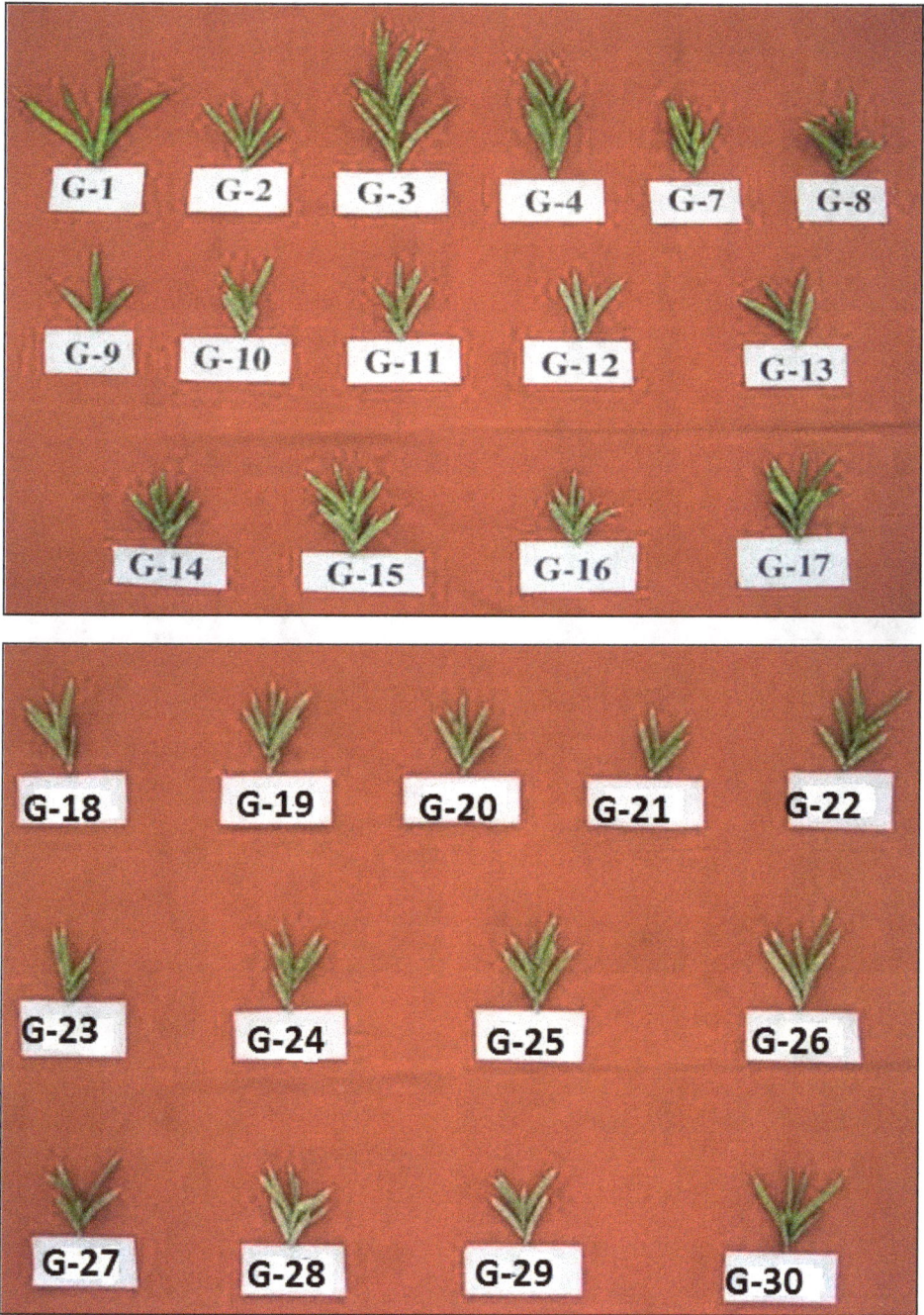

Plate 7.1: Variability in Cluster Characters of 30 Genotypes in Cluster Bean Observed under Study. (p. 182)

Plate 7.2: Variability in Pod Characters of 30 Genotypes in Cluster Bean Observed under Study. (p. 183)

Kale Cultivar 'KTK-64' under Testing in AICRP (VC) Trials.

Varieties availabe in Kale Crops. (p. 203)

Umble of Onion　　　　**Individual Flower of Onion**

Onion Seeds

Plate 9.1: Onion Flower and Seeds. (p. 214)

Plate 9.2: Photographs of Cultivated Allium Species.

Allium altaicum

Allium ampeloprasum

Allium fistulosum

Allium obliquum

Allium victorialis (p. 218)

Figure 11.1: Large cardamom (*Amomum subulatum* Roxb.). (p. 249)

Figure 12.2: *T. cacao*.

Figure 12.3: *T. grandiflorum*.

(p. 271)